1673063 2

16 m

Biophysical Aspects
of Cardiac Muscle

Members of the Organizing Committee

N. Cohanim
M. Edjehadi
T. Fakouhi
S. S. Jahromi
E. Meisami
A. Movassaghi
A. Nahapetian
G. Nayeri
M. Salimi
M. Tabatabai
M. Morad

Proceedings of the Cardiac Muscle Symposium
(May 14–16, 1977, Shiraz, Iran)

Sponsored by

Special Bureau of Her Imperial Majesty, the Shahbanou of Iran
Iranian Society of Physiology and Pharmacology
Pahlavi University
Iranian Ministry of Science and Higher Education

Biophysical Aspects
of Cardiac Muscle

Edited by

Martin Morad
Department of Physiology
School of Medicine
University of Pennsylvania
Philadelphia, Pennsylvania

Symposium Coorganizer

Mahmood Tabatabai
Department of Physiology
School of Medicine
Pahlavi University
Shiraz, Iran

Assistant Editor

Susan Smith
University of Pennsylvania
Philadelphia, Pennsylvania

Academic Press   *New York San Francisco London*   1978
A Subsidiary of Harcourt Brace Jovanovich, Publishers

ACADEMIC PRESS RAPID MANUSCRIPT REPRODUCTION

ACADEMIC PRESS, INC.
111 Fifth Avenue, New York, New York 10003

United Kingdom Edition published by
ACADEMIC PRESS, INC. (LONDON) LTD.
24/28 Oval Road, London NW1   7DX

Library of Congress Cataloging in Publication Data

Cardiac Muscle Symposium, Pahlavi University, 1977.
    Biophysical aspects of cardiac muscle.

    1.   Heart—Muscle—Congresses.   2.   Action
potentials (Electrophysiology—Congresses).   I.   Morad,
Martin.   II.   Tabatabai, Mahmood.   III.   Title. [DNLM:
1.   Myocardium—Congresses.   2.   Heart—Physiology—
Congresses. WG280 B615 1977]
QP113.2.C37 1977   596'.01'16   78-17025
ISBN 0-12-506150-1

PRINTED IN THE UNITED STATES OF AMERICA

Dedicated to the
Noble Persian Heritage

Contents

III. IONIC TRANSPORT MECHANISMS IN GENERATION OF CARDIAC ACTION POTENTIAL PLATEAU

IV. STRUCTURE AND FUNCTION OF THE SARCOTUBULAR SYSTEM

List of Contributors

Numbers in parentheses indicate the pages on which the authors' contributions begin.

Adrian, R. H. (45, 91), Physiological Laboratory, Cambridge CB2 3EG, England

Armstrong, C. M. (27, 75), Department of Physiology, University of Pennsylvania, Philadelphia, Pennsylvania

Baer, M. (129), Department of Pharmacology, University of Bern, Friedbuhlstrasse 49 30008 Bern, Switzerland

Baylor, S. M. (207), Department of Physiology, Yale University School of Medicine, New Haven, Connecticut

Best, P. M. (129), Department of Medicine, University of Chicago, Chicago, Illinois

Bezanilla, F. (229), Department of Physiology and the Brain Research Institute, University of California at Los Angeles, Los Angeles, California

Blood, B. E. (369, 379), University Laboratory of Physiology, Oxford University, Oxford, England

Carmeliet, E. (143), Laboratory of Physiology, Campus Grasthuisberg, 3000 Leuven, Belgium

Chandler, W. K. (31, 207), Department of Physiology, Yale University School of Medicine, New Haven, Connecticut

Cleemann, L. (153), Department of Physiology, University of Wisconsin, Madison, Wisconsin

Endo, M. (307), Department of Pharmacology, Tohoku University School of Medicine, Seiryo-machi, Sendai 980, Japan

Gilly, W. F. (31), Department of Physiology, Yale University School of Medicine, New Haven, Connecticut

Hille, B. (55), Department of Physiology and Biophysics, University of Washington, SJ-40, Seattle, Washington

Hui, C. S. (31), Department of Physiology, Yale University School of Medicine, New Haven, Connecticut

Huxley, A. F. (3), Department of Physiology, University College London, London, England

Ildefonse, M. (273), Laboratoire de Physiologie des Eléments Excitables, Université Claude Bernard, F69621, Villeurbanne, France

Kass, R. S. (345), Department of Physiology, University of Rochester School of Medicine and Dentistry, Rochester, New York

Kitazawa, T. (307), Department of Pharmacology, Tohoku University School of Medicine, Seiryo-machi, Sendai 980, Japan

Klitzner, T. (285), Department of Physiology, University of Pennsylvania, Philadelphia, Pennsylvania

McClellan, G. B. (329), Department of Physiology, University of Pennsylvania, Philadelphia, Pennsylvania

McNaughton, P. A. (107), Physiological Laboratory, Downing Street, Cambridge

Morad, M. (153, 285), Department of Physiology, University of Pennsylvania, Philadelphia, Pennsylvania

Noble, D. (369), University Laboratory of Physiology, Oxford University, Oxford, England

Page, S. (383), Department of Biophysics, University College London, Gower, Street, London WCIE 6BT, England

Peachey, L. D. (187, 191), Department of Biology, University of Pennsylvania, Philadelphia, Pennsylvania

Reuter, H. (129), Department of Pharmacology, University of Bern, Friedbuhlstrasse 49 3008 Bern, Switzerland

Roche, M. (273), Laboratoire de Physiologie des Eléments Excitables, Université Claude Bernard, F69621, Villeurbanne, France

Rougier, O. (273), Laboratoire de Physiologie des Eléments Excitables, Université Claude Bernard, F69621, Villeurbanne, France

Rüdel, R. (255), Physiologisches Institut der Technischen Universität, Biedersteiner Strasse 29, D-8000 Müchen 40, Federal Republic of Germany

Tsien, R. W. (345), Department of Physiology, Yale University School of Medicine, New Haven, Connecticut

Vergara, J. (229), Department of Physiology and the Brain Research Institute, University of California at Los Angeles, Los Angeles, California

Weingart, R. (345), Physiological Institute, University of Bern, Friedbuhlstrasse 49 3008 Bern, Switzerland

Winegrad, S. (329), Department of Physiology, University of Pennsylvania, Philadelphia, Pennsylvania

Foreword

It has been some 25 years since I had the privilege of visiting or contributing to the advancement of science in my native country. The idea for a cardiac muscle symposium was first discussed five years ago when Dr. Tabatabai visited my lab at the University of Pennsylvania. Though the idea for a scientific gathering was welcomed by all who were approached, the topic, the timing, and the financial sponsor for such a major undertaking ran into many obstacles. Finally, through the efforts of many individuals, the sponsoring organizations and the location were found so that the first cardiac muscle symposium could be held. The symposium was held in Shiraz under the auspices of her Imperial Majesty, the Shahbanou of Iran, the Ministry of Health and Higher Education, the Iranian Society of Physiology and Pharmacology and the Pahlavi University. I should like to express my personal gratitude to his excellency Dr. Motamedi, Chancellor Farhang Mehr, and Drs. Movassaghi, Meisami, and Rostami for their personal efforts on behalf of the symposium. It fell to Dr. Tabatabai, the co-organizer of the symposium and his colleagues at Pahlavi University to arrange for the taping of the proceedings and the accommodations of the visiting scientists. I shall always be grateful for his efforts to make the stay of our visiting scientists so joyous and unforgettable. Special thanks go to Miss Linda Loupe and Ms. Susan Smith, for transcribing and editing the tapes of the symposium proceedings, and to Ms. Linda Baird for typing the manuscripts.

There are two opposing thoughts as to whether the proceedings of a scientific symposium should be published. One holds that the freedom of the scientific presentations and discussions is stifled because the participants are aware that every word or idea uttered is being recorded. Another thought, though conceding to this criticism, maintains that the teaching benefits and wide exposure of published proceedings of the symposium far outweigh this minor reservation. In deciding to publish the proceedings of this symposium, I was persuaded by the latter argument and by the fact that young Iranian scientists and students who are trying to keep abreast of the new scientific development, despite fairly difficult academic conditions, would have an easily obtainable reference source for their scientific quests. On a more personal scale, the organization of this symposium and the publication of its proceedings represent my own confirmation of the love, honor, and gratitude that I have felt for about 25 years from afar for the country of my birth.

The high scientific standard of the participating scientists and high quality of their contributions were in keeping with the Persian scientific and literary heritage. The gathering of international scientists to discuss medical problems seems to have been an ancient Persian tradition. In fact, the first medical congress of Greek, Chinese, Indian, and Persian scholars gathered to discuss medical problems about the fourth century A.D. in Jundi Shapur, not very far from the location of this cardiac symposium.

In such a rich tradition of scientific and literary achievement, the proceedings of this symposium serve as a minor reminder to the continuance of the thread of scientific tradition and contribution throughout the last 2000 years of Persian history.

The romance of the Persian environment, the joy of newfound Iranian friends, the excitement of scientific discussions were best summed by Dr. Richard Adrian's concluding comments. I should like to quote from the last segment of his talk. Dr. Adrian reminded us of the trip of Sir Thomas Herbert who visited Shiraz in 1627, 60 years before the foundation of Harvard and a good many years before the foundation of the Royal Society of London.

> Herbert found Iran wise in civilization and in science, especially medicine. He described Shiraz as defended by nature, enriched by trade, and by art made lovely. He found here (Shiraz) a college wherein was read philosophy, medicine, chemistry, and mathematics. He noted also the poetry, the wine, the gardens, and the great joy of citizens of Shiraz. So well was Herbert entertained in Shiraz that he feared his virtue might be overcome by his pleasures. We too have been marvelously well looked after, and we too have found in Shiraz a flourishing medical school. And if I cannot find his poetic words to describe all our pleasures, I can at least echo Herbert when he said that on leaving Shiraz he felt as if he were leaving paradise. Speaking quite selfishly, I have learned an immense amount about what my colleagues are doing. Of course this is always the first thing that a scientific meeting should do. Beyond that, however, I believe that this meeting has been an extremely good example of the way progress in science is achieved. Sometimes that progress is not immediately visible. Our experiments and confusions show how far uncertainties and hesitations and even disagreements always precede the acquisition of sure knowledge. But perhaps more important than that, we have had the opportunity of meeting Iranian colleagues, physiologists, and pharmacologists. This is the first step which must precede scientific collaboration. Successful scientific endeavor begins and should end in friendship. To our many new friends in Iran, I would like to say from all of us thank you for asking us and thank you for listening to us. We all hope that these new friendships will lead to new scientific progress.

It is in the spirit of these comments that I prevailed upon my friends and colleagues to put this book together and collected all the detailed discussions and the photographic vignettes. I hope the book will serve to remind us and those who will follow us that we belong to an international community whose common bonds are scientific creativity and excellence and whose goal is to improve the quality of human life.

M. Morad
Blacksheep Farm
Newton Square, Pennsylvania

Preface

This book is based on the proceedings of the first cardiac muscle symposium held May 14–16, 1977, at the Pahlavi University in Shiraz, Iran. The symposium was organized in six sessions and dealt with the following topics: gating processes in excitable membranes; ionic transport mechanisms in the generation of cardiac action potential plateau; structure and function of the sarcotubular system; excitation–contraction coupling in heart muscle; and mechanisms of drug action in cardiac muscle. The symposium began with an opening lecture by Sir Andrew F. Huxley titled "On Arguing from one Kind of Muscle to Another."

Cardiac muscle physiology is an enormous field and it would be difficult to organize a symposium that would do justice to all fields of cardiac muscle research. It was therefore decided to limit the scope of the symposium to the biophysical aspects of cardiac muscle research. Speakers were chosen not only based on their international scientific contributions, but also because of their recent contributions to the topics to be discussed. This at times forced us to look beyond the cardiac muscle field and invite contributors whose primary work involved other excitable tissues such as nerve or skeletal muscle. It is our firm belief that molecular mechanisms operating in excitable tissue are often common to various biological tissues and therefore much can be learned about cardiac muscle from studies on nerve and skeletal muscle. In a limited symposium such as this, we painfully felt the strain of omitting many great scientific names who have contributed to the fields of cardiac muscle electrophysiology, mechanics, ionic transport, tissue growth, and pharmacodynamics. I should like to state to those who did not participate in this scientific gathering that the symposium and its proceedings sorely missed their contributions. We can only hope that money and time restrictions will be less stringent in the future so as to make the second cardiac symposium in Iran more comprehensive.

This book is in six sections. The first section deals with Sir Andrew F. Huxley's lecture on muscle, which traces the development of muscle physiology in the last century and discusses the pros and cons of the "unitary" contractile mechanism in biological tissues.

The second section deals with the "gating" mechanism in excitable membranes. Charge movement has been primarily studied in axons and skeletal muscle fibers. No experimental data are as yet available on cardiac muscle. Since the gating experiments provide rather basic information as to the mechanisms by which ionic channels operate, it is hoped that experimental data from nerve and muscle will provide some understanding as to how similar mechanisms may be operating in cardiac muscle.

The third section deals with transport mechanisms. In this respect the contributors to the session discuss Na–Ca exchange, inward rectification, and voltage dependence of TTX binding in nerve, skeletal, and cardiac muscle.

The section on the structure and function of the sarcotubular system deals not only with recent structural findings in muscle, but also with optical signals measured from internal membranes, which may reflect the function of the Ca^{2+} release and uptake system.

The next section discusses the mechanisms of Ca^{2+} transport systems across the cardiac membrane and the regulation of Ca sensitivity of the myofilament in response to hormonal or chemical manipulations.

The final section deals with some recent data on the action of adrenaline and cardiac glycosides in cardiac muscle and the possible implication of these data as to the molecular mechanisms responsible for the mode of action of these drugs on heart muscle.

At the end of each presentation, often a lively discussion occurred. The proceedings of the discussions have been in part edited for clarity. However, the humor and the critical nature of the question–answer period have been left intact. In the absence of a reviewing board, it is hoped that the discussion section provides the clarification and criticism often needed after a formal presentation. Some of the discussions are missing primarily due to the failure of the recording equipment.

For me personally, the proceedings of the symposium were a great learning experience. I am rewarded by the excellence of the contributed papers and the warmth and kindness that were so freely demonstrated by all of the participants toward the organizers of this symposium.

M. Morad

A picture at Pahlavi University. Symposium participants and members of the Iranian Society of Physiology and Pharmacology.

Fig. 1. Scientific poses at Persepolis a. "The Persian way." Lord Adrian, Drs. Peachey, Chandler, and Morad. b. Peachey—conscripted. c. Sir Andrew Huxley at the Gate of Nations—graffiti on the wall not his doing! d. Peachey, Chandler, and Morad trying to show off. e. The "Illinois Kid," S. Baylor, pocketing Persian artifacts. f. Morad pointing at graffiti made by more aggressive ancestors in the last century. g. Hille and Armstrong in the "Na$^+$ hole." h. Winegrad, Tsien, Rougier, and Bezanilla high above the Persepolis plane. i. McNaughton searching for the remains of the Na–Ca exchanger at the tomb of Artxerxes. j. & k. Armstrong and Sir Andrew Huxley searching for original results, i.e., photographic. l. Armstrong and Morad—"even in Texas they don't have this!" m. Reuter and Cleemann witness Peachey's "helicoid" approach. Ms. Bassange is amused. n. Tabatabai, pleased! o. Carmeliet a little sun struck and "inwardly rectifying!" p. Lord Adrian—just a family shot!

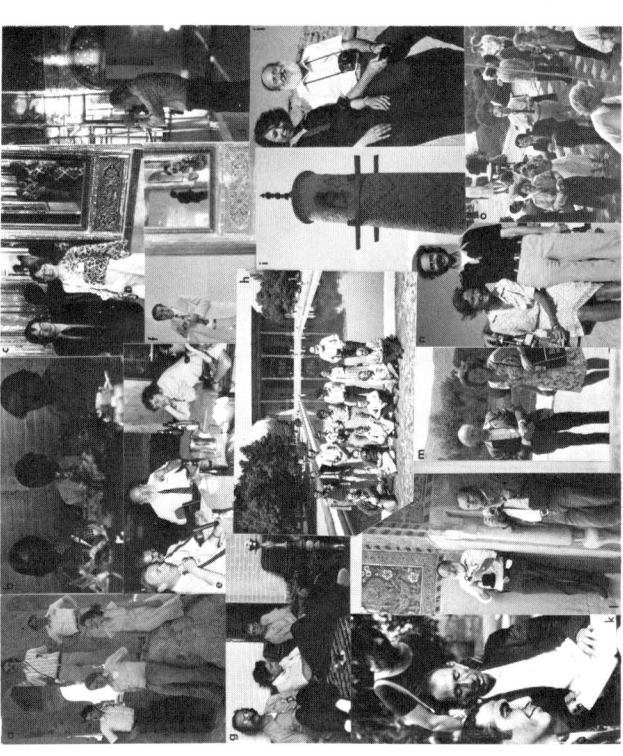

FIG. 2. *a. At the windswept Zoroashtrain Temple in Isfahan, Armstrong, Cleemann, and Tsien; Peachey and Reuter. b. Tsien, Endo, and Cleemann at a Persian feast—Does Ca-induced Ca release improve appetite? c. & d. "Reflections of self"—Tsien, Tsien-Shang, and Morad. e. The boys night out! Lord Adrian and Drs. Peachey, Chandler, Baylor, and Bezanilla. f. In search of the "module," Chandler. g. "Glazed" by the aromas of Isfahan—Peachey, Armstrong, and Hille. h. Some of the symposium participants at the Palace of 40 Columns in Isfahan. From left to right: Rougier, Carmeliet, Reuter, Hille, Cleemann, and Neil. Standing: Morad, Adrian, Peachey, Bezanilla, Tsien, Baylor, Page, and Chandler. i. Shaking the "shaking minaret" outside Isfahan—who else but a "crazy dane!" j. "The old man and the sea" or "papa" and an Iranian beauty. k. At the Tomb of Hafiz.—Morad's fortune told through poetry—Madame Bassange did not speak Farsi! l. "Standing on the corner" —Chandler and Rougier. m. Sir Andrew and Lady Huxley on the way to Persepolis. n. Peachey tries harder! See Fig. 1m. o. Loading up on cherries and cucumbers in a Persian garden—Baylor, Julie and Dick Tsien, Adrian, Cleemann, McNaughton, and Reuter. Also seen are Dr. Movassagi, the President of Iranian Physiological Society, Dr. Rostami, and Madame Carmeliet.*

I

The A. F. Huxley
Symposium Lecture

ON ARGUING FROM ONE KIND
OF MUSCLE TO ANOTHER

Andrew F. Huxley

Department of Physiology
University College London
London, England

It has been said for a very long time that cardiac muscle
is intermediate between skeletal muscle and smooth muscle.
Cardiac muscle is striated like skeletal muscle, and it is
moderately fast, although not as fast as most skeletal muscle.
But it also resembles smooth muscle in that it is composed of
many mononucleate cells between which there is some kind of
electrical continuity, but not protoplasmic continuity.
Cardiac muscle has a spontaneous rhythm like some sorts of
smooth muscle; there is a strong influence of length, and
indeed, of many other factors, on the force of contraction;
and it has nerve supplies which modulate the strength of contraction. Of course, in smooth muscle the nerve supply may
also initiate the contraction.

Both cardiac and smooth muscle are experimentally much
more difficult to investigate than skeletal muscle. For this
reason, much of the progress on cardiac and smooth muscle has
followed that on skeletal muscle. People do their experiments
on the easy tissue, and then they try to apply their results
to the more difficult types. Richard Adrian, a contributor to
this symposium, has allowed me to repeat a story which he
often tells. In his life he can only remember two occasions
when he received advice from his distinguished father, Lord
Adrian, the great neurophysiologist. He acted on both of
these pieces of advice. One of them was that somewhere in his
examination papers for his degree, he should include an
integral sign. The other was to keep off smooth muscle. And
he has kept off smooth muscle, and so have I, for the very
good reason that it is so difficult to work on. Those of us

who hope for clear-cut results choose an easy tissue and leave
it to bolder people to investigate cardiac and smooth muscle.

Arguments by analogy from one tissue to another are
valuable, and this is certainly a field in which they are
important; but they do raise some general and very difficult
problems. How far are these analogies valid? We may draw
analogies between different contractile systems, say, between
muscle and protoplasmic movement, ciliary movement, bacterial
motility, and the many forms of motility in protozoa. We also
draw analogies between the different sorts of muscle, skeletal,
cardiac, and smooth, and between different kinds of animals.
But how far can we assume that results from striated muscle in
a crab, a water-bug, or even a frog are applicable to man?
How similar are the numerous different fiber types of skeletal
muscle even in a single animal?

I. THE ASSUMPTION OF UNIFORMITY

In 1953, shortly after I started work on muscle, I was
looking at skeletal muscle fibers from the frog, a highly
specialized contractile tissue, with the interference
microscope that I had developed. I received a visit from Max
Delbrück, famous for his work on bacterial genetics, and he
suggested that I ought to look at primitive contractile
systems where "contractility" can be studied in its most basic
form. I did not take his advice, and I am sure I was right
not to take it. At that time there was the beginning of a
wave of what one might call "basicity," the argument that one
must go back to basic mechanisms which one assumes to be
similar in all kinds of living things. This outlook resulted,
no doubt, from the discovery of the double helix, and from the
finding that bacterial genetics is very much the same as
genetics in higher animals, and so forth. Indeed, I think
many of the molecular biologists of that time thought that
this was a new idea, that plants, animals and bacteria were
essentially the same. Of course, they were totally wrong.
Just before the Second World War, when I was an undergraduate,
all of us were tremendously struck by the way in which the
glycolytic cycle had been worked out partly on yeast, partly
on muscle, and partly on liver. Going back another thirty or
forty years, the rediscovery of Mendelian genetics was done
almost equally on animals and on plants. Back yet another
forty years, in each chapter of the *Origin of Species*, Darwin
brings forward examples both from the plant and the animal
kingdoms. Another thirty years before that, comparative
embryology was developed in von Baer's great book (1828-37):
as one traced animals back earlier in their development, they

became more and more similar. The cell theory itself was initiated by Schleiden, a botanist, and Schwann, an animal anatomist and physiologist: similarities between animal and plant tissues were what led them to postulate that all living things are made up of "cells."

Now, in relation to contractility, if one may use such a word, arguments based on the assumed similarity of motility or contractility in all types of organisms can be found in the literature at all times since the publication of Bowman's famous paper in 1840, the beginning of modern investigation of muscle. But there was a particular outburst of this type of reasoning which started in the middle 1880's and continued until about 1910. At this time the evolutionary theory had become universally accepted, and everybody had been brought up in the assumption of common ancestry and therefore of widespread uniformity. In 1885-88 three papers, widely known at that time, presented a reticular theory of muscle, in which contraction was done not by the fibrils but by a reticulum in between the fibrils, which could be stained with gold chloride (Melland, 1885; van Gehuchten, 1886; Ramón y Cajal, 1888). This idea was based on an analogy with a theory of protoplasmic movement, current at that time, which was due to Heitzmann and Carnoy, who believed that the protoplasm of an amoeboid cell was a network of contractile fibrils in a matrix of more or less uniform material. This reticular theory of muscle fortunately did not survive. Koelliker made it very clear that the fibrils are the contractile structures. In a famous book, *The Movement of Living Substance*, Verworn said that we should study not advanced and complicated contractile systems, but processes like protoplasmic movement. Bernstein, more famous for his investigations of biological electricity, also had theories of muscle contraction. About 1900 he forcefully put forward the view that it is no use to pay attention to the striations, because unstriated muscle also works perfectly well, and, as an ideal object for investigation, he named the contractile thread of Protozoa such as *Vorticella*.

This assumption that all contractile systems must be similar held up progress on muscle to a very substantial extent. The arguments that the striations cannot be important because smooth muscle works, and that contraction clearly is a chemical event and must therefore be at a far submicroscopic level, led physiologists and biochemists to disregard evidence obtained with the microscope. The striations were regarded as an irrelevance from about 1900 until the early 1950's, when structure came back into prominence.

The discoveries of the last quarter of a century have shown that arguments like Bernstein's are wrong in numerous ways. Nearly all of this recent progress has depended on experimental work on some of the most highly specialized

contractile systems: the skeletal muscle fibers of vertebrates, and the rather peculiar muscle by which certain of the
orders of insects fly. The two-winged flies, bees, beetles
and some other insects fly with an extraordinarily highly
developed mechanism in which the fibrillar structure of the
muscle fiber is even more regular than it is in the skeletal
muscle of vertebrates. Another thing that has been learned in
recent years is that although actomyosin systems are
surprisingly widespread in nonmuscle cells, there are several
other kinds of motility that have nothing to do with actin or
myosin. For example, the numerous systems depending on
microtubules (such as mitotic movement), and bacterial
flagella, seem to work in totally different ways, and in
particular, the contractile thread of *Vorticella* and related
animals seems to work by yet another totally different
contractile mechanism.

Until very recently, all of the new advances came from
study of the specialized muscle systems that I mentioned, but
two years ago, for the first time, a discovery applicable to
muscle resulted from study of a nonmuscle system. It had been
found that many of the proteins involved in contraction can be
phosphorylated, and of course one assumes that this has some
functional importance. But the first demonstration of a
change of function in one of these proteins as a result of
phosphorylation was made by Adelstein & Conti (1975) in the
United States, using actomyosin from the platelets of blood.
They showed that when the myosin was phosphorylated, its
ATPase activity was increased. This matter of phosphorylation
of muscle proteins is coming increasingly to the fore, and we
shall hear more about it in papers during this symposium.

II. THE ACTION POTENTIAL

I want to illustrate some of the difficulties of arguing
from one tissue to another in relation to various aspects of
contraction, particularly in connection with the heart.
Working inwards from the surface, the first event in the
contraction is a reduction or reversal of the potential
difference across the surface membrane. During the 19th
century it was recognized that when a muscle or a nerve propagates its event, there is a "depolarization," which about a
hundred years ago was identified as a rise in the internal
electric potential of a cylindrical cablelike structure. In
muscle, this change of potential lasts about a thousandth of a
second and is very much shorter in duration than the mechanical change, the twitch, that follows it. The time course of
the action potential was worked out just over a hundred years

ago by the same Bernstein (1868), using an ingenious strobo-scopic device. As regards the heart, of course, the contraction is a few times slower than in skeletal muscle, but the electrical event is the order of a thousand times slower. Instead of being a short event that is over before the contraction really begins, it has a duration comparable to that of the contraction itself. In case anybody still imagines that this was seen only after microelectrodes were developed, let me remind you that the long duration of the action potential was discovered just about a hundred years ago by Burdon-Sanderson (Burdon-Sanderson & Page, 1878; 1880), using a capillary electrometer.

In these respects, there are again analogies with smooth muscle. Long-lasting "heartlike" action potentials are given by some vertebrate smooth muscles, notably the ureter of the rat; furthermore, this was shown (by Bozler, 1942) with external electrodes on whole-muscle preparations long before intracellular recording was developed. Bozler, by the way, also showed that different smooth muscles, all from vertebrates, give a very wide range of different types of electrical response.

There are also differences in the ionic basis of the action potential. Hodgkin & Katz (1949) showed that the squid nerve action potential was generated by sodium entry, which then was also found in frog muscle and in frog myelinated nerve. At that point it looked as if sodium entry was the universal mechanism of producing excitation. Then Fatt & Ginsborg (1958) found that the excitation of crab muscle was worked by calcium entry. This mechanism was found to be widespread. Much vertebrate smooth muscle seems to work by calcium entry, not sodium entry, while heart seems again to be intermediate between skeletal and smooth muscle in this respect: both sodium and calcium entry appear to be important.

III. SLOW MUSCLE FIBERS IN VERTEBRATES

In the 1920's there was much evidence that some muscles of the frog were capable of very long-lasting contractures. If the familiar sartorius is placed in an acetylcholine or potassium solution, it gives a contracture but then it relaxes again in a matter of seconds, or a minute at the outside. But with the rectus abdominis or the iliofibularis or the palmaris, the contraction may last for hours so long as the contracture agent is kept on. Then in the early 1930's, Krüger found that certain muscles of the frog contain fibers of two types that are conspicuously different in appearance in ordinary cross-sections viewed with the light microscope [references and a review are given in Krüger (1952)]. One type, with small myofibrils, is similar to that in pure twitch muscles like the sartorius. The other fibers have what he referred to as "field structure," with the cross-sections of the myofibrils having a straplike appearance. Krüger found that the second type of fiber existed only in muscles which gave long-lasting contractures, and he quite correctly supposed that these fibers were responsible for the long-lasting responses.

Most physiologists paid no attention whatever to this work, and the principal reason was the argument that tonus-- long-lasting postural contractions--must be of the same nature in frogs as in mammals. In the late 1920's it was found that in mammals even a weak tonic contraction of a limb muscle consisted of twitch or tetanuslike activity, an unfused tetanus at low frequency involving a few units, but qualitatively just the same as twitch activity. So it was supposed that the same must be true in frogs, but this assumption was wrong.

Fig. 1 is from a key study, very little known in the West, which showed that frog muscles are different in this respect from those of mammals. This paper by Tasaki and a Japanese colleague (Tasaki & Mizutani, 1944) was published during the war in a Japanese journal. The curves show the tension developed by the gastrocnemius muscle of the frog on a time scale of 1-sec intervals. The frequency of stimulation applied was 10, 20, 50 or 100 per sec. In this experiment Tasaki is stimulating a single motor nerve fiber of small diameter, about 4 or 5 μm, as opposed to 15 μm or so of the fibers which give twitch responses. A single stimulus to this small motor nerve fiber produces no response. With repetitive stimulation at 10 per sec there is a slow contraction which one can just detect, rising over many sec; at 20 per sec it goes up faster, and at 50, faster still. This slow mode of contraction, produced by stimulating the small-diameter nerve fibers, is quite unlike the familiar twitch or tetanus. Stimulating a large nerve fiber at 50 per sec would have caused a contraction which reaches its plateau in a fraction of a second.

At that state it was uncertain whether there were two modes of contraction in the same muscle fibers, or whether, as Krüger had suggested, there were different fibers with different modes. It is not unreasonable to think of single fibers having two modes, for in crabs and other Crustacea, there are many fibers which can give either a slow contraction of this type or a rapid, twitchlike response. It was not until microelectrodes were used that in 1953, Kuffler and

Vaughan-Williams showed that in the frog, the two types of
contraction are carried out by separate muscle fibers.
Peachey and I confirmed what Krüger had surmised twenty or
thirty years before, that it is the fibers with the field-
structure which give this slow mode of contraction. At that
time, around 1960, it seemed very clear that frogs have this
type of fiber in their muscles, and mammals do not. Leksell,
for example, looked for this type of contraction when the small
nerve fibers in the motor nerve of mammals were stimulated, and
saw no mechanical response. It looked as if this was a clear
difference between amphibians and mammals.

Then, slow contractions turned up in muscles that are
among the fastest that are found anywhere, the external eye
muscles of mammals. This discovery was made in Russia by
Matiushkin (1961), although the authors who are most generally
credited are Hess and Pilar in the United States, who made the
same discovery independently just a year or so later.
Matiushkin's work was not widely known in the West until later,
in spite of the fact that it did appear in translation when
originally published. Matiushkin obtained intracellular
records from fibers of one of the extracellular eye muscles.
Some fibers give spikes of over a hundred millivolts amplitude,
while others give electrical changes of only a few tens of
millivolts and clearly are very similar to the slow fibers of
frogs. Fig. 2a from Matiushkin's paper is an extracellular
record with an appropriate stimulus, showing the quick twitch
response, and a delayed nonpropagating monophasic response
from the slow fibers. I think most people seeing this record
would guess that it is one that had been made by Kuffler and
Vaughan-Williams on a limb muscle of a frog. Fig. 2b shows
intracellular records.

IV. INWARD SPREAD OF ACTIVATION

The action potential, of course, is really an event in the
surface membrane itself; that is the only place where a
potential gradient exists which is large enough to produce
appreciable physicochemical effects. But how does this event
in the surface membrane cause the inside of the fiber actually
to contract? The existence of this problem became clear
shortly after the second war, from A.V. Hill's measurements of
the heat production of muscle. He made it quite clear that
the speed with which the contractile material of frog
sartorius muscle, for example, is activated after a stimulus
is much too great to be accounted for by any substance
spreading inwards from the surface membrane by ordinary
diffusion processes.

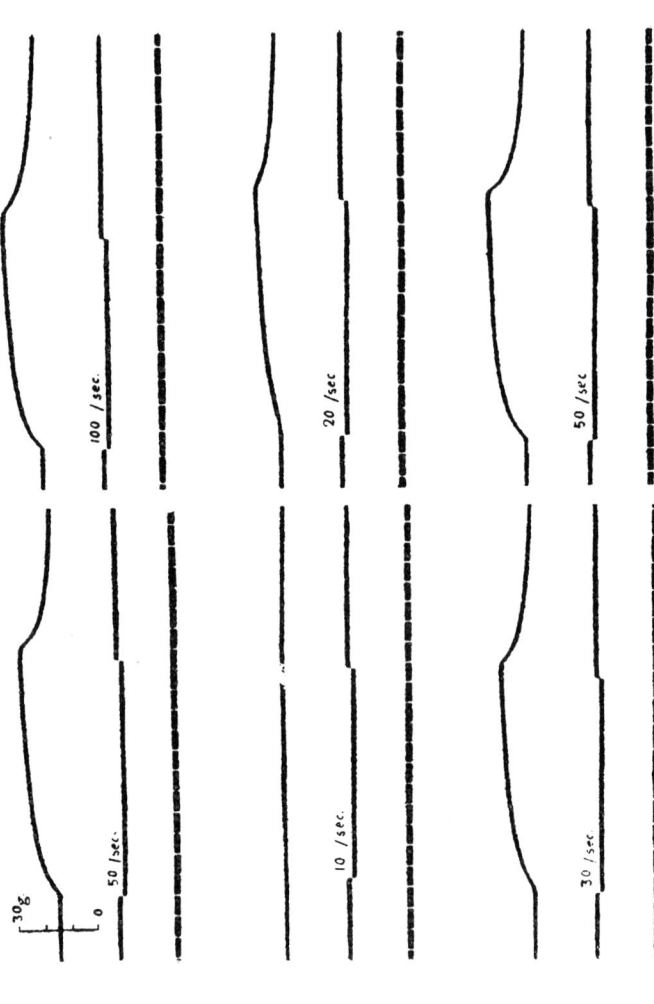

Fig. 1. Reproduction of Fig. 7 from Tasaki & Mizutani (1944). The upper trace of each section shows tension produced in a gastrocnemius muscle of a toad on stimulation of a single small diameter (5.3 μm) nerve fiber dissected from its motor nerve. This preparation developed "unusually great tensions" with a maximal value of about 14 g. Stimulus duration shown on middle trace, with statement of frequency; lower trace, 1-sec time marks. Temperature 22.4° C; at this temperature the tension developed by a fast unit on tetanic stimulation reaches its plateau in about 0.5 sec.

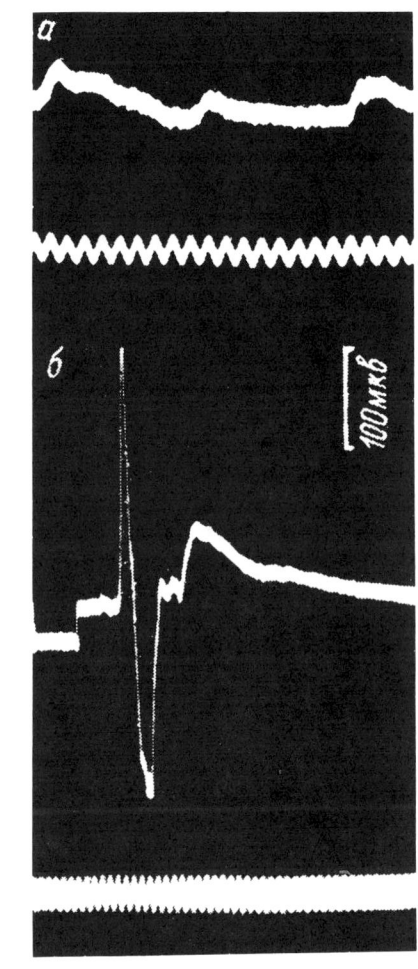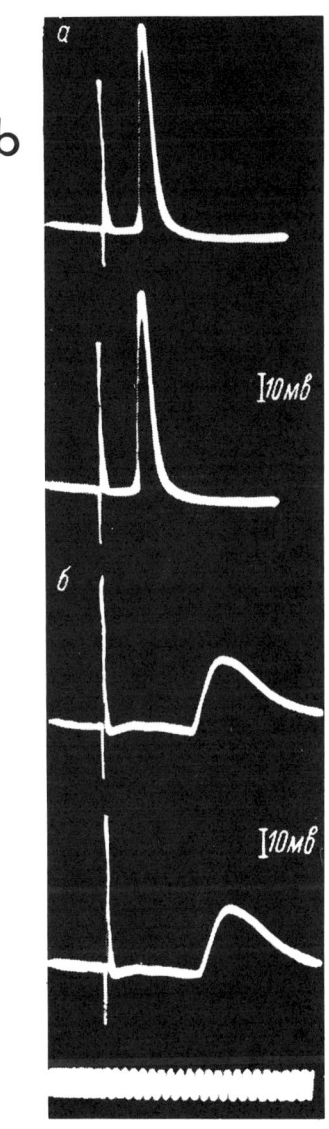

Fig. 2. Records of electrical activity of fast and slow
motor units in external eye muscles, from Matiushkin (1961).
Superior oblique muscle of rabbit. a. External recording
from the whole muscle. Upper section: spontaneous tonic
activity, time marker 500 Hz; lower section, response to
3-msec stimulus to nucleus of IV cranial nerve, time marker
2000 Hz. Scale bar, 100 μV. b. Intracellular records from a
fiber of a fast unit (two upper traces) and of a slow unit
(two lower traces). Stimulation of nucleus of IV cranial
nerve. Scale bar, 10 mV.

The problem was, how does it go so fast? One might expect that this mechanism might involve some transverse structure in the muscle. There are many suggestions in the old microscopic literature that the Z line, the dense line that separates each sarcomere of each fibril from the next, is in some sense a continuous structure across the whole muscle fiber. Each Z line of course exists separately in each fibril, but the whole muscle fiber is made up of a great many of these fibrils lined up with the striations in register, and there are many old indications that the Z lines within the fibrils are connected together across the fiber to form the structure that was referred to as Krause's membrane. Thus it seemed plausible that this structure, the Z lines plus their transverse connections, might be the thing which somehow conducted activity inwards.

Dr. R.E. Taylor and I thought we would test this by the experiment which is shown diagrammatically in Fig. 3. The potential difference across a small patch of the surface membrane of a frog muscle fiber is reduced by applying a negative electric pulse to the interior of a pipette whose tip is pressed against the fiber surface. The membrane potential

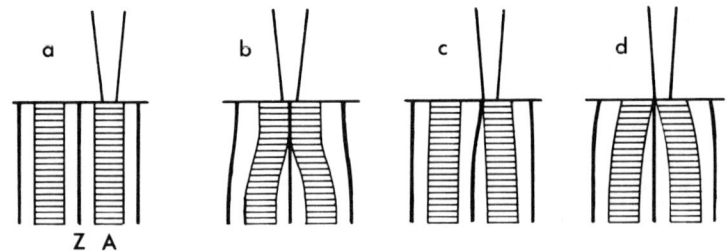

Fig. 3. *Diagrams illustrating local-activation experiments. An isolated muscle fiber is surrounded by Ringer solution and is observed with a light microscope using either polarized light or interference microscopy. The glass pipette in contact with the fiber surface is also filled with Ringer solution, and is connected to a circuit by means of which its interior can be made to go electrically negative by about 100 mV relative to the surrounding fluid. In all the types of muscle discussed here, there is no response to the electric change if the pipette is placed over the middle of an A band (a). b. The response given by a frog twitch fiber when the pipette is placed over a Z line (middle of I band). c,d. Responses when the pipette is slightly to one side of a Z line; c is the type of response given by fibers from crab and from lizard, and d is the type of response given by a frog twitch fiber.*

change is similar to that which occurs during an action potential, but it is confined to the area of membrane under the tip of the pipette, whose diameter is a fraction of the repeat distance of the striation pattern. A local contraction occurred only if the tip of the pipette covered the point where a Z line reached the fiber surface (Fig. 3*b*). We published this result in *Nature* with the title, "Function of Krause's Membrane" (Huxley & Taylor, 1955). It was, I think, a genuine indication that there is some inward conducting structure at that position, and we assumed that the structure in question actually was the Z lines and the connections between them, since no other transverse structures were known at that time.

We showed this evidence to the Physiological Society in 1955. Dr. J.D. Robertson, the electron microscopist, was then working at University College, London, where we gave this paper, and he produced from his pocket a slide showing an electron micrograph of a longitudinal section of vertebrate skeletal muscle [the same micrograph is published as Fig. 10 of a paper that appeared the next year (Robertson, 1956)]. The slide very clearly showed tubules running transversely in the spaces between the myofibrils. Now, tubes going across the fiber are very promising things for conducting something inwards from the surface membrane. But there was not just one at the Z line; there were two, in pairs, flanking each Z line. Robertson suggested that perhaps our pipettes were not small enough to resolve the two sets of tubes in this position.

To get better resolution, we repeated our local stimulation experiment on a crab muscle because it has much broader striations. When we applied a negative pulse with the pipette just to one side of the Z line, a contraction was produced, but it was a contraction of just half of the I band. The Z line was pulled across into contact with the A band, as shown diagrammatically in Figure 3*c*. Thus the contractile unit was half a sarcomere, just as though there were tubes going in, like those in Robertson's electron micrograph.

Of course, our study was of a crab, while Robertson's was of a lizard. Figure 3*d* illustrates what we found when we looked more carefully at the frog. We could not get the two halves of the I band to work separately, even if we put the edge of the pipette just over the Z line so as to activate one half much more than the other. The result was always symmetrical: the Z line stayed in the middle, implying a single inward conducting system at the level of Z, unlike the double set of tubes seen in Robertson's electron micrograph and suggested in our experiment on crab muscle. Now, as regards phylogenetic relationships, it is rather odd that a lizard should be more like a crab than like a frog.

More evidence became available in 1957, when Porter and
Palade (1957) published the first good electron microscope
study of the reticulum of muscle. In amphibian muscle, at
the level of Z, they found two large "lateral vesicles"
flanking a row of "intermediary vesicles," and they named this
composite structure the "triad." These triads were thus at
the right position to be responsible for the inward conduction
of activation that we had observed in frog muscle, but they
did not appear to be continuous across the fiber. The middle
element was a row of vesicles, and the outer vesicles did not
appear to be continuous. According to Porter and Palade's
description, there was continuity between another set of
vesicles that they found midway between successive Z lines,
but this was just the position where we did not find inward
spread in any kind of muscle.

This was partially clarified in 1959 (Huxley, 1959;
Andersson-Cedergren, 1959), when people began to see that the
"intermediary vesicles" were really a continuous tube which
either became very convoluted or was broken up into vesicles
on fixation with osmium tetroxide. With the introduction of
glutaraldehyde as a fixative, it became a matter of routine to
show the middle element of the triad as a continuous structure.
In frog muscle, and so far as is known, in all Amphibia, there
is a single set of tubes at this level, and there is no doubt
that these tubes do the inward conduction. Our suggestion
that "Krause's membrane" was the conducting structure turned
out to be a mistake, although in the frog the tubes are at
the same position in the striation pattern as Krause's
membrane. In many other animals, however, the tubes occur in
pairs on either side of the Z line, as Robertson had already
seen in a lizard.

In mammals, the tubes always seem to turn up in pairs, and
around 1960 it was believed that in each of the classes of
vertebrates there was one or the other arrangement, mammals
and reptiles having pairs, and amphibians having only single
sets. But recently it has been found that birds have both:
Dr. Page, one of the contributors to this symposium, showed
that even in the same chick, one muscle will have a single
set, and another will have pairs. Both arrangements are also
found in fishes.

Yet another variation was found by Peachey and Porter in
the muscle of *Amphioxus*, a primitive chordate, which consists
not of round fibers but of very thin sheets, each about 1 μm
thick. This muscle does not have a tubular system at all; it
does not need it because the distance from surface membrane
to contractile material is only a fraction of a micrometer,
short enough for diffusion to be sufficiently fast.

Although all our knowledge of the transverse tubular system has been gained with the electron microscope since about 1957, beginning with Porter and Palade, no one had any excuse for not knowing about it in advance. Professor H.S. Bennett, who with Porter took some of the earliest electron micrographs of this reticulum, rediscovered a paper of 1902 by Veratti, who applied the Golgi method to many types of skeletal muscle. A translation of this paper, with excellent reproductions of its very beautiful illustrations, was printed in the *Journal of Biophysical and Biochemical Cytology* in 1961. In every case (except for one figure, which I think must be a mistake) this paper shows transverse networks at the positions where the electron microscope shows transverse tubules in the same type of muscle.

About 1959, Peachey and I examined crab muscle with the electron microscope in the hope of finding a tubular system that would account for the separate activation of half-sarcomeres. We were surprised and puzzled to find a conspicuous set of tubules at the level of each Z line. A few years later, Peachey found additional sets of tubules on either side of Z; i.e. at the positions required for explaining the local-activation results in this animal, and these tubules (but not the ones at Z) made contact with vesicles of the sarcoplasmic reticulum. All three of these transverse networks in each sarcomere are clearly shown in Veratti's pictures. The function of the tubules at the level of the Z line is still completely unknown.

It is now fairly clear that wherever there are tubules which form contacts with vesicles of the reticulum, there is inward spread from the surface, and a local contraction occurs if just one of the tubules is activated at a time. This story now appears very straight-forward and consistent. But at the time when it was being worked out, it was a thoroughly confusing business. The differences between different kinds of animals were quite unexpected.

The nature of the connection between these tubules and the surface membrane was another question that caused a great deal of difficulty. The electron micrographs that showed the existence of continuous transverse tubules in frog muscle did not clear up this aspect of the matter. The fact that the tubules are open to the extracellular space was demonstrated in 1963 in several sets of experiments in different laboratories. Marker substances were added to the extracellular fluid, and with the appropriate type of microscope--fluroescence microscopy of living fibers in one case, and electron microscopy in the others--it was shown that the marker had entered the tubule systems. Even now, there are only very few published electron micrographs of

adult amphibian muscle (or of mammalian muscle) that directly
show the tubules as open to the extracellular space, although
the openings are readily seen in cardiac muscle and in muscles
from some fishes and from several kinds of arthropods. Here
again, however, the answer was lying unsuspected in the old
literature. A paper of 1897 by a Swede, Nyström, showed that
carbon particles could enter transverse structures in cardiac
muscle, and the famous histologist Holmgren confirmed this
and showed that the structures were at the level of Z. The
relevant illustration in Nyström's paper is reproduced in a
review article of mine (Huxley, 1971).

V. THE ACTION OF CALCIUM

After electrical change has spread inwards along these
tubules, the next step, of course, is the release of calcium,
which acts as the internal stimulus. The question of the
control of the contraction also went through many periods of
uncertainty. Well before World War II, Heilbrunn and others
had indications that calcium might be the key factor. But
after the existence of a relaxing factor had been recognized,
in Marsh's famous work of 1951 (Marsh, 1951; 1952), one
theory after another identified different soluble factors.
It was not until 1960 that Ebashi, and soon after, Hasselbach,
Weber, and others demonstrated the action of the reticulum
in reabsorbing calcium and thereby producing relaxation.
Again, these phenomena now have a very clear-cut explanation
after being extremely obscure for a period of ten years or so.
Another great surprise came in 1963, when Ebashi found
that he could isolate a factor from muscle that conferred
this sensitivity to calcium on actomyosin preparations which
had lost their sensitivity during purification or after
storage. This finding opened up another tremendous field,
and again everything looked very clear-cut: the troponin-
tropomyosin system, on the thin filaments, is sensitive to
calcium and regulates the interaction with the thick
filaments. Again, there followed a period of a year or two
when everybody thought that all muscles were controlled in
this way. Then Andrew Szent-Györgyi found that in *Pecten*, a
mollusc, the contraction is indeed regulated by calcium, but
troponin is not present, and the action of calcium is not on
the thin filament at all, but on the thick filament. He
showed that in many other groups of animals both types of
control exist. It is still controversial whether, in addition
to the control through troponin and tropomyosin on the thin

filament, there may be another control working on the myosin
in the thick filament of vertebrate muscle. This has not
been proved, but also it has not yet been definitely excluded.
    Recently there was yet another surprise in this field, in
smooth muscle. It has been clear for a good many years that
calcium is the activator. Just a little over a year ago,
scientists in three different laboratories found that the
ATPase activity of smooth muscle actomyosin is greatly
increased if one phosphorylates the myosin--an effect similar
to that already found by Adelstein & Conti (1975) with
platelet myosin (Adelstein, Chacko, Barylko, Scordilis &
Conti, 1976; Chacko, Conti & Adelstein, 1977; Aksoy, Williams,
Sharkey & Hartshorne, 1976; Gorecka, Aksoy & Hartshorne, 1976;
Sobieszek & Small, 1976). A specific enzyme for this
phosphorylation is itself highly sensitive to calcium. There
is a strong hint that in smooth muscle, the action of
calcium on the contractile system is not direct but, rather,
the calcium activates this protein kinase, which
phosphorylates the myosin. When the myosin is phosphorylated,
its interaction with actin is stimulated, and this causes
contraction. The myosin is not the only contractile protein
that can be phosphorylated; two of the components of
troponin, troponin-I and troponin-C, can also be
phosphorylated by sepcific kinases, and we shall be hearing
more about these phosphorylation processes in the regulation
of cardiac muscle during this symposium. Thus calcium
activates contraction in all these different tissues, but in
three totally different ways.
    There is yet a fourth totally different calcium
mechanism. I mentioned earlier that at the turn of the
century, the contractile thread of *Vorticella* appeared to be
an ideal model for muscle contraction, but that in the last
twenty years it had been shown, first by Hoffmann-Berling
(1958), and then in much more detail by Weis-Fogh & Amos
(1972), that the contraction mechanism is totally unlike
that of muscle. It is not a sliding filament system at all,
but it works by a mechanism that was popular as a theory of
muscle around 1950. The contractile material is a protein
with rubbery elasticity. When the protein is activated, it
is plasticized; that is to say, it becomes free to shorten
by thermal agitation more than it can when it is in the
resting state. This plasticization is achieved by binding of
calcium ions. So again, calcium causes contraction, but not
by activating an interaction between two proteins, which
requires ATP, as it seems to do, directly or indirectly in
all muscles. In this case, however, the calcium itself does
the work. Its binding to the contractile protein becomes
stronger when the protein shortens, so that dissociation of

the calcium, and therefore relaxation, only occurs if the
free calcium concentration in the surrounding fluid is
reduced below the level needed to activate the system. The
net effect therefore is that some calcium is transferred from
a more concentrated to a more dilute solution, and this is
the source of free energy which drives the contraction--a
purely entropic system. This example is yet another warning
against assuming that things that look similar at first
sight are really essentially the same.

VI.   THE ARRANGEMENT OF FILAMENTS

     After the calcium action is completed, the actual sliding
process takes place. Fig. 4*a*, based on Hugh Huxley's work
of 1953, shows the arrangement of filaments that has been
found in all vertebrate muscles; there is a thin filament
at the center of each triangle formed by three of the thick
filaments. At that time, probably everyone involved in
muscle would have supposed that this filament arrangement was
the same in all striated muscles. But it was found around
1960 that in the highly specialized flight muscles of insects,
the arrangement is different, as shown in Fig. 4*b*. Here there
is a thin filament midway between each two thick filaments.
Each thin filament is not at the center of one of the

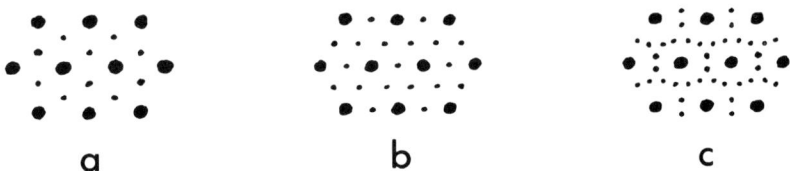

a   b   c

Fig. 4. Diagrams of cross-sections of zones of overlap
of thick and thin filaments in various types of muscle.
a.   In all vertebrate muscles so far investigated, there is
one thin filament at the center of each triangle formed by
thick filaments.   b.   Arrangement with a thin filament at the
center of each line joining two adjacent thick filaments,
found in the asynchronous flight muscles of certain insects,
and also in fast fibers of some crustaceans.   c.   Arrangement
with 12 thin filaments around each thick filament, found in
many slow fibers of insects and crustaceans; in other cases a
less regular arrangement with less than 12 thin filaments
around each thick filament is found.

triangles, but at the center of the line between two thick
filaments. This arrangement is also found in fast muscles of
crustaceans, as shown by Jahromi & Atwood (1969). Fig. 4c
shows a different arrangement again: in the slow muscles of
arthropods, each thick filament is surrounded by up to a
dozen thin filaments. So here are three widely different
filament arrangements that are found in different parts of
the animal kingdom.

Among vertebrates, the lengths of these filaments are
remarkably uniform. There are small differences as regards
the length of the thin filament, but the thick filament, as
far as is known, has the same length in all muscles and in
all species of all the vertebrate classes. But crabs are
very different. Dr. Clara Franzini-Armstrong (1970) found a
surprising degree of variability in the lengths of thick
filaments in muscle from a crab. Even within a single fiber
there was variation between the filaments of a single *A* band
in one myofibril, between adjacent *A* bands in the same
myofibril, and between adjacent myofibrils, as well as the
long-known variation between fibers and between different
muscles. There is even a strong suggestion (Dewey, Levine &
Colflesh, 1973) that in the muscles of *Limulus*, the horseshoe
crab, the length of the thick filaments diminishes
substantially during shortening of the muscle, in contrast
to the constancy of length of the *A* bands found in vertebrate
muscle.

VII. CONCLUSION

I have gone through these different aspects of
contractility in order to emphasize that it has turned out
that there is a quite unforeseen degree of variation between
different contractile systems, between muscles of different
animals, and even between different muscles of a single
animal, in many different respects: the action potential,
the speed of the contraction, the mechanism of inward spread,
the regulatory processes, the filament arrangement, and even
the contraction mechanism itself.

There is an expectation of uniformity between different
animals, based on evolution. Common ancestry is undoubtedly
the reason why similar mechanisms are found in different
animals and different organisms. But there is another aspect
to evolution, which is that it has produced the
extraordinarily great diversity of creatures that exist.
Sometimes there is a tendency to forget this second aspect

of evolution. An *a priori* argument from evolution can be
used in either direction: one can say either that things
will be the same because of common ancestry, or that they
will be different because they have adapted in some way to
the needs of the situation by natural selection.

There is a tendency to assume that if a feature is
fundamental, it must be the same everywhere. I regard this
as a statement which is true but has zero information content.
It is not a statement of fact; it is a definition of the word
"fundamental." Suppose that we found that in all places where
motility exists, it works by a sliding filament mechanism.
We should then say that sliding filaments are fundamental to
all biological contractility. We have seen, however, that
this is not so, so we cannot say that they are "fundamental."

There is perhaps also a second criterion for calling some
process or structure fundamental, namely, a judgment that it
is important for all, or at least many, kinds of organisms.
But whichever definition one uses, there are very severe
limitations on the usefulness of the concept "fundamental."
At the stage of an investigation where one would really like
to know whether something that has been discovered in one
contractile situation is true of others, of course one does
not know whether this feature is universal. So, if one wants
to judge whether it is truly fundamental and to be expected
everywhere, the first criterion cannot be used because other
places have not yet been investigated. The second criterion,
being a matter of personal judgment, is totally unreliable.
The proposition of uniformity of nature has a sort of
half-truth quality, and investigating the degree to which it
is true is valuable because it provides evidence about the
course of evolution: the way in which evolutionary
relationships are traced is by finding the extent to which
mechanisms and structures are similar in different organisms.
But the predictive value of propositions based on an
expectation of uniformity is very small indeed. I think there
is a close parallel between arguments of this kind, which
depend on the common ancestry aspect of evolution, and
teleological arguments, which are really based on the
adaptive and selective aspect of evolution. From the point
of view of an experimentalist, both are *a priori* arguments,
and neither is any substitute for actual experimental
evidence. In each case we must find what really is the
mechanism. When we know, we may be able to look back and
decide whether this is something that evolved for a particular
purpose, or whether it exists because it derived from some
more primitive organism. Nevertheless, these arguments are
valuable for the experimentalist. They cannot tell us what
the answers will be, but what they can and do suggest, in a
most valuable way, is what experiments we ought to do next.

REFERENCES

1. Adelstein, R.S., Chacko, S., Barylko, B., Scordilis, S.P. & Conti, M.A. (1976). In *Contractile Systems in Non-Muscle Tissues*, ed. Perry, S.V. *et al.*, pp. 153-163. Elsevier/North Holland Biomedical Press.

2. Adelstein, R.S. & Conti, M.A. (1975). Phosphorylation of platelet myosin increases actin-activated ATPase activity. *Nature, Lond. 256*, 597.

3. Aksoy, M.O., Williams, D., Sharkey, E.M. & Hartshorne, D.J. (1976). A relationship between Ca^{2+} sensitivity and phosphorylation of gizzard actmyosin. *Biochem. biophys. Res. Commun. 69*, 35.

4. Andersson-Cedergren, E. (1959). Ultrastructure of motor end plate and sarcoplasmic components of mouse skeletal muscle fiber as revealed by three-dimensional reconstructions from serial sections. *J. Ustrastr. Res., Suppl. 1.*

5. v. Baer, K.E. (1828-37). *Ueber Entwicklungsgeschichte der Thiere.*

6. Bernstein, J. (1868). Ueber den zeitlichen Verlauf der negativen Schwankung des Nervenstroms. *Arch. ges. Physiol. 1*, 179.

7. Bozler, E. (1942). The action potentials accompanying conducted responses in visceral smooth muscles. *Am. J. Physiol. 136*, 553.

8. Burdon-Sanderson, J. & Page, F.J.M. (1878). *Proc. R. Soc. 27*, 410.

9. Burdon-Sanderson, J. & Page, F.J.M. (1880). On the time-relations of the excitatory process in the ventricle of the heart of the frog. *J. Physiol. 2*, 384.

10. Chacko, S., Conti, M.A. & Adelstein, R.S. (1977). Effect of phosphorylation of smooth muscle myosin in actin activation and Ca^{2+} regulation. *Proc. natn. Acad. Sci. U.S.A. 74*, 129.

11. Dewey, M.M., Levine, R.J.C. & Colflesh, D.E. (1973). Structure of *Limulus* striated muscle. *J. cell Biol. 58*, 574.

12. Ebashi, S. (1960). Calcium binding and relaxation in the actomyosin system. *J. Biochim. 48*, 150.

13. Ebashi, S. (1963). Third component participating in the superprecipitation of "natural actomyosin." *Nature, Lond. 200*, 1010.

14. Fatt, P. & Ginsborg, B.L. (1958). The ionic
     requirements for the production of action potentials
     in crustacean muscle fibres. *J. Physiol. 142*, 516.
15. Franzini-Armstrong, C. (1970). Natural variability in
     the length of thin and thick filaments in single fibers
     from a crab, *Portunus depurator*. *J. cell. Sci. 6*, 559.
16. van Gehuchten, A. (1886). *La Cellule 2*, 289.
17. Gorecka, A., Aksoy, M.O. & Hartshorne, D.J. (1976).
     The effect of phosphorylation of gizzard myosin on
     actin activation. *Biochem. biophys. Res. Commun. 71*,
     325.
18. Hodgkin, A.L. & Katz, B. (1949). The effect of sodium
     ions on the electrical activity of the giant axon of
     the squid. *J. Physiol. 108*, 37.
19. Hoffmann-Berling, H. (1958). Der Mechanismus eines
     neuen, von der Muskelkuntraktion verschiedemen
     Kontraktionszyklus. *Biochim. biophys. Acta 27*, 247.
20. Huxley, A.F. & Taylor, R.E. (1955). Function of
     Krause's membrane. *Nature, Lond. 176*, 1068.
21. Huxley, A.F. (1959). Local activation of muscle.
     Ann. N.Y. Acad. Sci. 81, art 2, 446.
22. Huxley, A.F. (1971). The activation of striated muscle
     and its mechanical response. *Proc. R. Soc. B, 178*, 1.
23. Huxley, H.E. (1953). Electron microscope studies of
     the organization of filaments in striated muscle.
     Biochim. biophys. Acta 12, 387.
24. Jahromi, S.S. & Atwood, H.L. (1969). Correlation of
     structure, speed of contraction, and total tension in
     fast and slow abdominal muscle fibers of the lobster
     (*Homarus americanus*). *J. exp. Zool. 171*, 25.
25. Krüger, P. (1952). *Tetanus und Tonus der quergestreiften
     Skelettmuskeln der Wirbeltiere und des Menschen*.
26. Marsh, B.B. (1951). A factor modifying muscle fibre
     syneresis. *Nature, Lond. 167*, 1065.
27. Marsh, B.B. (1952). The effects of adenosine
     triphosphate on the fibre volume of a muscle homogenate.
     Biochem. biophys. Acta 9, 247.
28. Matiushkin, D.P. (1961). Phasic and tonic neuro-motor
     units in the oculomotor system of the rabbit
     (experiments with intracellular potential derivation).
     Fiziol. Zh. SSSR. 47, 878; translated in *Sechenov
     Physiological Journal of the USSR*, Jan. 1962, pp. 960-965,
     New York: Pergamon.
29. Melland, B. (1885). *Q. Jl. microsc. Sci. 25*, 371.

30. Nyström, G. (1897). Ueber die Lymphbahnen des Herzens. *Arch. anat. Physiol. (Anat. Abt.)*, 361.

31. Porter, K.E. & Palade, G.E. (1957). Studies on the endoplasmic reticulum. III. Its form and distribution in striated muscle cells. *J. biophys. biochem. Cytol.* *3*, 269.

32. Ramón y Cajal, S. (1888). *Int. J. Anat. Physiol.* *5*, 205, 253.

33. Robertson, J.D. (1956). Some features of the ultrastructure of reptilian skeletal muscle. *J. biophys. biochem. Cytol.* *2*, 369.

34. Sobieszek, A. & Small, J.V. (1976). Myosin-linked calcium regulation in vertebrate smooth muscle. *J. molec. Biol.* *102*, 75.

35. Tasaki, I. & Mizutani, K. (1944). Comparative studies on the activities of the muscle evoked by two kinds of motor nerve fibers. Part I, Myographic studies. *Jap. J. med. Sci., III. Biophys.*, *10*, 237.

36. Veratti, E. (1961). Investigations on the fine structure of striated muscle fiber. *J. biophys. biochem. Cytol. Suppl. 10.*

37. Weis-Fogh, T. & Amos, W.V. (1972). Evidence for a new mechanism of all motility. *Nature, Lond.* *236*, 301.

II

Gating Processes in Excitable Membranes

INTRAMEMBRANOUS CHARGE MOVEMENT AND CONTROL
OF CELLULAR FUNCTIONS BY MEMBRANE VOLTAGE

Clay Armstrong

Department of Physiology
University of Pennsylvania
Philadelphia, Pennsylvania

INTRODUCTORY REMARKS

Membrane voltage is known to control a number of cellular functions, and two of the best examples are provided by nerve and muscle. In both tissues it is the influence of voltage on ionic pores that underlies transmission of action potentials. In muscle, membrane voltage has the additional function of controlling the contractile machinery, which is activated when voltage exceeds a threshold of about −50 mV.

It was clearly realized by Hodgkin & Huxley (1952) that a change of membrane voltage might exert its influence on ionic permeability by causing rearrangement of charged or dipolar gating structures, and that movement of gating charge would generate a detectable "gating current" as does any charge moving through the membrane. If the gating charge were confined to the membrane, the gating current that followed a change of membrane voltage would be transient and would cease when the gating structures reached an equilibrium at the new voltage. Gating current would thus have properties of a capacitative current: a transient flow following a voltage change. Viewed in this way, the gating structures would be a slow and nonlinear component of the membrane dialectric; slow because rearrangement of the gating structures requires an appreciable fraction of a millisecond, and nonlinear because they move primarily when membrane voltage is altered within the gating range from roughly −70 to +20 mV.

The gating structures would thus perform the dual function of sensing the membrane voltage and controlling ionic movements. There are two other ways worth mentioning in which voltage sensing might occur. The first is that detection might be electromagnetic. A rapid change of the membrane field produces a magnetic field within the membrane and conceivably some membrane element might detect it. Such an element, however, would be sensitive not to membrane voltage but to its derivative with respect to time. There is much evidence that voltage rather than its time derivative is sensed, so this possibility can be safely discarded. The second possibility is exemplified by the Ca plug theory which was examined by Frankenhauser & Hodgkin (1956). According to this theory, Ca ion from the external solution can move part way through the membrane to a site where it blocks ion movement. The equilibrium between Ca in solution (at fixed concentration) and Ca in the blocking sites would be sensitive to membrane voltage, such that making voltage more positive would move Ca out of the membrane and increase permeability. This theory was discarded by Frankenhauser and Hodgkin for quantitative reasons, and it is now thought that calcium exerts its effect by neutralizing negative charges that cover the outer surface of the membrane (suggestion of A.F. Huxley, quoted in Frankenhauser & Hodgkin, 1956; Hille, Woodhull & Shapiro, 1975). But it is worth noting that the charge carried by Ca moving to and from its blocking site would be a form of intramembranous charge movement, and Ca movement would generate a current with many of the properties of "gating current" as described above.

Whatever the details of mechanism, it seemed clear that gating current or something must exist in all cases where a cellular function is under control of membrane voltage. The first experimental evidence for such a current came from muscle fibers (Schneider & Chandler, 1973), and it seems likely that this current is associated with a step in excitation-contraction coupling. Dr. Chandler describes similar studies performed on slow (non-twitch) muscles from the frog. Observation of gating current in nerve fibers was made soon after (Armstrong & Bezanilla, 1973) and quickly confirmed (Keynes & Rojas, 1973). My paper will describe some of the properties of the gating current associated with Na channels in nerve membrane, and the information that gating current gives about the organization of the channels. No gating current associated with K channels has yet been observed in nerve fibers, but Dr. Adrian discusses evidence for a K channel gating current in skeletal muscle fibers.

Dr. Hille's paper is concerned not with gating current but with interaction between antiarrythmic agents and the gates of Na channels. The observations he described provide an excellent basis for understanding how these drugs aid in controlling irregularities of the heartbeat.

REFERENCES

1. Armstrong, C.M. & Bezanilla, F. (1973). Currents related to movement of the gating particles of the sodium channels. *Nature 242*, 459-461.
2. Frankenhaeuser, B. & Hodgkin, A.L. (1957). The action of calcium on the electrical properties of squid axons. *J. Physiol. 137*, 218-244.
3. Hille, B., Woodhull, A.M. & Shapiro, B. (1975). Negative surface charge near sodium channels of nerve: divalent ions, monovalentions and pH. *Phil. Trans. R. Soc. Lond. B 270*, 301-318.
4. Hodgkin, A.L. & Huxley, A.F. (1952). A quantitative description of membrane current and its application to conduction and excitation in nerve. *J. Physiol. 117*, 500-544.
5. Keynes, R.D. & Rojas, E. (1973). Characteristics of the sodium gating current in the squid giant axon. *J. Physiol. 233*, 28-30P.
6. Schneider, M.F. & Chandler, W.K. (1973). Voltage dependent charge movement in skeletal muscle: A possible step in excitation-contraction coupling. *Nature 242*, 244-246.

ELECTRICAL PROPERTIES
OF AMPHIBIAN SLOW MUSCLE FIBERS

W.K. Chandler
W.F. Gilly
C.S. Hui

Department of Physiology
Yale University School of Medicine
New Haven, Connecticut

INTRODUCTION

The frog skeletal muscle system contains two distinct
fiber types, twitch and slow, which differ qualitatively in
many ways (see reviews by Peachey, 1968; Hess, 1970;
Lännergren, 1975). For example, a twitch fiber responds to
nerve input at one or two end-plate regions by the
propagation of an action potential, leading to a brief
contraction or twitch.

A frog slow fiber, on the other hand, will not conduct
an action potential. Instead, depolarization spreads
electrotonically from multiple synapses distributed along
the fiber length (Kuffler & Vaughan Williams, 1953a;
Burke & Ginsborg, 1956a,b). Sodium channels, if present at
all, are not abundant. The membrane resistance, and, hence,
time constant are large (Adrian & Peachey, 1965; Stefani &
Steinbach, 1969); and this would increase the duration of
the postsynaptic response. In this way the strength of
contraction can be regulated efficiently, though somewhat
slowly (Kuffler & Vaughan Williams, 1953b).

These examples are only a few of the many distinctions
that can be made between twitch and slow fiber types, and
some interesting comparative physiological questions can

be asked. For example, what sort of ionic channels are present, and are they similar to those seen in twitch fibers? Do brief depolarizations, sufficiently large to activate contraction in twitch fibers, activate the slow fiber contractile response, or are long-lasting depolarizations required? Are charge movements, present in twitch fibers and possibly involved in excitation-contraction coupling, also present in slow fibers? Experiments using the voltage clamp technique have been carried out by two of us (W.F.G. & C.S.H.) in an attempt to answer questions such as these. Preliminary reports have appeared (Hui & Gilly, 1977; Gilly & Hui, 1977a,b; 1978).

METHODS

Pyriformis muscles were dissected from English frogs maintained at room temperature. The anterior edge of the muscle's dorsal surface, containing both slow and twitch fibers, was positioned under a water immersion objective. Slow fibers could usually be recognized microscopically by their zig-zag Z lines and by the absence of lipid droplets (see also Smith & Ovalle, 1973). Positive identification was based on the presence of high input resistance (usually greater than 10 MΩ) and the absence of early inward current. Membrane currents were measured using the three microelectrode voltage clamp technique (Adrian, Chandler & Hodgkin, 1970); contractile activation was studied with the two microelectrode voltage clamp (Adrian, Chandler & Hodgkin, 1969; Costantin, 1974). Strength duration measurements were carried out in normal Ringer: NaCl = 115 mM, KCl = 2.5 mM, $CaCl_2$ = 1.8 mM, pH = 7.0 buffered with 1.5 mM phosphate. In other experiments changes in solution were made as indicated.

MEMBRANE CURRENTS

Fig. 1a shows records of membrane current under voltage clamp. For these experiments movement was minimized by using D_2O instead of H_2O and adding sucrose to double the osmolality. The currents associated with the hyperpolarization (-150 mV) and the smallest depolarization

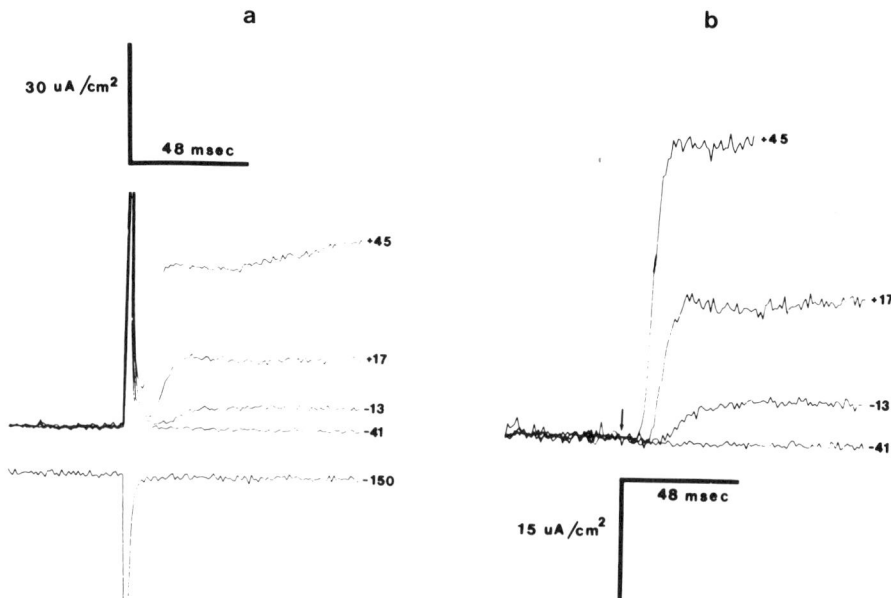

Fig. 1. *Membrane currents in slow fibers. (a) currents recorded using the three microelectrode voltage clamp technique. (b) currents corrected for capacitative and leakage components, obtained by appropriately scaling the current at -150 mV and adding to the currents recorded during depolarizing steps. The records in b have been terminated at the end of the plateau and before the rise of the slower phase, most prominantly shown in the record of +45 mV in part a. The bathing solution contained D_2O instead of H_2O, was made hypertonic by adding 230 mM sucrose, contained 11.8 mM Ca, and was buffered with 1 mM PIPES, 10° C. Electrode spacing $l_1 = l_2 = 156$ μm; fiber capacitance (C_m) 6.4 μF/cm^2; 1 μA/cm^2 corresponds to 0.100 mV on the ΔV trace.*

(-41 mV) were similar and showed an early capacitative transient followed by a small maintained current. The capacitative component corresponds to the current required to charge the capacitance of the surface and tubular membranes; the maintained component represents the membrane ionic current plus any shunt current around the V_1 electrode (Appendix A in Schneider & Chandler, 1976).

Fig. 1b shows records obtained after subtracting the capacitative transient and leakage current. None of the records shows any indication of inward ionic current (Stefani & Steinbach, 1969; Miledi, Stefani & Steinbach, 1971; Stefani & Uchitel, 1976). Records taken at -13 mV and more positive voltages show evidence of a time dependent outward current which develops during depolarization. This current resembles delayed rectification seen in twitch fibers or squid axons in that it develops along a sigmoid time course (Fig. 1b); it has a reversal potential which is close to V_K (Stefani & Steinbach, 1969) and is sensitive in the expected manner to external potassium concentration (not shown); it is blocked by TEA (tetraethylammonium) ions (Fig. 5), in a manner previously reported for twitch fibers (Stanfield, 1970).

Fig. 2a shows an *I-V* plot of the final currents from the experiment in Fig. 1b. The currents can be converted to conductances by dividing by the driving force, $V-V_K$, where V_K is the reversal potential for the channel. Fig. 2b shows the results. The curved line was calculated from equations appropriate for twitch fibers (Adrian, Chandler & Hodgkin, 1970) but using a smaller value of $\bar{g}_K$, 0.6 mS/cm^2. It is clear that g_K varies less steeply with voltage in slow fibers than in twitch. The value for $\bar{g}_K$ was chosen somewhat arbitrarily to agree with the data around +80 mV. The average value of $\bar{g}_K$ in 7 slow fibers was 0.5 mS/cm^2 (range 0.2 - 0.9 mS/cm^2). The main point in comparing potassium conductances in slow and twitch fibers is that $\bar{g}_K$ is an order of magnitude smaller in slow fibers. Values found in twitch fibers are 8.5 - 20 mS/cm^2 (Almers, 1976), and 6 - 20 mS/cm^2 (Gilly & Hui, unpublished, obtained using the same hypertonic D$_2$O solution).

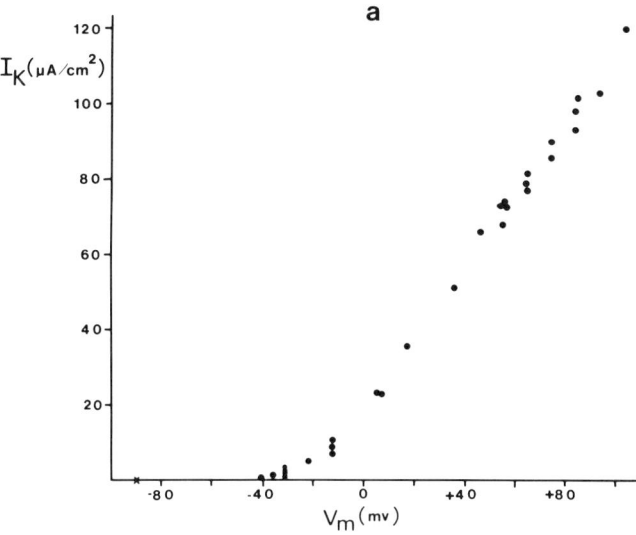

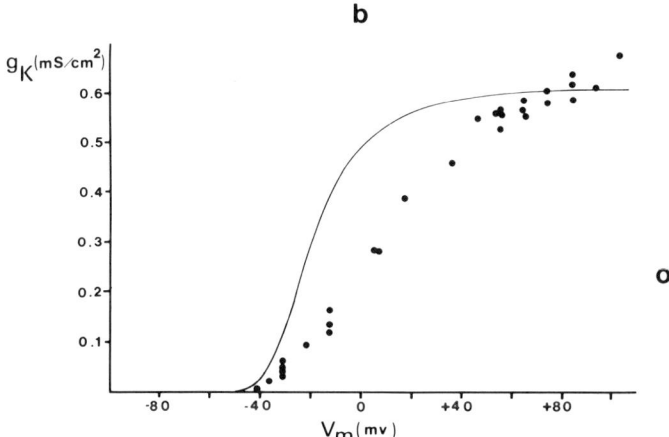

Fig. 2. *Current and conductance as a function of membrane potential. (a) final currents from the experiment in Fig. 1b. (b) conductance vs. voltage, obtained from a using V_K = -75 mV. The curve was calculated using the equation for twitch fibers (Adrian, Chandler & Hodgkin, 1970) with $\bar{g}_K$ = 0.6 mS/cm^2, but shifted 7 mV to the right to allow for the elevated Ca (Costantin, 1968).*

MECHANICAL ACTIVATION

   In an attempt to learn whether short depolarizations can
activate contraction in slow fibers, as they do in twitch
(Adrian, Chandler & Hodgkin, 1969), strength-duration curves
were measured in both fiber types in the manner described by
Costantin (1974). Fig. 3 shows mean results at 9° and 20° C.
The open symbols are from the twitch fibers, the filled
symbols from slow. It is clear that brief depolarizing steps
are equally effective in eliciting threshold contractions in
both fiber types.
   Another important observation is that both twitch and slow
fibers show similar strength-duration curves when external
calcium is removed and 10.8 mM $MgCl_2$ plus 2.5 mM EGTA is
added (not shown).

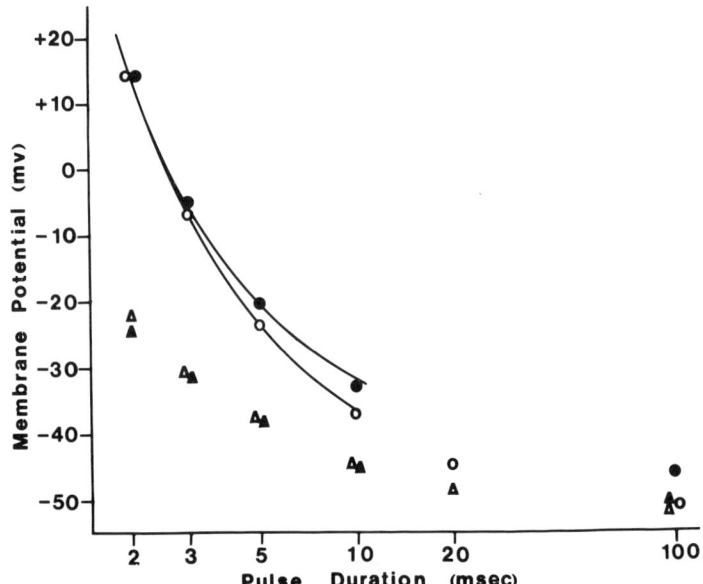

*Fig. 3. Strength-duration curves for twitch and slow
muscle fibers. For a given pulse duration, the threshold
membrane potential was taken as the minimum value at which
discrete sarcomere shortening was visually detected. Solid
circles (twitch fibers) and open circles (slow fibers) taken
at 9° C; solid triangles (twitch fibers) and open triangles
(slow fibers) taken at 20° C. From Gilly & Hui (1977a).*

Although the requirements for mechanical activation by
brief depolarizations are similar in slow and twitch fibers,
the decay of excitability is different. This is illustrated
by the two-pulse experiment in Fig. 4. A brief subthreshold
pulse V_1 was followed by a second pulse V_2 after a period Δt
of repolarization (see inset in upper right of Fig. 4). The
magnitude of the voltage V_2 which was required to elicit a
just visible contraction was used to estimate the level of
mechanical excitability which remained after the period Δt.
Fig. 4a shows results from one twitch fiber (open circles)
and two slow fibers (filled symbols). In the twitch fiber
the decay of excitability was rapid, being 97% complete by
10 msec (see also Costantin, 1974). Slow fibers also showed
an initial rapid decay, but the process was incomplete. The
final fraction, 0.1-0.2, decayed with a time constant
greater than 0.3 sec, Fig. 4b. This final time constant is
similar to that associated with the relaxation of tension,
determined in a voltage-clamped short cruralis slow fiber
(not shown).

These findings are consistent with the idea that the
early stages in activation leading to calcium release may be
similar in both fiber types, but that the uptake of calcium
by the SR (sarcoplasmic reticulum) proceeds at a greatly
reduced rate in slow fibers, possibly because slow fibers
have less SR than twitch fibers (Page, 1965).

CHARGE MOVEMENT

Intra-membranous charge movements have been described in
twitch fibers and are possibly involved in the activation of
contraction (Schneider & Chandler, 1973) or the gating of
potassium channels (Adrian & Perez, 1977). In either case
it is of some interest to find out whether similar currents
are present in slow fibers. Fig. 5 shows an experiment
designed to examine this point. In this experiment movement
was blocked with tetracaine (Almers & Best, 1976).

Records on the left show membrane current arising from
charge movements. This component of current was measured in
the manner described by Chandler, Rakowski & Schneider (1976),
and the transient currents resemble those seen in twitch
fibers. The "on" and "off" areas are similar in magnitude,
and a plot of charge against voltage follows a sigmoid curve
(right side of Fig. 5) drawn according to

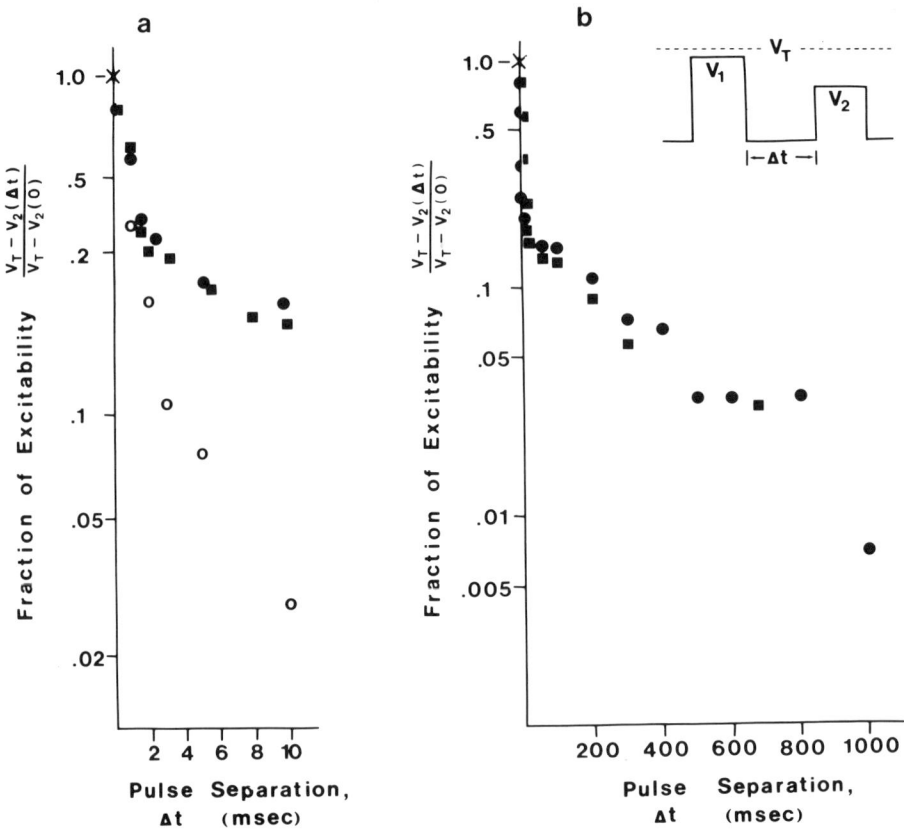

Fig. 4. *Decay of mechanical excitability after a brief*
subthreshold pulse. Inset at upper right shows the protocol.
A just subthreshold pulse V_1 *(3-5 msec duration) was followed*
by a period Δt *of repolarization. The voltage* V_2 *during a*
test pulse (same duration as V_1*) was adjusted to give a just*
visible contraction. The fraction of mechanical excitability
remaining after the period Δt *was estimated according to the*
relation $[V_T - V_2(\Delta t)]/[V_T - V_2(0)]$. V_T *is the usual*
threshold membrane potential for a single pulse of the same
duration as V_1 *and* V_2*. Open circles, twitch fiber; solid*
circles and squares, slow fibers; crosses, points at $\Delta t = 0$
shared by all fibers; $9°$ *C. Parts* a *and* b *are plotted on*
different time scales. From Gilly & Hui (1977a).

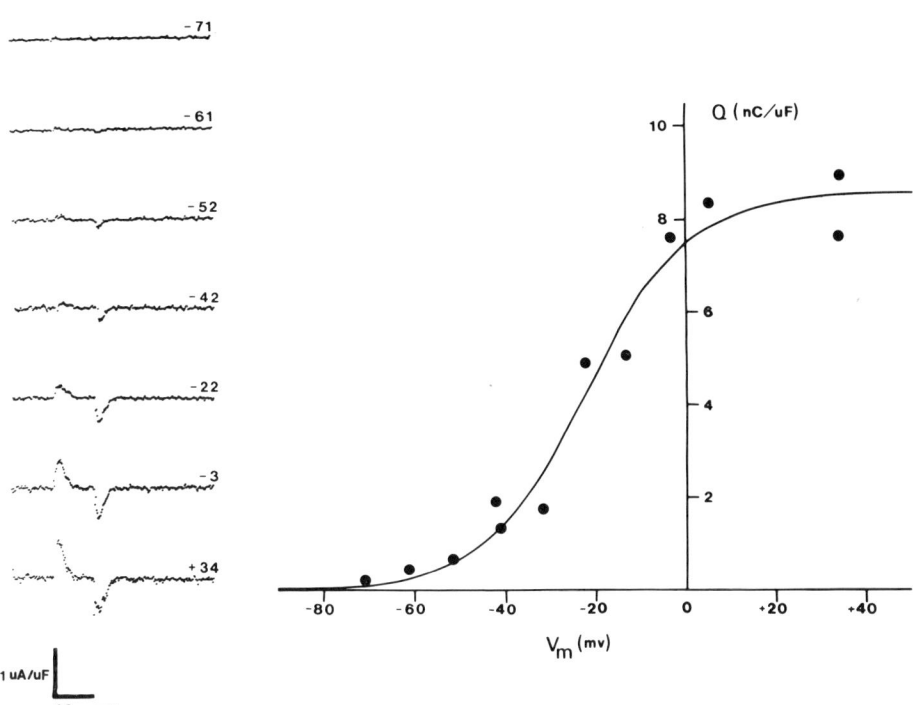

Fig. 5. *Charge movement currents in slow fibers. The records on the left show charge movement currents (averages of 4-16 traces) recorded at the potentials indicated. The bathing solution contained 11.8 mM $CaCl_2$ (to reduce leakage around electrodes), 2 mM tetracaine (to block movement), and TEA-Cl instead of NaCl (to block potassium currents); 4.5° C. Electrode spacings $l_1 = l_2 = 273$ µm; fiber diameter = 72µ; $C_m = 6.3$ µF/cm^2; 1 µA/µF corresponds to 1.05 mV ΔV. The plot on the right shows charge Q plotted against membrane potential V_m. The curve is drawn according to equation (1) with $Q_{max} = 8.6$ nC/µF, $\bar{V} = -22$ mV, k = 11.5 mV.*

$$Q = \frac{Q_{max}}{1 + \exp - \dfrac{(V - \bar{V})}{k}} \tag{1}$$

an equation which has been used for twitch fibers (Schneider
& Chandler, 1973; Adrian & Almers, 1976; Chandler *et al*,
1976). Q_{max} is the total amount of charge, $\bar{V}$ is the voltage
for $Q = 0.5 \, Q_{max}$, and k is the steepness factor.

Table 1 lists values of Q_{max}, $\bar{V}$, and k from slow and
twitch fibers. The values of $\bar{V}$ and k for the two fiber types
are similar, but Q_{max} in slow fibers is about 0.3 times the
value in twitch (see also Adrian & Almers, 1976). Charge
movement currents are therefore present in slow fibers, but
are reduced in amplitude.

CONCLUSIONS

Ionic channels in slow fibers seem fewer in number than
in twitch fibers. The resting resistance is high, of the
order of 10^5 Ωcm^2 or more, and there is no evidence of
ingoing rectification. Sodium channels, if present at all,
are sparsely distributed and the delayed potassium
conductance is probably at most 0.1 times the value in
twitch fibers. These properties are consistent with the
absence of an action potential and the reliance on passive
electrical spread to activate contraction.

The finding that contraction can be activated by a
brief depolarization and in the absence of external calcium
is compatible with the somewhat tentative idea that calcium
release from the sarcoplasmic reticulum may occur in similar
ways in frog slow and twitch muscle. The mechanism in slow
fibers for coupling tubular potential to the regulation of
calcium release may involve intra-membranous charge
movements but, as for twitch fibers, the idea is only
plausible and needs evidence of a more direct kind.

DISCUSSION

Cleemann: Is there any inactivation of the charge movement in the slow fibers with maintained depolarization?

Chandler: This is what Gilly and Hui are studying now, and so far, their preliminary findings are rather confusing. If they depolarize the membrane potential to −40 mV for a period of a half hour or so there is very little change in the amount of charge. If they depolarize to −20 mV for a half hour, the value of Q_{max} typically falls by about one-half. Charge movements are present and the fiber still gives tension. The slightly confusing aspect of this is that although the amount of charge does not fully inactivate, the halfway point shifts to the left, by 20 or 30 mV. More studies should clearly be made on this.

Morad: Did you use tetracaine on twitch fibers in the experiments in your lab? How does the total Q value in your lab compare to those obtained by others?

Chandler: Almers and Best are the only people who have published results on tetracaine on twitch fibers. We did not do that. The total amount of Q in our lab has only been obtained on fibers in hypertonic solution. Adrian and Almers' Q's are slightly higher.

REFERENCES

1. Adrian, R.H. & Almers, W. (1976). Charge movement in the membrane of striated muscle. *J. Physiol. 254*, 339-360.

2. Adrian, R.H., Chandler, W.K. & Hodgkin, A.L. (1969).
    The kinetics of mechanical activation in frog muscle.
    J. Physiol. *204*, 207-230.
3. Adrian, R.H., Chandler, W.K. & Hodgkin, A.L. (1970).
    Voltage clamp experiments in striated muscle fibres.
    J. Physiol. *208*, 607-644.
4. Adrian, R.H. & Peachey, L.D. (1965). The membrane
    capacity of frog twitch and slow muscle fibres.
    J. Physiol. *181*, 324-336.
5. Adrian, R.H. & Peres, A.R. (1977). A gating signal for
    the potassium channel? *Nature 267*, 800-804.
6. Almers, W. (1976). Differential effects of tetracaine
    on delayed potassium channels and displacement currents
    in frog skeletal muscle. *J. Physiol. 262*, 613-637.
7. Almers, W. & Best, P.M. (1976). Effects of tetracaine
    on displacement currents and contraction of frog
    skeletal muscle. *J. Physiol. 262*, 583-611.
8. Burke, W. & Ginsborg, B.L. (1956a). The electrical
    properties of the slow muscle fibre membrane.
    J. Physiol. 132, 586-598.
9. Burke, W. & Ginsborg, B.L. (1956b). The action of the
    neuromuscular transmitter on the slow fibre membrane.
    J. Physiol. 132, 599-610.
10. Costantin, L.L. (1968). The effect of calcium on
    contraction and conductance thresholds in frog skeletal
    muscle. *J. Physiol. 195*, 119-132.
11. Costantin, L.L. (1974). Contractile activation in frog
    skeletal muscle. *J. gen. Physiol. 63*, 657-674.
12. Chandler, W.K., Rakowski, R.F. & Schneider, M.F. (1976).
    A non-linear voltage dependent charge movement in frog
    skeletal muscle. *J. Physiol. 254*, 245-283.
13. Gilly, W.F. & Hui, C.S. (1977a). Contractile activation
    in slow and twitch muscle fibres of the frog. *Nature
    266*, 186-188.
14. Gilly, W.F. & Hui, C.S. (1977b). External calcium ion
    and mechanical activation in frog slow fibers.
    Biophys. J. 17, 6a.
15. Gilly, W.F. & Hui, C.S. (1978). Charge movement in frog
    slow and twitch muscle fibers. Soc. Abstracts of
    Biophys. J. 18, (in press).
16. Hess, A. (1970). Vertebrate slow muscle fibers.
    Physiol. Rev. 50, 40-62.
17. Hui, C.S. & Gilly, W.F. (1977). Membrane electrical
    properties of frog slow muscle fibers. *Biophys. J.
    17*, 4a.

18. Kuffler, S.W. & Vaughan Williams, E.M. (1953a). Small-nerve junctional potentials. The distribution of small motor nerves to frog skeletal muscle, and the membrane characteristics of the fibres they innervate. *J. Physiol. 121*, 289-317.

19. Kuffler, S.W. & Vaughan Williams, E.M. (1953b). Properties of the slow skeletal muscle fibres of the frog. *J. Physiol. 121*, 318-340.

20. Lännergren, J. (1975). Structure and function of twitch and slow fibers in amphibian skeletal muscle, in *Basic Mechanisms of Ocular Motility and Their Clinical Implications*, G. Lennerstrand & P. Bach-Y-Rita, eds. Pergamon Press, New York, pp. 63-84.

21. Miledi, R., Stefani, E. & Steinbach, A.B. (1971). Induction of the action potential mechanism in slow muscle fibres of the frog. *J. Physiol. 217*, 737-754.

22. Page, S.G. (1965). A comparison of the fine structure of frog slow and twitch muscle fibres. *J. cell Biol. 26*, 477-497.

23. Peachey, L.D. (1968). Muscle. *Ann. Rev. Physiol. 30*, 401-440.

24. Schneider, M.F. & Chandler, W.K. (1973). Voltage dependent charge movement in skeletal muscle: a possible step in excitation contraction coupling. *Nature 242*, 244-246.

25. Schneider, M.F. & Chandler, W.K. (1976). Effects of membrane potential on the capacitance of skeletal muscle fibers. *J. gen. Physiol. 67*, 125-163.

26. Smith, R.S. & Ovalle, W.K., Jr. (1973). Varieties of fast and slow extrafusal muscle fibres in amphibian hind limb muscles. *J. Anat. 116*, 1-24.

27. Stanfield, P.R. (1970). The effect of the tetraethylammonium ion on the delayed currents of frog skeletal muscle. *J. Physiol. 209*, 209-229.

28. Stefani, E. & Steinbach, A.B. (1969). Resting potential and electrical properties of frog slow muscle fibres. Effect of different external solutions. *J. Physiol. 203*, 383-401.

29. Stefani, E. & Uchitel, O.D. (1976). Potassium and calcium conductance in slow muscle fibres of the toad. *J. Physiol. 255*, 435-448.

TABLE 1. PROPERTIES OF CHARGE MOVEMENT IN ISOSMOTIC SOLUTIONS

Fiber type	Solution	$\bar{V}$ (mV)	k (mV)	Q_{max} ($nC/\mu F$)	Reference
Slow	Na,Cl*	-23	14	8 [3]	Gilly & Hui
Slow	TEA,Cl* 11.8mM Ca	-25	13	7 [3]	Gilly & Hui
Twitch	TEA,Cl* 11.8mM Ca	-24	8	29+	Gilly & Hui
Twitch	TEA,SO$_4$*	-27	13	29 [4]	Almers & Best (1976)
Twitch	TEA,Cl	-30	12	25‡[5]	Schneider & Chandler (1976)

*2mM Tetracaine
+$\bar{V}$ & k from 2 fibers, Q_{max} from 4 fibers
‡Q_{max} assumed from the average value obtained in hypertonic solution

CHARGE MOVEMENT IN THE MEMBRANE
OF STRIATED MUSCLE

R.H. Adrian

Physiological Laboratory
Cambridge CB2 3EG England

In the study of excitable membranes interest has
recently concentrated on the small signals which may be
related to the opening and closing of the systems which
regulate the ionic currents (Almers, in press). Armstrong
& Bezanilla (1974) and Keynes & Rojas (1974) have studied
the asymmetrical displacement current in squid axon which
shows some of the characteristics of m particle movement
originally suggested by Hodgkin & Huxley (1952). Schneider
& Chandler (1973) have detected a similar, but slower,
transient charge movement in striated muscle which they
suggested might be related to the activation of contraction.

The signals in question are only detectable when the
much bigger ionic currents are prevented, usually by
tetraethyl ammonium (TEA) and tetrodotoxin (TTX), and they
are currents which are essentially capacitative in their
behavior. To impose an instantaneous change in potential on
the membrane, a certain quantity of charge must be transferred
from one side to the other in addition to any current which
flows through resistive pathways in the membrane. If the
membrane behaved as a perfect condenser, this quantity of
charge would be independent of the initial and final
potentials and would depend only on their difference. Also,
an abrupt potential change would require an equally abrupt
transfer of charge. Despite the fact that for many years
electrophysiologists have assumed this ideal behavior for
the membrane capacity, neither of the predictions is true
experimentally. The charge transfer associated with equal
potential steps from different initial potentials is not
equal: the charge transfer continues for some time after the
potential has changed. In other words, the membrane

dielectric saturates and is lossy; polarization is neither linear nor instantaneous. Behavior of this kind is to be expected if the membrane contains charges or permanent dipoles within it which can move under the influence of the field but which do not move instantaneously and cannot leave the membrane. If there are numerous species of charges or dipoles, one can predict that the dielectric constant of the membrane will be a complexly nonlinear function of the membrane potential.

Several authors have examined the dependence of membrane capacity on membrane potential in striated muscle. Eliminating the complexities introduced by the transverse tubular system, Schneider & Chandler (1976) first showed that there was a conspicuous rise in capacity for potentials positive to the resting potential. In fibers depolarized to -20 mV, however, the capacity appeared to be about 10% larger at -90 mV than at -20 mV. Adrian & Almers (1976a) confirmed these findings.

Working with A.R. Peres, we have tried to extend the measurement of membrane capacity as a function of membrane potential, since it seemed to us that it presents a method of separating possibly multiple charge or dipole movements in the membrane (Adrian & Peres, 1977). We have used the method of Adrian & Almers (1976a), which is based on the three-electrode clamp at the end of a sartorius muscle fiber (Adrian, Chandler & Hodgkin, 1970). We imposed 10 mV measuring steps of potential: one at a control potential, usually -90 mV, and the other at some other potential in the range -90 to +20 mV. The capacity may be estimated as the integral of the transient membrane current at the beginning and end of the voltage steps. Records of the time course of the charge movement are given by subtracting the membrane currents for control and test potential steps. The necessary numerical operations are carried out automatically by a PDP 11 computer which controls the experiment and takes in digitized records of membrane potential and membrane current.

Fig. 1 illustrates both the design of these experiments and the principal result. On the left is shown the control potential step at -90 mV and membrane current; above is the test step of the same size but superimposed on a potential step to some other membrane potential -48 mV. The computer examines only the two 10 mV steps and displays the results of the subtracted membrane currents. These are shown in the middle column for a range of membrane potentials between -80 and -20 mV. It is at once apparent that the charge movement during the step, and especially in the potential

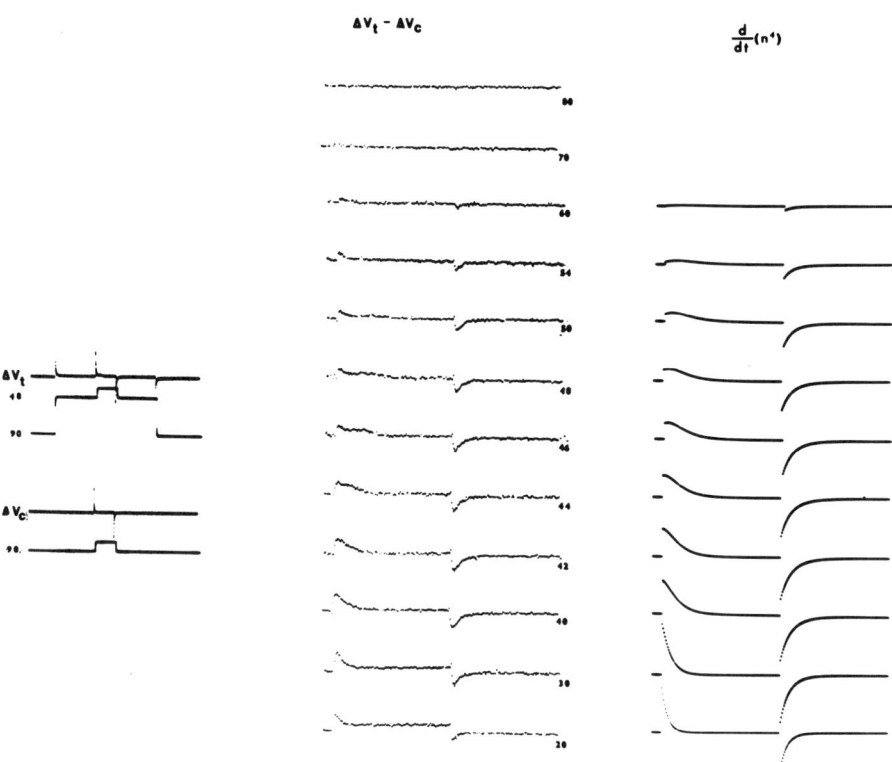

Fig. 1. The central column of records is the result of
subtracting the membrane currents for two 10 mV steps which
start from different potentials. On the left are records
of membrane potential and membrane current ($\propto\Delta V$) for a
control step from -90 mV and a test step from -48 mV. The
right-hand column shows the rates of change of potassium
conductance calculated from the measurements of Adrian et al.
(1970) for 10 mV steps of potential from the various starting
potentials. Calibration lines (which do not apply to the
left-hand records) are 50 msec, 3.3 mV for ΔV_t - ΔV_c and
0.01 msec^{-1} for $d(n^4)/dt$. The figure is reproduced from
Adrian & Peres (1977).

range -54 to -44 mV, is complicated and cannot be described
by any simple function such as an exponential. On the other
hand, the charge movement after the end of the step might be
so described. The configuration of the charge movement
during the step suggests that there may be two charges moving

with quite different behavior, one whose rate of movement is
initially large and then declines, and one whose initial
movement is low and increases before declining at least at
some values of the potential. For comparison, the rate of
change of the potassium conductance is shown on the right of
Fig. 1. Potassium currents were suppressed in this
experiment by tetraethylammonium ion (Stanfield, 1970), but
the measurements and analysis of Adrian *et al.* (1970) allow
one to calculate what the potassium conductance changes would
have been for the voltage steps involved in Fig. 1, and from
them, the rates of change of conductance. The right-hand
column of Fig. 1 shows $d(n^4)/dt$ which, ignoring any
inactivation, can be considered proportional to $d(g_K)/dt$.
A striking similarity exists between the differential of the
conductance change which would have taken place in the
absence of TEA and the delayed component of the charge
movement in the experimental records. This similarity
certainly suggests a connection between the charge movement
and the delayed potassium system. But it may be worth noting
a difference between this charge movement and the charge
movement associated with the sodium conductance change.
Whereas here the charge movement appears to resemble $d(g_K)/dt$,
in the squid it resembles $d(g_{Na}^{1/3})/dt$ (ignoring inactivation
in both cases) at least for increases in sodium conductance.
We do not know the reasons for this difference.

   An experiment of the kind shown in Fig. 1 can be
examined in another way, and we can plot the membrane
capacity against membrane potential. The clearest method is
to determine the ratio of the capacity at the test potential
to the capacity at the control potential. C_T/C_C therefore
gives the ratio of the total charge required for control and
test potential steps. (The integral of the transient parts
of the charge movement records in Fig. 1 gives the *difference*
between the charge required for control and test steps.)

   The uppermost points (half-filled circles) in Fig. 2
show the way C_T/C_C varies when the membrane potential is
held at -90 mV at all times other than during the capacity
measurements. The peak in the capacity can be seen to
correspond to the potential at which the integral of the
charge movement in Fig. 1 is greatest. If, however, the
fiber is permanently depolarized to -40 mV, C_T/C_C has a
different shape (open circles), and depolarization to -20 mV
increases the difference (filled circles). Prolonged
depolarizations to -50 mV or more negative potentials do not
alter the large peak in C_T/C_C seen when the permanent
potential is -90 mV. Prolonged depolarizations positive to

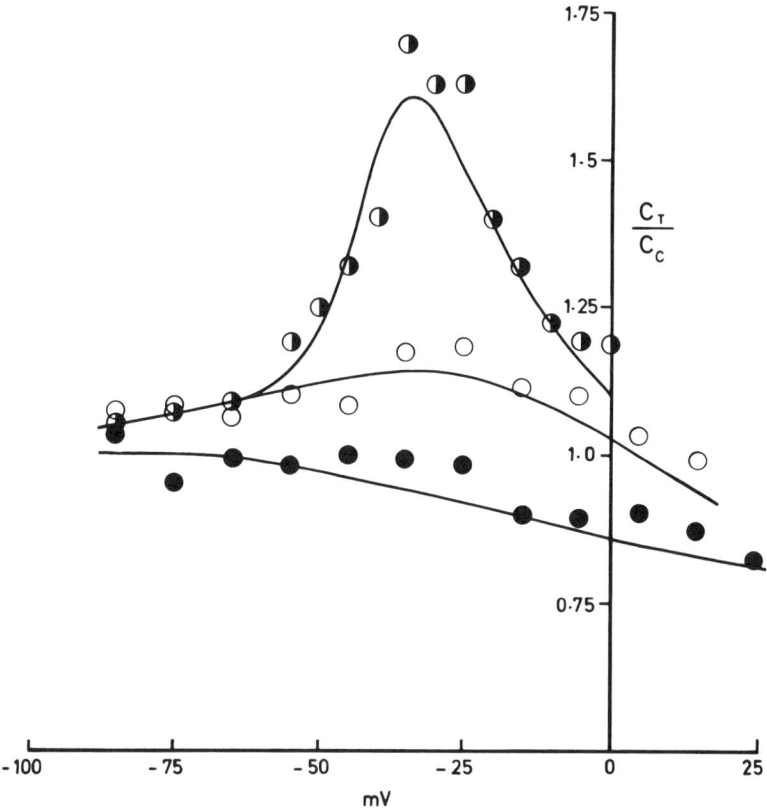

Fig. 2. The experimental points are the ratio of the membrane capacity at a control potential to the membrane capacity at the potential given by the abscissa. For the half-filled circles the potential was held at -90 mV (except during the measuring steps), for the open circles at -40 mV, and the filled circles at -20 mV. Control capacities were measured at the holding potential. The two upper sets of points have been scaled vertically to coincide with the curve between -90 and -75 mV. The curves from below up are:

$$C_T/C_C = (1 + C_\alpha)/(1 + C_\alpha^*)$$
$$C_T/C_C = (1 + C_\alpha + C_\beta)/(1 + C_\alpha^*)$$
$$C_T/C_C = (1 + C_\alpha + C_\beta + C_\gamma)/(1 + C_\alpha^*)$$

where $C_\alpha = 35 \ nC\mu F^{-1}/[45(1 + \cosh(-(V + 80)/45))]$
$C_\beta = 13 \ nC\mu F^{-1}/[22(1 + \cosh(-(V + 22)/22))]$
$C_\gamma = 33 \ nC\mu F^{-1} \cdot d(n_\infty^4)/dt$
and C_α^* *is* C_α *at -90 mV.*

-50 mV are known to inactivate both the delayed potassium
current (Adrian *et al*, 1970) and the ability to contract
(Hodgkin & Horowicz, 1959). An attractive interpretation of
Fig. 2 would be that a potential of -40 mV immobilizes the
charge, which in some way regulates the potassium current,
but that holding at -20 mV immobilizes in addition a charge
involved in regulating contraction--perhaps the charge which
moves first in Fig. 1 before the slowly moving charge which
seems to be associated with the potassium channel. The lines
in Fig. 2 are drawn on the basis of this suggestion. The
lowest line would be the capacity voltage relation if the
membrane contained only a single mobile charge with the
characteristics of Charge 2 described by Adrian, Chandler &
Rakowski (1976) and Adrian & Almers (1976*b*). The middle line
supposes the addition of a charge with characteristics like
the reprimed charge seen in Adrian *et al*. (1976). The upper
line supposes the further addition of a charge whose voltage
distribution is given by n_{∞}^{4}.

At the moment, this interpretation must remain only a
working hypothesis. Its confirmation will depend on finding
other ways to separate the supposedly different charge
species. But the results clearly show that the dielectric
behavior of the membrane is much more complex than has been
previously assumed, and that we must be prepared for the
possibility of several species of intramembrane charge
movement.

DISCUSSION

Armstrong: We now have evidence from skeletal muscle
for intramembranous charge movement regulating the process
of excitation-contraction coupling and for controlling the
potassium permeability which is involved in conduction of the
action potential. The type of charge movement that Dr. Adrian
has described was not observed in the experiments by
Schneider and Chandler, and I wonder if I might ask Dr.
Chandler to comment.

Chandler: That's quite right, the original experiments
which Martin Schneider and I published didn't show this, but
we certainly didn't look for it nearly so carefully as Dr.
Adrian has. I think it's quite possible that we overlooked
it. His records certainly are very convincing. The only
thing close to it that we found was in the very first
experiments that we did, when we used the same solution
tonicity that Dr. Adrian used. When we looked in the
microscope under these conditions we saw fiber movement. I

don't mean to imply that I think this is fiber movement, but
we were a little worried that we were getting mechanical
artifacts, and we went to higher tonicity. The fully
published analysis by Schneider and Rakowski was done on
fibers that had a somewhat higher tonicity. That could be
part of the reason why we missed it.

Adrian: I suspect that the charge movement that we've
seen up to now is a mixture of two charges. One is perhaps
contraction charge and the other is perhaps potassium charge
or something that has highly complex kinetics which by
coincidence may resemble those of a potassium system, but I
certainly think that the normally detected charge in the
muscle fiber is by no means only that connected to the
potassium channel. My present hunch is that quite a large
amount of the charge movement which I described as being the
initial charge movement is probably related to the
contraction, although I suspect that the later part of it
may be more related to the potassium channel.

Reuter: For the first charge, you see the same results
as Dr. Chandler--that means that $\bar{Q}$ in each case is the same as
is attributed to excitation-contraction coupling. Secondly,
as I understand it the reason why delayed rectification was
rejected as being related to the activation of contraction
was on the basis of pharmacological tests which shifted the
inward rectification in one direction and the activation of
contraction in the other direction. Have you done such
experiments on these two charges?

Adrian: The short answer to both these questions is no.
We haven't done the pharmacology, but we certainly should and
we intend to do so. For your first question, however, I'm
not quite sure whether it's fair to say no to that. If we
take the amount of charge that is necessary to account for
the lowest curve in Fig. 2, we can describe that with a
function that defines an amount of charge, that is Charge 2.
We can then add another charge which adds to make the middle
curve in Fig. 2, and I would tentatively suggest that it
represents the contraction charge. Then we can add to those
two charges a third charge, which is what we would expect if
it were related to the potassium system. In this case, the
total amount of charge for the lowest one is something like
35 nanocoulombs, which is not unlike the magnitude of Charge
2 that was determined by Chandler and ourselves. The second
amount of charge is about 13 nanocoulombs, and the third
amount of charge is also about 33 nanocoulombs. The
difficulty is that here we have 35 + 13 + 33, which is nearly
80 nanocoulombs of charge. Whether you see that or not
depends entirely on what pulse structure you use to detect

it, because there is quite a large amount of Charge 2 moving
in the control pulse that we use at -100, but it is not
moving at 0 mV. Thus an amount of charge actually subtracts
from the amount of charge that we tend to see when we do a
normal charge movement determination. The problem, then, is
that there are a large number of charges. One must know
over what voltage range they move in order to be able to
interpret the difference between the control pulse and the
test pulse charge movement. Although I don't want to press
it very hard, I think there is a charge movement for
contraction, a charge movement for potassium gating and a
separate Charge 2 movement. I don't know what the function
of Charge 2 is. There probably is also a sodium-gating
current, but for experimental reasons we wouldn't expect to
detect that.

   Peachey: It's convenient to think that the first charge
movement, the earlier one, is related to *E-C* coupling and
the second one perhaps to potassium currents, but just how
strong is the evidence that the first one is related to
E-C coupling, and that the second is not related to *E-C*
coupling? How strongly should we feel that potassium
currents themselves could not be related to *E-C* coupling?

   Adrian: The evidence that the first charge is related to
E-C coupling is not strong, and I would also say that the
evidence that the humped charge movement is related to
potassium current is not strong. The kinetics suggest a
relation to potassium conductance, but then there may be
other systems which have complicated kinetics of that kind.
I think that the evidence that the excitation-contraction
coupling is not related to potassium conductance change is in
fact fairly good, and it is of the kind that was referred to
by Dr. Reuter earlier. To summarize, I would not like
anyone here thinking that I was strongly committed to the
interpretation that I've put upon these data this afternoon,
it seems to me at this point to be suggestive but certainly
no more than that.

   Hille: I thought Wolf Almers published a paper in
November* in which he applied tetracaine to the fiber in the
same kind of experiment, and found (1) that tetracaine
shifted the voltage dependence of the potassium conductance
by possibly 30 mV in the positive direction, and (2), that
whatever component of the charge movement he was looking at
was not changed by this treatment. Therefore he concluded

*Almers, W. (1976). Differential effects of tetracaine
on delayed potassium channels and displacement currents in
frog skeletal muscle. J. Physiol. 262, 613-637.

that there is no relationship between g_K and the charge movement. Was he looking at the wrong charge movement?

Adrian: No, I'm sure he wasn't looking at the wrong charge movement, but I'm not sure what charge movement he was looking at. I think that this is the difficulty. Certainly we know that tetracaine shifts the activation potential of g_K 30 or 40 mV to the right. We don't know that it doesn't alter some other charge movement. I think that you have put your finger on a rather difficult matter for the interpretation that I've put upon the results here. It's not inconceivable that tetracaine affects one of the other charge movements, and therefore the particular mix of charge movement that was seen by Almers in those experiments is not necessarily the same as the mix that is seen in the absence of tetracaine. Until we can be really sure that we are actually affecting one and only one charge, we should be rather careful in the interpretation of any of these results.

REFERENCES

1.   Adrian, R.H. & Almers, W. (1976a). The voltage dependence of membrane capacity. *J. Physiol. 254*, 317-338.
2.   Adrian, R.H. & Almers, W. (1976b). Charge movement in the membrane of striated muscle. *J. Physiol. 254*, 339-360.
3.   Adrian, R.H., Chandler, W.K. & Hodgkin, A.L. (1970). Voltage clamp experiments in striated muscle fibres. *J. Physiol. 208*, 607-644.
4.   Adrian, R.H., Chandler, W.K. & Rakowski, R.F. (1976). Charge movement and mechanical repriming in skeletal muscle. *J. Physiol. 254*, 361-388.
5.   Adrian, R.H. & Peres, A.R. (1977). Charge movement associated with the opening and closing of the activation gates of the Na channels. *Nature, Lond. 267*, 800-804.
6.   Almers, W. Gating current and charge movement in excitable membranes. *Reviews of Physiology, Biochemistry and Pharmacology*. In press.
7.   Armstrong, C.M. & Bezanilla, F. (1974). Charge movement associated with the opening and closing of the activation gates of the sodium channels. *J. gen. Physiol. 63*, 533-552.

8.   Hodgkin, A.L. & Horowicz, P. (1959). Potassium
     contractures in single muscle fibres. *J. Physiol. 153*,
     386-403.

9.   Hodgkin, A.L. & Huxley, A.F. (1952). Quantitative
     description of membrane current and its application to
     conduction and excitation in nerve. *J. Physiol. 117*,
     500-544.

10. Keynes, R.D. & Rojas, E. (1974). Kinetics and steady
     state properties of the charged system controlling
     sodium conductance in the squid giant axon. *J. Physiol.
     239*, 393-434.

11. Schneider, M.F. & Chandler, W.K. (1973). Voltage
     dependent charge movement in skeletal muscle; a possible
     step in excitation-contraction coupling. *Nature, Lond.
     242*, 244-246.

12. Schneider, M.F. & Chandler, W.K. (1976). Effects of
     membrane potential on the capacitance of skeletal muscle
     fibres. *J. gen. Physiol. 67*, 125-163.

13. Stanfield, P.R. (1970). The effect of the tetraethyl
     ammonium ion on the delayed currents of frog skeletal
     muscle. *J. Physiol. 209*, 209-229.

LOCAL ANESTHETIC ACTION ON INACTIVATION OF THE
Na CHANNEL IN NERVE AND SKELETAL MUSCLE:
POSSIBLE MECHANISMS FOR ANTIARRHYTHMIC AGENTS

Bertil Hille

Department of Physiology and Biophysics
University of Washington SJ-40
Seattle, Washington, 98195

I will discuss work on local anesthetic drugs done in
Seattle by Gary Strichartz, Ken Courtney, Wolfgang Schwarz,
Philip Palade, and myself over the last several years.
Although this is a symposium on cardiac muscle, I do not ever
deal with the heart. In my laboratory, we primarily study
myelinated nerve fibers and skeletal muscle from the frog.
But as it happens, the phenomena which we see with local
anesthetics on nerve and muscle seem to me quite convincingly
relevant to the problem of the use of the same molecules,
including lidocaine, as antiarrhythmic agents on the heart.
Much of the work described here closely parallels results
coming from the laboratories of T. Narahashi and B.I.
Khodorov, both working on axons, and naturally many similar
observations have already been made with cardiac tissues,
which I shall not discuss (c.f., Weidmann, 1955; Heistracher,
1971; Chen, Gettes & Katzung, 1975; Weld & Bigger, 1975).
    The primary target of local anesthetics is the sodium
permeability mechanism in the axon or muscle membrane. Fig. 1
shows how we monitor the changes of sodium permeability in a
single fiber of frog semitendinosus muscle. This is a
voltage clamp experiment of the same type as was originally
used by Hodgkin and Huxley. The membrane potential is
stepped from resting potential to various other values in
test pulses lasting 6 msec. The potassium currents are
blocked pharmacologically, and leakage and some capacity
currents are subtracted before the signal reaches the
oscilloscope, so in this case sodium currents may be

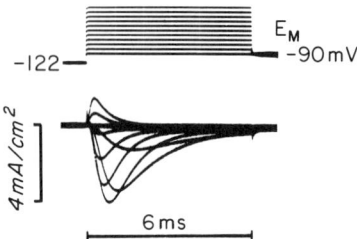

Fig. 1. Sodium currents recorded from a semitendinosus muscle fiber under voltage clamp at 5° C. The upper traces show the 11 potential steps spaced at 16-mV intervals from -90 to +70 mV; the lower traces show the corresponding family of membrane currents after subtraction of leakage and the slow components of capacity current. The fiber contains nearly isotonic CsF, and there are no potassium currents. Inward flow of Na^+ ions is seen as a transient downward deflection of the current trace. (After Hille & Campbell, 1976.)

photographed directly off the oscilloscope. It is because the sodium currents may be studied so directly and reliably that axon or skeletal muscle preparations offer some advantages over cardiac tissue in studying antiarrhythmic agents.

Although this example happens to be a skeletal muscle, these sodium currents have the same form in the squid giant axon or in myelinated nerve fibers, and indeed, in every axon which has been studied. The important feature here is that the currents turn on and off in response to the single depolarization. We say that Na channels activate and inactivate, or go from a resting state to an open state to an inactivated state. When the fiber is repolarized, inactivated channels gradually return to the resting pool, and then after what might be considered a refractory period, we can repeat the experiment again and see sodium currents.

Fig. 2 shows diagrammatically a modern conception of an ionic channel that should not violate the general ideas of any of the participants in this symposium. The lipid bilayer, which is relatively oily and relatively fluid, allows motion. Within the bilayer there is a macromolecule which is primarily protein, with a molecular weight of perhaps 250,000. A property of this protein is that it can

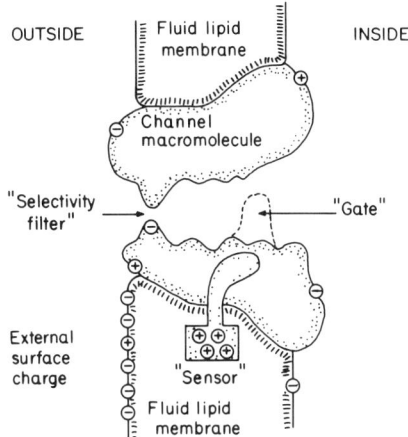

Fig. 2. Schematic diagram of an ionic channel like the Na or K channel of nerve membranes. The channel is a protein macromolecule forming a pore through the lipid-bilayer membrane. The pore has a narrow selectivity filter near the outside and a gate near the inside. A voltage sensor extending into the lipid moves under the influence of the intramembrane field and controls the opening of the gate. Surface charges and the ionic double layer associated with them modify the field within the membrane. The drawing is fanciful and the dimensions and shapes of the components are not known.

have a hole through it, a pore. The aqueous pore is the region through which ions like sodium pass. This figure might serve as a diagram of a sodium, potassium, or calcium ionic channel, or even of other kinds of voltage-dependent channels about which we do not know much.

An important feature of such channels is that they have ionic selectivity; there is some narrow place where the channel decides "Yes, this is a sodium ion." That narrow region need not be the whole channel. In fact, it would be quite inefficient to have the entire channel narrow, because then the ions would be stopping all the way across. There need only be one difficult place to go across, which I call the "selectivity filter." Another important feature is a device in the membrane called the sensor, which senses the electric field in the membrane. This kind of device is necessary for any voltage sensitive process, such as the types discussed earlier in this meeting: those that determine the contraction threshold in muscle, those that determine the

turning on of potassium conductance, and those that determine
the turning of sodium conductance. This voltage sensor may
be regarded as a collection of charge or of dipoles which
experiences forces caused by the electric field across the
membrane. The pulling back and forth, or the movement, of
the sensor is seen as gating current. In an ionic channel
the effect is in some way to open and close a gate. When the
gate is closed, the channel does not allow ions to go
through, and when it is open, it does.

I.   BLOCK OF Na CHANNELS

   The following experiments focus on the question of what
local anesthetics do. Fig. 3 shows voltage clamp currents of
a single node of Ranvier from a myelinated nerve fiber. The
figure shows traces for many different voltages recorded at
15° C. The early currents in (A) are sodium currents. At
later times outward potassium currents begin to develop, but
they have been cut off to make the picture. They would
continue to grow, however, because potassium channels have

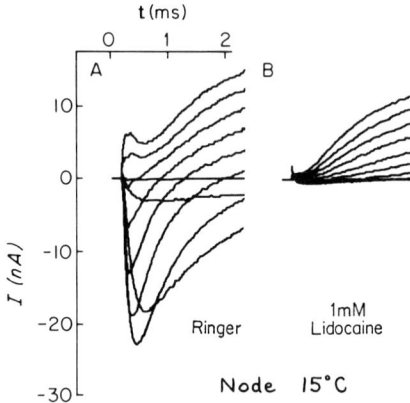

*Fig. 3. Block of sodium current by 1 mM lidocaine in the
external solution. A family of voltage clamp currents
recorded from a single node of Ranvier at 15° C and corrected
for leakage. (A) Before treatment with drug. (B) A couple
of minutes after application of lidocaine. Holding
potential -83 mV. No prepulse. Test pulses applied at
1/sec. (After Hille, 1977.)*

not been blocked in this fiber. After 1 mM lidocaine is added in (B), the sodium currents are blocked and the potassium currents are slightly reduced. This block of Na channels is the mechanism of local anesthesia; it is why one does not feel pain, for example, after receiving an injection of lidocaine or procaine. Without Na channels there can be no excitation. The drug concentration here may be thought of in the following terms: The dentist might typically use a 2% solution of lidocaine, and that contains 72 mM lidocaine, so 1 mM is clearly within the range that might be experienced in local anesthesia, although it is well above the 5-10 μM range effective in treatment of cardiac arrhythmias.

It is now believed that local anesthetic molecules may approach the Na channel from the intracellular side and also from the membrane and may find a receptor which is partly within the aqueous region of the pore. The binding site is neither on the external membrane surface, as it is for tetrodotoxin, nor entirely in the fluid lipid membrane. We believe that the drug may sit with its hydrophobic part buried in the hydrophobic wall of the channel and with its hydrophilic part sticking out in the ion pathway as is shown in the diagram in Fig. 4. The most direct evidence for placing the receptor within the pore comes from the work of Strichartz (1973) with an N-ethyl quaternary derivative of lidocaine. Such permanently charged anesthetic analogs cannot

Drug-Receptor Transitions

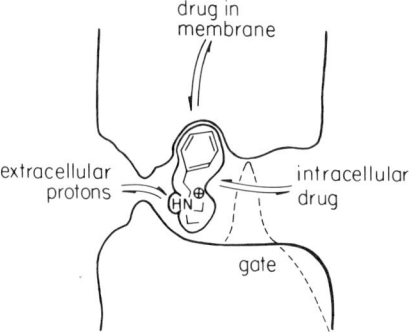

Fig. 4. Diagrammatic view of a local anesthetic molecule on its receptor in the sodium channel. The molecule can gain access to the receptor from the lipid membrane and from the intracellular medium if the "gate" is open. Extracellular protons can change the ionization state of the bound anesthetic. (From Schwarz et al., 1977.)

cross the cell membrane. In agreement with the work from
Narahashi's laboratory (Frazier, Narahashi & Yamada, 1970) on
squid giant axons, Strichartz showed with myelinated nerve
that the quaternary molecule has no anesthetic effect when
applied outside the cell, but blocks strongly when applied
internally. In addition, however, he found that the drug
block develops only if nerve is stimulated to open the gates
of the channel, and then, when the gates are open, the drug
binding equilibrium is similar to that for a charged molecule
which must move across more than half of the electric field
of the membrane to reach its binding site. He therefore
suggested that, proceeding from the inside out, the Na
channel has gates, drug receptor, and then the narrow
selectivity filter as in Fig. 4.

II. FOUR MODULATING FACTORS

     Now let us turn to experiments with lidocaine of possible
relevance to antiarrhythmic action. Some of the figures
show results with frog myelinated nerve and some with frog
semitendinosus, but the results do not depend on which
preparation is used. The block by local anesthetics is
complicated by many factors. I will limit this discussion to
four major factors which change the degree of block of Na
channels in the presence of a local anesthetic: membrane
potential, calcium, stimulation, and pH.
     The first modulating factor is the membrane potential.
For example, in Fig. 5A where the potassium channels are
blocked with tetraethylammonium ion and the nerve fiber is
held at the normal resting potential, we do not obtain
sodium currents because of the presence of 1 mM lidocaine.
But when the fiber is hyperpolarized for a minute by 15 or
60 mV (Fig. 5B, C), then the sodium currents return to almost
half the value they had in the untreated fiber. If this
experiment is done with a much lower concentration of local
anesthetic, some sodium currents are evident when holding at
the resting potential, but they are very strongly blocked if
the holding potential is made 5-10 mV more inside positive
(Khodorov, Shishkova & Peganov, 1974, 1976).
     Another manifestation of the effect of membrane potential
is seen in Fig. 6 from experiments of the conventional type
to measure inactivation of sodium channels. The nerve is
hyperpolarized by 50-msec prepulses to different values and
then a test pulse to a fixed value is applied to measure the
fraction of available Na channels. In the absence of drug

(open circles), the amplitude of the current in the test pulse
is large for all prepulses more negative than -90 mV. The
experiment is then repeated with 1 mM lidocaine or 1 mM
neutral benzocaine (filled circles). Both drugs reduce
sodium currents even with the most hyperpolarizing prepulse

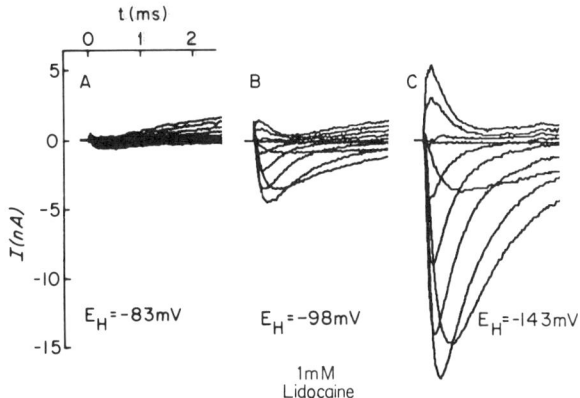

Fig. 5. Reversal of lidocaine block by hyperpolarization.
Voltage clamp currents corrected for leakage and with
potassium currents largely suppressed by 3.6 mM
tetraethylammonium ion. Same fiber as in Figs. 3 and 7,
treated with 1 mM lidocaine and held for at least 1 min at the
stated holding potential, E_H, before measuring the voltage
clamp currents. (After Hille, 1977.)

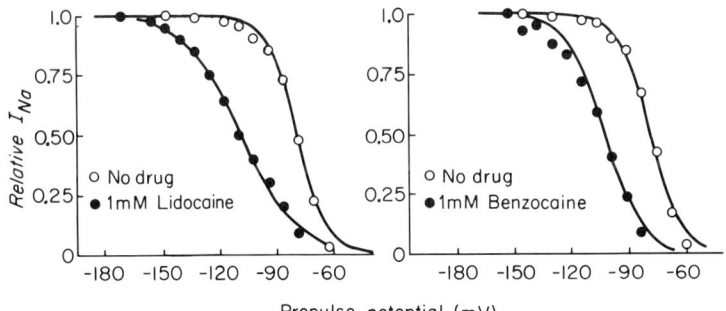

Fig. 6. Effect of 50-msec conditioning prepulses on peak
sodium currents with lidocaine and benzocaine. Single node of
Ranvier held at -90 mV and tested with test pulses to -15 mV.
For each treatment the current at the most negative prepulse
potential is normalized to 1.0. (From Hille, 1977.)

used, but the points are normalized to eliminate this block. The significant result is that the prepulse dependence of the remaining current is shifted to more negative potentials, so only a small fraction of the current is seen with a prepulse to -90 mV. For both drugs, the inactivation curve measured with 50-msec prepulses is shifted and its slope is reduced. In the later discussion, we shall see that this type of measurement is complex, combining the kinetics of the drug-receptor reaction and of the inactivation gating processes without clear distinction. Similar shifts of inactivation with a change of slope were first seen by Weidmann (1955) in Purkinje fibers and have since been confirmed numerous times for various drugs acting on cardiac tissues, but a clear demonstration of the same effect in axons and skeletal muscle is relatively recent, although quite straightforward (Khodorov *et al.* 1974, Khodorov, Shishkova, Peganov & Revenko, 1976; Courtney, 1975; Hille, 1977; Schwarz, Hille & Palade, 1977).

The first modulating factor, then, is the membrane potential. It could be quite important in the use of lidocaine and other drugs, including quinidine and propanolol, in controlling cardiac arrhythmias, because some of the tissues that need to be controlled are damaged regions of the heart, which are already depolarized. The depolarization would potentiate block of Na channels by the drug and thus would selectively lower the excitability in these regions.

The second modulating factor is the external calcium concentration. Sodium currents are blocked by 1 mM lidocaine at normal calcium concentration (Fig. 7), but if we increase the calcium concentration, the sodium currents can be restored. Thus, calcium ions decrease the potency of local anesthetics.

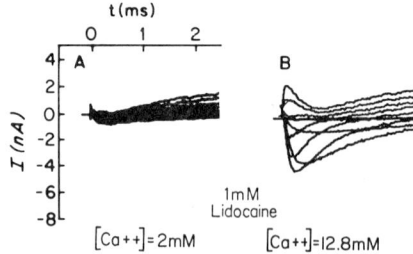

Fig. 7. Reversal of lidocaine block by elevated external Ca^{++} ions. Voltage clamp currents as in Fig. 5. Same fiber as in Figs. 3 and 5, held at a holding potential of -83 mV with either 2 or 12.8 mM Ca^{++}. (After Hille, 1977.)

Still another factor is stimulation. Fig. 8, from the work of Courtney (1975), shows an experiment on a nerve fiber in the presence of 0.5 mM lidocaine. The fiber has been rested for some time. Voltage clamp pulses are begun at a constant pulse potential, and if the preparation is stimulated at 1/sec, the sodium current remains more or less constant, although it is smaller than in a fiber that has not been treated with drug. If the fiber is stimulated at 8/sec, we see an increasing block with each stimulus. In the presence of local anesthetic, the nerve remembers the previous stimulus, and the block is cumulative. If the stimulation is stopped, the extra block is gradually lost. In each of the four runs, the first stimulus drives some local anesthetic into the receptor, and we can use the second stimulus in the train as an assay to determine how much local anesthetic is left on the receptor after a given interval. We see that after 125 msec, quite a bit is left, and with time it decays with a time constant on the order of a few hundred msec. Thus, increasing the rate of stimulation of the fiber increases the block of Na channels. This has been called "frequency-dependent" or "use-dependent" block.

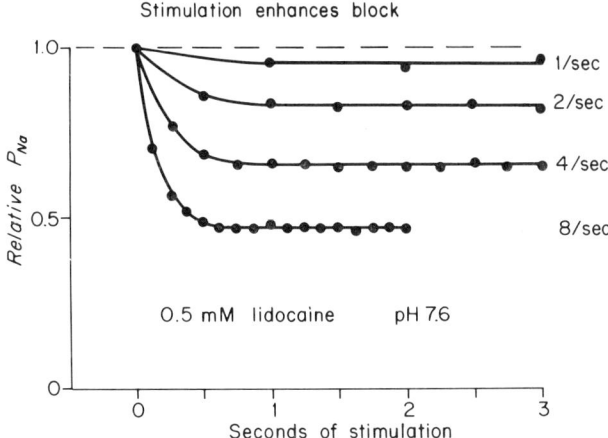

Fig. 8. *Development of use-dependent block in a node of Ranvier equilibrated with 0.5 mM lidocaine. Stimulation is applied at each of four different frequencies following rest periods. Points plot the size of peak sodium current relative to the first pulse for 7-msec test pulses to -20 mV from a holding potential near -80 mV. Temperature 10° C. (From Hille et al., 1975; based on Courtney, 1975.)*

     This factor could be important clinically. It is quite
conceivable that the kinds of activities we would like to
block with anesthetics are high-frequency activities. It is
possible that pain comes up as high-frequency firing in a
nerve fiber, whereas other sensations or motor activities are
not such high-frequency activities, and, from the clinical
point of view, we may care less about blocking them.
Consider also cardiac arrhythmias. A cardiac action potential
would drive some of the drug into Na channels, and then, when
the cell repolarizes, it takes a while before the system is
ready to fire again because the drug has to leave first.
This factor could obviously have importance in preventing
extra systoles as a reentrant excitation shortly after a
previous action potential. In effect, it could lengthen the
minimum interval between the end of the last action potential
and the beginning of the next.
     Finally, block by anesthetics can be modulated by a
fourth factor, which is pH. Fig. 9B shows the same
experiment as before, with pH 7.2, except now it is with a
skeletal muscle fiber exposed to 200 µM lidocaine. With
rapid stimulation, much extra block develops, and at a low
frequency of stimulation there is not much extra block.
Again, this result shows that the channel "remembers" the
previous stimulus for a short time. If the external pH is
lowered to 6 (Fig. 9A), the channel remembers the previous
stimulus much longer. We can go to intervals longer than
1 sec and still accumulate drug. On the other hand, at pH 8.8
(Fig. 9D), the sodium channels forget the previous stimulus
very rapidly. Evidently the drug leaves quickly, and even
when stimulating every 400 msec, we have hardly any
cumulative effect. As will be discussed later, we believe
we understand the effect of external pH in terms of a change
in the charge form of the ionizable drug molecule bound to
the receptor.
     The effect of pH has possible clinical significance in
cases where ischemia is associated with regions supporting
arrhythmias. The pH in such ischemic tissue is abnormally
acid, and therefore local anesthetics can reduce the rate of
firing of action potentials even more than if the tissue
were at the normal pH. In this way, both the reduction of
membrane potential and the reduction of pH in diseased
cardiac tissue might potentiate the therapeutic effect of
preventing high-frequency or premature firing.

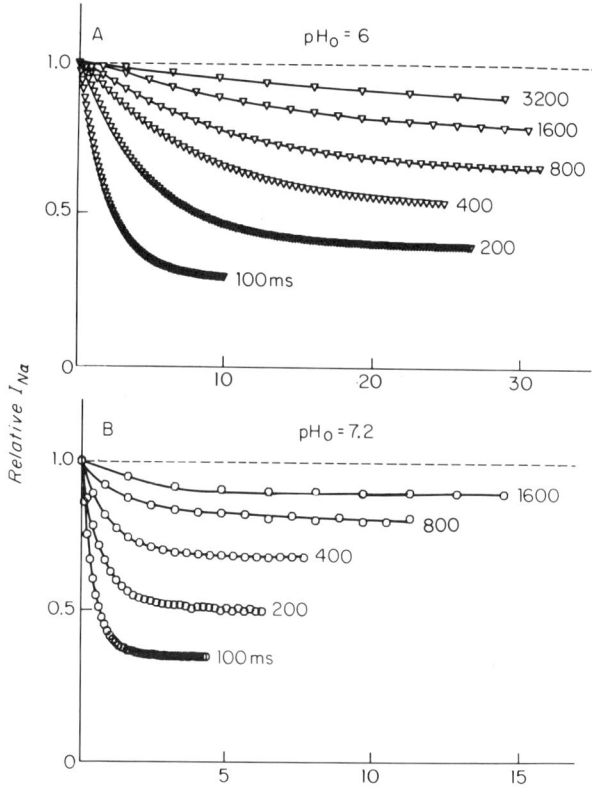

Fig. 9. Use-dependent block with 0.2 mM lidocaine at
four values of external pH. Measurements similar to Fig. 8
but with a single semitendinosus muscle fiber held at
-100 mV and depolarized with 1.5-msec test pulses to -10 mV.
The numbers on the curves indicate the interval in msec
between test pulses. Temperature 12.5° C. (From Schwarz
et al. 1977)

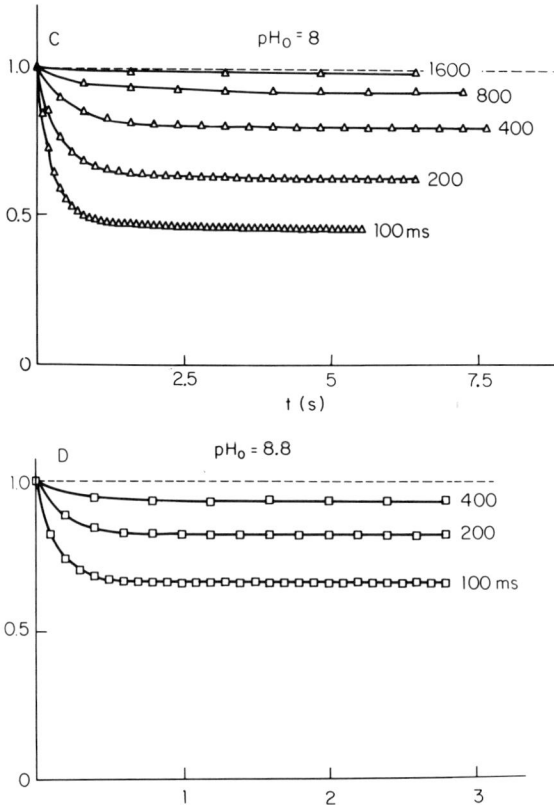

Fig. 9c, d

III.   INACTIVATION, MEMBRANE POTENTIAL, AND STIMULATION

   I now turn to our interpretation of the origin of the
modulatory effects I have spoken about. Fig. 10 is a
simplified model summarizing some of the ideas we use in my
laboratory. Normal Na channels are shown going from rest (R)
to open (O) to inactivated (I) in the upper part. We
envision the possibility that each form of the channel can
react with drug (*) to give the corresponding blocked
channel form, but the reactions are not all equally probable.
Following the suggestions of Khodorov *et al.* (1974, 1976)
and Courtney (1975), working with myelinated nerve, we
believe that local anesthetics bind most strongly to channels
in the inactivated conformation, I.   In depolarized
membrane (Fig. 10), resting channels become inactivated
and then may bind a drug molecule, reaching the quite stable
blocked-inactivated state, I*. In hyperpolarized membrane
or even at the resting potential, the blocked-inactivated
channel may undergo a conformational change that produces a
blocked-resting state (R*) with a much lower affinity for

POTENTIAL – DEPENDENT BINDING OF LOCAL ANESTHETICS

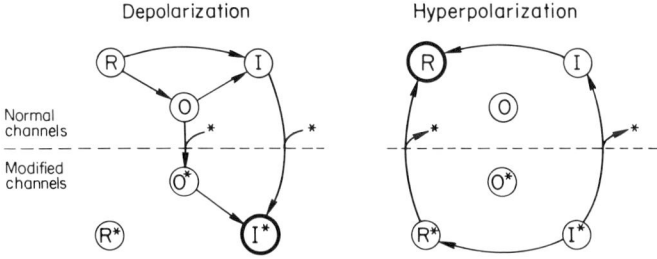

R = Resting channel
O = Open channel
I = Inactivated channel
* = Local anesthetic molecule

*Fig. 10.   Sequence of steps for local anesthetic binding
to depolarized excitable membrane and unbinding from
hyperpolarized membrane.   The sodium channel is shown in
three conformational states, resting, open, and inactivated.
Drug-blocked forms have an asterisk (*).   Depolarization
favors the blocked-inactivated form (I*), and
hyperpolarization favors the unblocked-resting form (R)
(Courtney, 1975; Hille, 1977).   The drug-receptor reactions
shown are those for neutral form of drug.   The charged form
may only react via the O-O* pathway (Strichartz, 1973).*

drug. The drug comes off and we recover a resting channel
without drug. Thus, repetitive activity inactivates channels
and tends to build up a pool of inactivated channels with
drug on them, and in the intervals between each
depolarization, those inactivated channels are slowly
returning to the resting state and losing their drug
molecules. The observed amount of block in any measurement
reflects the balance of these processes. The model of
Fig. 10, therefore, accounts at least qualitatively for the
effects of the holding potential and of repetitive
stimulation.

Let us examine more closely the necessary
interrelationships between preferential binding of drug to
inactivated channels and alterations of the gating process
by bound drug. Consider Fig. 11, which is an enlargement
of the R-I-I*-R*-R cycle of Fig. 10 with Hodgkin-Huxley
rate constants α_h and β_h for opening and closing of
inactivation gates and with equilibrium dissociation
constants K^* for the drug-channel complex. The notation
distinguishes the K's for the R and I forms of the channel
and the α's and β's for drug-free and drug-blocked forms of
the channel. A principle of statistical physics called
"detailed balances" now allows us to write down a
relationship among all these constants. Provided that there
is no external source of energy injected continuously to
power indefinite net cycling of channels around the scheme,

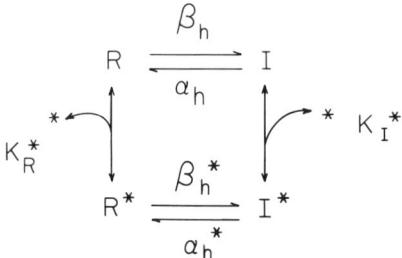

*Fig. 11. Possible kinetic basis for the interactions
between drug (*) binding and inactivation of Na channels.
States of the channel as in Fig. 10. The normal inactivation
process has the conventional rate constants α_h and β_h.
Inactivation in drug-blocked channels has modified rate
constants α_h^* and β_h^*. The drug dissociation constants are K_R^*
from the resting state and K_I^* from the inactivated state.*

the product of all equilibrium constants proceeding in one direction around the cycle must be 1.0. For example, proceeding clockwise in Fig. 10 we get:

$$(\beta_h/\alpha_h) \cdot (1/K_I^*) \cdot (\alpha_h^*/\beta_h^*) \cdot K_R^* = 1$$

and rearranging gives:

$$\frac{(\beta_h^*/\alpha_h^*)}{(\beta_h \; \alpha_h)} = \frac{K_R^*}{K_I^*}$$

In words, the result is that an increased tendency (β_h^*/α_h^*) for blocked channels to become inactivated goes hand in hand with stronger binding of drug to inactivated channels. Thus, while we believe that gating continues to occur in blocked channels, at least one of the rate constants α_h^* or β_h^* must be different from the normal, stabilizing the blocked inactivated state as is seen in the inactivation experiments of Fig. 6. Such considerations must be included in any quantitative model of inactivation in the presence of drugs.

IV. CALCIUM AND GATING

We can now understand the effect of calcium ions. In the absence of drug, raising the external calcium concentration decreases the fraction of sodium channels in the inactivated state at the resting potential. Indeed, the entire voltage dependence of sodium inactivation is shifted to more positive potentials (Weidmann, 1955; Frankenhaeuser & Hodgkin, 1957). The shift is now understood to result from an electrostatic effect of divalent cations attracted to negative external surface charges on the membrane (Fig. 2; see also references in Hille, Woodhull & Shapiro, 1975). The local field set up by these cations fools the sensor into reporting that the membrane is hyperpolarized, and, therefore, the inactivation gates open. In the presence of local anesthetic, adding calcium increases the tendency for inactivation gates to be opened and decreases anesthetic binding. Hence, calcium ions probably act quite indirectly to antagonize local anesthetic block by an effect on the gating which secondarily modulates drug binding (Hille, 1977).

V. DRUG CHARGE AND pH

Strichartz's (1973) study of block with quaternary lidocaine inside the node of Ranvier closely paralleled Heistracher's (1971) observations with a quaternary derivative of ajmaline on cardiac Purkinje fibers. Both found strong use dependence of the block and a slow recovery at the resting potential, taking minutes. In terms of the diagram in Fig. 10, it is likely that lipid-insoluble charged drug molecules are constrained to enter or leave the receptor by the O-O* pathway involving open gates. This probably corresponds to the aqueous path by the infrequently open gates in Fig. 4. In effects, the lifetime of the drug-receptor complex is greatly prolonged because closed gates trap the cationic drug molecule for a long time. On the other hand, neutral drug or amine molecules capable of being neutral can react with and leave Na channels even when the gates are closed (pathways R-R* and I-I*), making their equilibration rates much faster. This reaction probably involves diffusion of neutral drug through the hydrophobic phase of the membrane, the second major path indicated in Fig. 4.

We have already seen (Fig. 9) that lowering the external pH slows the on and off reactions of amine molecules like lidocaine, transforming their fast use dependence, characteristic of lipophilic neutral molecules, into the slow use dependence and slow recovery more characteristic of charged drug forms. The pH effect is due to a change in the degree of ionization of the drug rather than to a change in the receptor, since we do not find an effect of external pH on the action of permanently neutral benzocaine or of permanently charged quaternary lidocaine (Schwarz *et al.* 1977). However, we also obtained the suprising result that the external (and not the internal) pH is the important variable, although external charged drug molecules have no access to their receptor. This paradox may be resolved by postulating a direct pathway for protons to exchange between the external medium and drug already bound to the receptor. This new postulated exchange is shown as an arrow through the external selectivity filter in Fig. 4. Low external pH lengthens the lifetime of the drug-receptor complex by protonating the bound drug molecule, thus causing it to be trapped by the gates of the channel. A kinetic model of this process is given by Schwarz *et al.* (1977).

VI.   CONCLUSION

    Most experimental analyses of the action of lidocaine-like
drugs on the heart use the maximum upstroke velocity as a
measure of inward sodium current (e.g., Weidmann, 1955;
Heistracher, 1971; Chen *et al.*, 1975; Weld & Bigger, 1975;
and many others). Such experiments led to the well-known,
plausible explanation for antiarrhythmic action based on
modifications of conduction velocity, action potential
duration, wavelength refractory period, excitability, etc.,
found in most textbooks. The recent work on local
anesthetics from our laboratory and others on axons and
skeletal muscle has rediscovered many phenomena already
established in the cardiac literature, but because axon and
muscle preparations can be voltage clamped more reliably, it
has been possible to carry the analysis further and suggest
underlying mechanisms in more detail. The action of
lidocaine-like drugs is quite complex, as it is closely tied
in with the gating processes of the Na channels which are not
yet understood in any preparation. The considerable overlap
of phenomena seen with heart, axon, and skeletal muscle
suggests that the Na channels of the latter two preparations
can serve as a useful model for Na channels of the heart in
analyzing the molecular mechanism of these phenomena.

DISCUSSION

    Tsien: To add to the possible points of relevance
between your results and those of heart, you mentioned the
fact that ischemic tissue in heart becomes acidotic, and you
didn't specify whether you meant intracellular or
extracellular pH. One of the first things that happens when
you cut off the oxygen supply is that the intracellular pH
falls. If that were the case, then the local anesthetic
might get trapped inside the cell purely on the basis of there
being more protons inside the cell.
    Hille: Do you know whether the extracellular pH changes
and how fast? I'm interested in that.
    Tsien: I don't think the extracellular pH changes
nearly as much or as fast when you put on mild events of
anoxia. Perhaps one problem with the experiments that you
do is that you change the extracellular pH. It's not clear
to me whether you actually clamp intracellular pH or that
you know the intracellular pH. That might be an important

factor in interpreting the experiments. Do you think that the protons are going through the sodium channel, or through some other pathway? The question I had was whether any molecule, charged or uncharged, which biases the membrane toward the inactivated state (by binding to the "I state" forming the "I* state") should also be expected to change in steepness of the voltage dependence of the apparent inactivation curve. Lidocaine and procaine seem not to change the steepness. Cocaine, which is the drug that Weidmann used, did not change the steepness, but some of the other drugs that he used did change the steepness.

Hille: The answer to that might lie in the following consideration: In the Hodgkin-Huxley scheme for the opening and closing of inactivation gates, there is an α_h and a β_h. You could derive the kinds of changes that I showed, that is, shifts together with a decrease of slope, as with lidocaine, by lowering the opening rate, α_h, and not changing β_h. In fact, to fit our results you lower the opening rate to 1/50 the normal rate. This is also in answer partly to that question about voltage shift; in order to derive a shift of this type, it's not necessary to postulate that all parameters are shifted. You could just say that the rate of opening has slowed down, i.e., a drug bound to the channel slows its rate of opening. This might also account for the slowness of the recovery from block in addition to the shift.

Morad: I wonder whether you considered the possibility that in these rapid beatings or in the case of anoxia, internal sodium concentration rises, and therefore driving force for Na across the membrane decreases. Have you done an experiment where you clamp the membrane above E_{Na} or where you drive sodium out basically, and then see whether in fact repetitive stimulation will or will not decrease this type of local anesthetic effect?

Hille: If you make fast repetitive depolarizations beyond E_{Na} in the presence of drug, then the drug reduces the conductance of sodium channels. But I think that may not be what you meant. You may have meant that you wanted some pulse structure to be used which changed the internal sodium concentration first and then to test the drug. In our experiments we would do that a different way; we would just change the sodium concentration inside by cutting the ends of the fiber in whatever we wanted to, which takes 5 min. I have never studied the interactions between sodium concentration and drug effects.

Armstrong: I wonder if you might summarize briefly the mechanism that you postulate for antiarrhythmic drugs in the heart?

Hille: These kinds of compounds bind preferentially to the inactivated state of sodium channels. Therefore, they will bind preferentially to any channel which is depolarized or has been depolarized for more than a few msec. This will include those channels which have gone through a normal action potential and those channels which happen to be at a low "resting potential" because that part of the tissue is at low potential. The next action potential cannot follow in some kind of cyclic reentry or extrasystole situation until the slow recovery or the slow loss of this drug from channels has occurred. And this will only occur after repolarization is sufficient. Already one reads in textbooks that many of these drugs prolong the refractory period, etc., so the kind of explanation I have made, which comes from our experiments, is really not very different from what is generally thought from less specific experiments that were done previously.

REFERENCES

1. Chen, C.-M., Gettes, L.S., & Katzung, B.G. (1975). Effect of lidocaine and quinidine on steady-state characteristics and recovery kinetics of $(dV/dt)_{max}$ in guinea pig ventricular myocardium. *Circulation Res.* *37*, 20-29.

2. Courtney, K.R. (1975). Mechanism of frequency-dependent inhibition of sodium currents in frog myelinated nerve by the lidocaine derivative GEA 968. *J. Pharmac. exp. Ther.* *195*, 225-236.

3. Frankenhaeuser, B. & Hodgkin, A.L. (1957). The action of calcium on the electrical properties of squid axons. *J. Physiol.* *137*, 217-243.

4. Frazier, D.T., Narahashi, T. & Yamada, M. (1970). The site of action and active form of local anesthetics. II. Experiments with quaternary compounds. *J. Pharmac. exp. Ther.* *171*, 45-51.

5. Heistracher, P. (1971). Mechanism of action of antifibrillatory drugs. *Nauyn-Schmiedebergs Arch. Pharmac.* *269*, 199-212.

6. Hille, B. (1977). Local anesthetics: hydrophilic and hydrophobic pathways for the drug-receptor reaction. *J. gen. Physiol.* *69*, 497-515.

7. Hille, B. & Campbell, D.T. (1976). An improved vaseline gap voltage clamp for skeletal muscle fibers. *J. gen. Physiol.* *67*, 265-293.

8. Hille, B., Woodhull, A.M. & Shapiro, B.L. (1975). Negative surface charge near sodium channels of nerve: Divalent ions, monovalent ions, and pH. *Phil. Trans. R. Soc. B. 270*, 301-318.

9. Khodorov, B.I., Shishkova, L.D. & Peganov, E.M. (1974). The effect of procaine and calcium ions on slow sodium inactivation in the membrane of Ranvier's node of frog. *Bull. exp. Biol. Med. 3*, 13-14.

10. Khodorov, B., Shishkova, L., Peganov, E. & Revenko, S. (1976). Inhibition of sodium currents in frog Ranvier node treated with local anesthetics: Role of slow sodium inactivation. *Biochim. biophys. Acta 433*, 409-435.

11. Schwarz, W., Hille, B. & Palade, P.T. (1977). Local anesthetics: Effect of pH on use-dependent block of Na channels in frog muscle. *Biophys. J. 20*, 343-368.

12. Strichartz, G.R. (1973). The inhibition of sodium currents in myelinated nerve by quaternary derivatives of lidocaine. *J. gen. Physiol. 62*, 37-57.

13. Weidmann, S. (1955). The effects of calcium ions and local anesthetics on electrical properties of Purkinje fibres. *J. Physiol. 129*, 568-582.

14. Weld, F.M. & Bigger, Jr., J.T. (1975). Effect of lidocaine on the early inward transient current in sheep cardiac Purkinje fibers. *Circulation Res. 37*, 630-639.

MODELS OF GATING CURRENT
AND SODIUM CONDUCTANCE INACTIVATION

Clay M. Armstrong

Department of Physiology
University of Pennsylvania
Philadelphia, Pennsylvania

It is now widely accepted that sodium and potassium ions
pass through nerve cell membrane by means of pores that can
open and close in a fraction of a millisecond. The opening
and closing is often thought to result from movement of a
specialized "gating" region of the pore, and the remainder of
the pore is viewed as fixed in conformation. (An alternate
view is that gating instead is actually the aggregation of
several subunits to form a pore, where none pre-exists.)
A sodium channel has two gates, called activation and
inactivation, since one opens or activates the pore following
depolarization, and the other closes or inactivates it after
it has conducted for a time. The two gates are distinct both
behaviorally (Hodgkin & Huxley, 1952*b*) and biochemically,
since inactivation can be destroyed by perfusing an axon
internally with the enzyme mixture pronase (Armstrong,
Bezanilla & Rojas, 1973). In the Hodgkin & Huxley
formulation (1952*a*) activation and inactivation are
completely independent of each other (each gate can open and
close regardless of the other) but more recent information
(Goldman & Schauf, 1972), particularly from gating current
studies (Armstrong & Bezanilla, 1977), indicates that the two
gates interact with each other. Gating current is a small
nonlinear component of capacitative current that is produced
by movement of charged structures that give the activation
gate its sensitivity to membrane voltage: the structures

move in response to a change of the membrane field and open
or close the activation gate. In theory, any voltage-
sensitive gating process must have gating current associated
with it.

This paper briefly reviews the gating current evidence
regarding coupling of the activation and inactivation gates,
and discusses three models of increasing complexity to
explain the coupling.

I. METHODS

In order to see gating current it is necessary to
suppress ionic current through the membrane pores. This is
done by substituting the impermanent ions, e.g. Tris
(tris-hydroxymethyl-aminomethane) or TMA (tetramethylammonium)
for Na^+ and K^+ both internally and externally. As a further
precaution, tetrodotoxin (TTX) is used to eliminate any
residual current through the Na channels. TTX eliminates
I_{Na} without affecting gating current (I_g), and this has been
rationalized by saying that the gating structures are at
the inner surface of the membrane while the binding site for
TTX is at the outer surface.

A second important procedure is the suppression of the
large linear portion of capacitative current. Gating current
is a small part of capacitative current, and is seen only in
the voltage range where gating occurs, from roughly -70 mV
where all the gates are closed, to +20 mV where almost all
are open. Gating current is determined experimentally
by measuring total capacitative current during a voltage step
into the gating range, and subtracting the current from one
or a series of steps to a voltage outside the gating range.
The records shown here were obtained with what is called the
$P/4$ procedure; total current was recorded for a pulse of
amplitude P beginning at -70 mV, and from this was subtracted
the current for four control steps each of amplitude $P/4$,
beginning at -150 mV (see Armstrong & Bezanilla, 1975, for
details).

Sodium current was determined by recording total current
in a solution that contained only 33% of the normal sodium
concentration, and subtracting from it gating current, which
was determined after removing all Na ions and adding TTX.
A low sodium concentration was used to reduce the difficulties
inherent in attempting to control the voltage of a membrane
with a negative resistance characteristic.

II. RESULTS AND DISCUSSION

There are two major pieces of evidence for coupling of
activation to inactivation, one of them negative, the other
positive. The negative evidence is that no gating current
directly associated with inactivation has been detected.
In the Hodgkin and Huxley formulation (1952), the inactivation
gate of each channel has associated with it approximately
three electronic charges, while the activation system has a
total of six to nine. Inactivation gates should thus
contribute one-third to one-fourth of the total gating
charge movement, but this component has not been detected so
far. If, as seems likely, there is little or no inactivation
gating charge, inactivation must gain its voltage
dependence from coupling to the activation process.

The positive evidence for coupling is that inactivation
immobilizes part of the gating charge that is associated
with the activation process. This effect is illustrated in
Fig. 1. In part (a) the upward trace is I_g recorded
during a pulse from -70 to +80 mV, with control pulses from
-150 to -112.5 mV. This outward current is associated with
opening of the Na channels, and will be referred to as ON
gating current. The downward traces are inward, OFF gating
currents, recorded when V_m was returned to -70 mV after the
interval indicated. The tail after 0.3 msec at +80 mV is
larger than the one after 5 msec, as is shown clearly in
Fig. 1 (upper right trace), where the two traces are
superimposed. This has been interpreted to mean that a
portion of gating charge is immobilized by inactivation and
returns to OFF position too slowly to be detected (Armstrong
& Bezanilla, 1977). Fig. 1 (low right trace) was taken from
an axon in which approximately 70% of inactivation had been
removed by internal perfusion with the enzyme mixture
pronase. In this case, the 0.3 and 5 msec tails are more
similar, for gating charge is immobilized only in the
channels with inactivation still intact.

At -140 mV the recovery from inactivation is very rapid,
and has a time constant of about 0.5 msec (Armstrong &
Bezanilla, 1977). At this voltage the immobilized charge
is released so rapidly that it forms a distinct slow
component (Armstrong & Bezanilla, 1977), which is illustrated
in Fig. 2a. The traces are recorded at -140 mV after a
pulse to +10 mV for the indicated interval. After 0.3 msec
inactivation has not occurred to an appreciable extent, and
the current tail decays approximately as a single exponential.

After 5 msec, however, inactivation has reached a steady state, and the gating current tail is seen to have reduced amplitude and a prominent slow component. Removing about 70% of inactivation with pronase almost erases the difference between the 0.3 and the 5 msec tail, as shown in Fig. 2b. Again the interpretation is that inactivation immobilizes some of the gating charge associated with activation of the channels, and that pronase, which destroys inactivation, prevents charge immobilization. The gating current evidence thus shows that activation and inactivation are not independent, for inactivation affects gating current generated by the activation process.

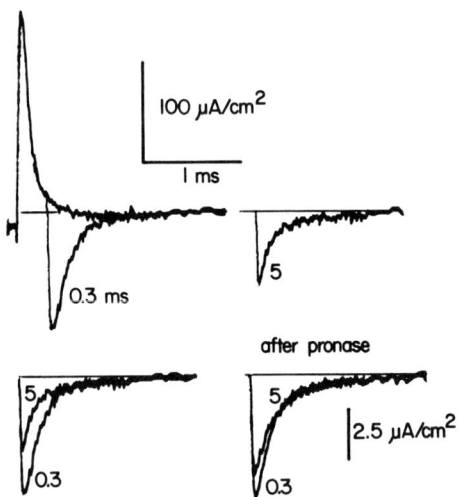

Fig. 1. Immobilization of gating charge by inactivation. The upper traces show gating current that flows following a depolarization from -70 to +80 mV (upward transient, ON gating current) and the inward OFF gating current (down) that follows return of the voltage to -70 mV after 0.3 or 5 msec at +80 mV. Total charge movement in the 0.3 msec tail is substantially larger than after 5 msec, as can be seen when the two tails are superimposed (lower left), because inactivation immobilizes a fraction of the gating charge. After destruction of about two-thirds of inactivation by internal perfusion with pronase, immobilization is much less noticeable (lower right).

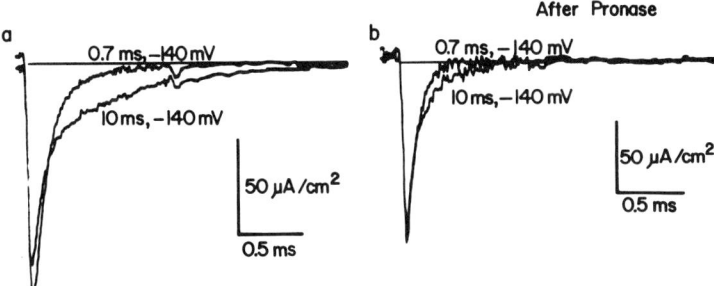

Fig. 2. At -140 mV, recovery from inactivation is sufficiently rapid that the gating charge immobilized by inactivation can be seen as a distinct slow component of OFF current in the 10 msec tail. During a 0.7 msec pulse (to +10 mV) inactivation does not occur to a significant degree, and there is no slow component on return to -140 mV. (b) after destruction of about 70% of inactivation by pronase the slow component formed by the immobilized charge is largely eliminated. (From Armstrong & Bazanilla, 1977).

The first model which was used in an attempt to explain this behavior was the following:

$$\text{Model 1} \quad x_5 \underset{b_4}{\overset{a_4}{\rightleftarrows}} x_4 \underset{b_3}{\overset{a_3}{\rightleftarrows}} x_3 \underset{b_2}{\overset{a_2}{\rightleftarrows}} x_2 \underset{b_1}{\overset{a_1}{\rightleftarrows}} x_1^* \underset{\lambda}{\overset{K}{\rightleftarrows}} x_1 z$$

In this state diagram, a sodium channel is imagined to have four closed states, (x_5, x_4, x_3, x_2) one open state (x_1^*) and one inactivated state $(x_1 z)$. The rate constants for each step are named in the diagram. Q_1, Q_2, Q_3, and Q_4 are the charge movement associated with the first four steps. The last step, inactivation, is postulated to involve no charge movement; i.e., the equilibrium and rate constants for this step are not voltage dependent. A postulate with regard to the relative values of the Q's that is being explored is the following:

$$Q_4 = Q_3 = Q_2 = 1; \quad Q_1 = 2$$

A major interest of this postulate is that from it can be derived realistic predictions of the time course of gating current tails, as described below.

The g_{Na} and I_g predictions of this scheme when the channels are opening are shown in Fig. 3 for progressively larger depolarizations (see traces 1 to 3). The lower traces are normalized g_{Na} calculated for several depolarizations, using rate constants given in Table 1. Of particular interest is the fact that the time course of inactivation as measured by τ_h (Hodgkin & Huxley, 1952a) varies widely with voltage, as it does experimentally, even though the rate constants of the inactivation step ($x_i^* \rightleftharpoons x_1 z$) are fixed. The reason for this is that the overall rate of inactivation depends on the voltage sensitive rate constants of the first four steps, $x_5 \rightleftharpoons x_1^*$.

The corresponding gating current predictions are shown in the upper traces of Fig. 3. Interestingly, the form of the predicted gating current transient changes as the voltage is made higher. The current falls monotonically for a small depolarization (1), while with larger depolarizations it has either an initial plateau (2), or a distinct rising phase (3). Experimental traces usually show similar behavior.

Predictions for tail conductance are shown in Fig. 4a,c (see Table 1 for rate constants). After a pulse of 0.3 msec (Fig. 4a) the theoretical conductance decays exponentially,

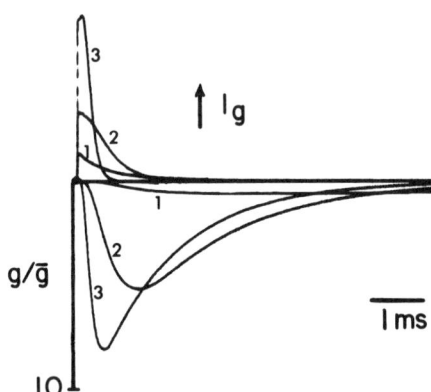

Fig. 3. Gating current and conductance predictions of the first model given in the text as channels open. $\bar{g}$ is the maximum theoretical conductance, the value of the conductance with all pores open simultaneously. Gating current is in arbitrary units.

TABLE I. Parameters of Calculations (rate Constants are in $msec^{-1}$)

	a_1	b_1	a_2	b_2	a_3	b_3	a_4	b_4	κ_1	λ_1	κ_2	λ_2	x_{50}^*	x_{10}^*	$x_1z_0^*$
Fig. 3 trace 1	3	3	3	3	3	3	1	2.5	0.7	0	–	–	1.0	0	0
trace 2	7	1	7	1	7	1	4.6	1	0.7	0	–	–	1.0	0	0
trace 3	16	0	16	0	16	0	16	0	0.7	0	–	–	1.0	0	0
Fig. 4 0.3 msec	0	20	0	30	0	50	0	8	0.7	0.09	–	–	0	1.0	0
5 msec	0	20	0	30	0	50	0	8	0.7	0.09	–	–	0	0.11	0.89
Fig. 5 0.3 msec	0	20	0	30	0	40	0	8	0.7	0.09	–	–	0	1.0	0
5 msec	0	20	0	30	0	40	0	8	0.7	0.09	–	–	–	0.11	0.89

with a time constant of 125 msec, which is the time constant
of the $x_1^* \longrightarrow x_2$ step. Gating current is predicted to have
an initial rising phase, and its final decay is approximately
exponential with a time constant similar to that for I_{Na}.
The predicted tails resemble the experimental ones
illustrated in Fig. 4b. The I_{Na} (experimental) begins to
decline approximately exponentially after 30 or 40 msec.
The conservative interpretation is that the first 40 msec of
the traces are unreliable because the clamp requires this
time to settle fully. The gating current trace continues to
rise in amplitude for some time after 40 msec, peaks, and
then decays with a time constant of about 175 msec. The
rising phase of tail gating current has in the past been
viewed cautiously, but there seems no valid experimental
reason for doubting that it exists. Thus Model 1 correctly
predicts similar decay time constants for I_{Na} and I_g, and
predicts an early time course for I_g that resembles
experimental traces. It also correctly predicts that at
more negative voltages the ratio of τ_{I_g} and $\tau_{I_{Na}}$ becomes
larger (Neumcke, Nonner & Stämpfli, 1976) a consequence of
the greater voltage sensitivity of the $x_1^* \rightleftarrows x_2$ step.

Predictions of the model for g_{Na} and I_g following a 5 msec
pulse during which inactivation goes to completion are shown
in Fig. 4c. Both currents are very small and I_{Na} has been
plotted at expanded scale in Fig. 4d, to emphasize the
prediction that it has a finite value after the fast phase
of decay; i.e., during recovery from inactivation. The
corresponding experimental records are shown in Fig. 4e, and
they are clearly different from the prediction. Experimental
I_g is much larger than predicted, while I_{Na} (and consequently
g_{Na}) are immeasurably small after the phase of rapid decay.

The absence of I_{Na} in the period after the step back to
-70 mV can be confirmed in another way. It can be shown
that for Model 1 the time integral of g_{Na} after the 10 msec
pulse should be equal to or larger than (if not all channels
have opened in 0.3 msec) the same integration performed after
the 0.3 msec pulse. The reason is that the inactivated
channels must pass through state x_1^* as they are closing, and
the actual closing rate is the product of g_{Na} and b_1. The
number closed is then the time integral of this rate, and
is the same regardless of whether all channels start in state
x_1^* or in state z. In the latter case, however, the period
of integration must be longer--long enough to allow complete
recovery from inactivation. As an empirical test of this,
I_{Na} (which is proportional to g_{Na}) has been integrated both
after the 0.3 and the 10 msec tail. In both cases the
period of integration was for 7 msec following the step,

sufficient time to allow recovery of a large fraction of
inactivated Na channels. The integrals from the experiment
had the ratio 0.03 (10 msec tail:0.3 msec tail) rather than
the predicted ratio of about one. It is clear then, that
Na channels do not conduct transiently during recovery from
inactivation, and that the predictions of this model are thus
incorrect.

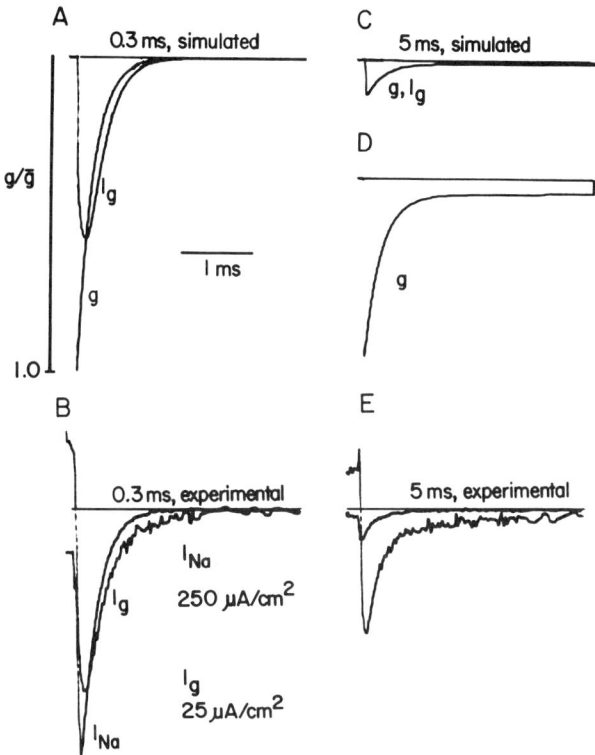

*Fig. 4. (A,C,D) Gating current and conductance
simulations of channel closing after depolarizing pulses of
0.3 and 5 msec duration. After 5 msec the (theoretical)
membrane is inactivated, and both conductance and gating
current are very small. There is, however, a detectable level
of conductance during recovery from inactivation, which is
illustrated in (D), where the g curve from (C) has been
amplified by a factor of 5. (B,E) Comparable experimental
traces, recorded at -70 mV after a 0.3 or 5 msec pulse to
+80 mV.*

Model 1 thus has two serious defects. It predicts a measurable amount of I_{Na} during recovery but none is observed, and it predicts that all gating charge is immobilized by inactivation, while in fact only about two thirds of it is. Both of these problems are overcome by the following model:

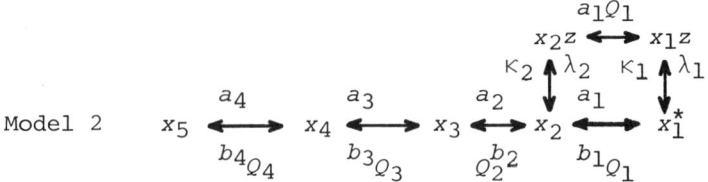

In this model x_5, x_4, x_3, and x_2 are closed states; x_1^* is the conducting state; $x_1 z$ is an open-inactivated state, meaning that the activation gate is open but the inactivation gate is closed; and $x_2 z$ is a closed-inactivated state, meaning that both the activation and the inactivation gates are closed. The steps that are voltage dependent have a Q next to them, and it is assumed that Q_1 is twice as large as the other Q's. The κ_1, λ_1, κ_2, and λ_2 are voltage-independent rate constants that govern inactivation and recovery. In this model the activation gate can partially close (to state $x_2 z$ while the channel is inactivated and the component of gating current that does not inactivate is associated with this step ($x_1 z \rightleftharpoons x_2 z$). This component is thus generated by a step that closes the channel so that it does not conduct during recovery from inactivation.

Calculations from this model are shown in Fig. 5 (parameters are in Table 1). The conductance and gating current predictions after a 0.3 msec step are almost identical to those for Model 1, which were described above. After a 5 msec step the predicted conductance is largely inactivated, and the portion that is not decays very quickly (Fig. 5B). Unlike Model 1, there is no detectable conductance after this fast phase of decay, for the inactivated channels have entered state $x_2 z$, and recovery follows the path $x_2 z \longrightarrow x_2 \dashrightarrow x_5$. x_1^* is not on the recovery path, and the channels thus do not conduct during recovery, as illustrated in Fig. 5C where the g prediction has been multiplied by a factor of 5. Most of the gating current in Fig. 5B is generated by the step $x_1 z \longrightarrow x_2 z$, but there is a small contribution from closing of those channels that have not inactivated ($x_1^* \dashrightarrow x_5$). The amplitude of this current is in reasonable agreement with experiment (Fig. 1).

Model 2 is thus in much better agreement with the data than is Model 1, but it nonetheless fails to reproduce two experimental observations. The first problem is that the rate of recovery from inactivation is voltage dependent. As noted above, the time constant of recovery changes from about 7 msec at -70 mV to 0.5 msec at -140 mV (e-fold/27 mV), and Models 1 and 2 cannot account for this, since both κ's and λ's are postulated to be independent of voltage. The experimental basis for this postulate is that no charge movement that is directly associated with inactivation has been observed. The second problem is that neither of the models provides an explanation for the slow component of ON gating current that has been reported (Armstrong & Bezanilla, 1977; Bezanilla & Armstrong, 1975).

Both of these defects have been remedied by a more complex model (Armstrong & Bezanilla, 1977) that can account for both voltage dependent recovery kinetics and the slow

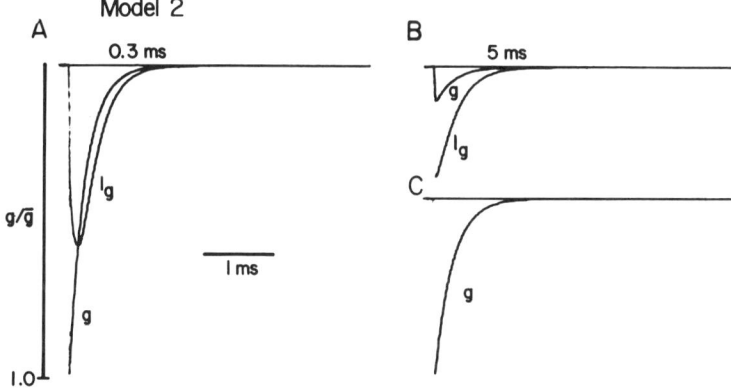

Fig. 5. Simulations of OFF gating current and conductance using the second model in the text. The simulated tails after 0.3 msec (A) are very similar to those in Fig. 4A. After inactivation has occurred (B), there is a relatively large gating current tail that is associated with the step from open inactivated to closed inactivated ($x_1z \longleftrightarrow x_2z$). This step closes the channel so there is no detectable conductance after about a millisecond, as can be seen in (C) where the g trace from (B) is amplified by a factor of 5. (C) should be compared with part (C) of Fig. 4.

component of ON gating current. This model has been
presented in full in the reference cited, and will not be
described here except to say that it is an extension of
Model 2, and like that model, it predicts partial
immobilization of charge and the absence of conductance
during recovery from inactivation.

In summary, this paper has presented two initial stages
in the development of a model for Na inactivation.
Observations that must be accounted for are the apparent
voltage dependence of inactivation, the absence of gating
current directly associated with inactivation, coupling of
activation and inactivation, partial immobilization of gating
charge by inactivation, absence of measurable Na conductance
during recovery from inactivation, and voltage dependence
of the recovery rate. The second model presented here falls
short of the complexity required to explain all of these
findings. A more complex model has been presented elsewhere
(Armstrong & Bezanilla, 1977), but requires further
experimental verification.

DISCUSSION

Huxley: A purely verbal point. I wonder whether it
isn't reasonable on the basis of the scheme that you put
up to call the slow component an inactivation gating
current. In your model the slow component is controlling
inactivation, but it is not simultaneous with it. But I
think that is not suggested by the verbal statement that
inactivation gating current does not exist.

Armstrong: Yes, well you're quite right. I suppose
I've leaned over backwards for the sake of provocation.
Certainly in this model the slow component is associated
with inactivation, but not directly.

Hille: You showed apparent immobilization of gating
charge during depolarization itself. Does immobilization
occur only after a delay, as you might expect if
inactivation itself only occurs after a delay? Secondly,
it was interesting to note that when you plotted
immobilization on the same scale as the sodium current,
the immobilization was a much longer trace.

Armstrong: Oh well, that is simply a question of
scaling.

Hille: Charge movement decreased a lot in here (before
the conductance rose) which seems to imply that a lot was
happening before any inactivation should occur according
to your model. That's one of the big issues, because the

Hodgkin-Huxley scheme says that inactivation starts
immediately from zero and goes all the way to completion.
Apparently in your slide gating charge has inactivated or
immobilized quite a bit before the conductance rises.

Armstrong: Well the data are really not good enough to
detect the small lag which is in question here, which is a
few hundred μsec. However, the other side of the slide was
a semilogrhythmic plot which I didn't talk about, but it
showed that in at least one experiment there is a lag in
immobilization of the charge which is consistent with the
lag in the onset of inactivation.

Tsien: I'd like you to clarify a point that you didn't
have very much time to speak about during your talk. It's
the extra OFF charge that appears after you depolarize the
fiber for a long period of time, the 20% extra OFF charge;
that is, OFF charge exceeds ON by 20%. If that's
associated with potassium gates, one would like to get rid
of it in some way, i.e., subtract it off. Clearly you can't
really subtract it off in a neat way. Is it possible, then,
that this will foul-up all the analysis of the earlier
gating current?

Armstrong: Well, there's no experimental way I know of
for getting rid of that component. There's no evidence,
nothing worthy of the name, as to its origin. A reasonable
guess is that it's associated with potassium channels and
an almost equally reasonable guess is that it's associated
with inactivation. As for subtracting it off, I don't know
how to do it. It certainly fouled up the tails after long
depolarizations in the sense that there's a component
which one cannot identify.

Adrian: I wonder whether expressing the activation
processes in four serial steps means that we can now abandon
the expectation that the gating current should obey dm/dt
precisely and start off initially at maximal amplitude.

Armstrong: Yes, by writing the rate constants in a
scheme of this sort appropriately, it mimmicks Hodgkin and
Huxley kinetics exactly.

Adrian: Yes, but the chances of it doing that might be
rather slim, and instead of getting a gating current which
started at the maximum amplitude, it might be able to
produce a gating current which has a rising phase which is
what the records show so much more often.

Armstrong: That's exactly the subject of the talk that
disappeared with the power failure. If rather than writing
the rate constants 3 alpha, 2 alpha, alpha, which is the

Hodgkin and Huxley sequence, one writes alpha, alpha, alpha, then the conductance looks almost exactly the same. That is, for the sequence alpha, alpha, alpha or 3 alpha, 2 alpha, alpha or even for alpha, 2 alpha, 3 alpha, the conductance time course looks almost exactly the same for all. The predicted gating current, on the other hand, looks quite different. In the Hodgkin and Huxley sequence it decays exponentially, while with this one (alpha, alpha, alpha) it has an initial plateau, and for alpha, 2 alpha, 3 alpha it has a very distinct rising phase. Whether there is in fact a rising phase or not, we don't know, but in theory there could well be one.

REFERENCES

1. Armstrong, C.M. & Bezanilla, F. (1975). Currents associated with ionic gating structures in nerve membrane. *Ann. N.Y. Acad. Sci. 264*, 265-277.
2. Armstrong, C.M. & Bezanilla, F. (1977). Inactivation of the sodium channel. II Gating current experiments. *J. gen. Physiol. 70*, 567-590.
3. Armstrong, C.M., Bezanilla, F. & Rojas, E. (1973). Destruction of sodium conductance inactivation in squid axons perfused with pronase. *J. gen. Physiol. 62*, 375-391.
4. Goldman, L. & Schauf, C.F. (1972). Inactivation of the sodium current in *Myxicola* giant axons. Evidence for coupling to the activation process. *J. gen. Physiol. 59*, 659-675.
5. Hodgkin, A.L. & Huxley, A.F. (1952a). The dual effect of membrane potential on sodium conductance in the giant axon of *Loligo*. *J. Physiol. 116*, 497-506.
6. Hodgkin, A.L. & Huxley, A.F. (1952b). A quantitative description of membrane current and its application to conduction and excitation in nerve. *J. Physiol. 117*, 500-544.
7. Neumecke, B., Nonner, W. & Stämpfli, R. (1976). Asymmetrical displacement current and its relation with the activation of sodium current in the membrane of frog myelinated nerve. *Pflügers Arch. 363*, 193-203.

III

Ionic Transport Mechanisms in Generation of Cardiac Action Potential Plateau

INWARD RECTIFIER: CARRIERS VS. CHANNELS

R.H. Adrian

Physiological Laboratory
Cambridge CB2 3EG England

It is a property of both cardiac and striated muscle that in the resting state, when they are not generating an action potential, the current-voltage relation of the membrane is markedly nonlinear. Normally, one assumes that the current-voltage relation of the leak current is linear. Hodgkin & Huxley (1952) for instance, assumed that the leak current in the squid axon was a linear function of the voltage, and indeed it probably is to a reasonable degree of approximation. But in striated muscle and cardiac muscle, there are several systems where one cannot make that assumption; that is, the membrane current rectifies very strongly, in what was originally called an anomalous way. These systems are now referred to as inward rectifying systems.

Originally these systems were called anomalous rectification systems, for the following reason. If one supposes that the membrane is a region of a certain width in which an ion is operated upon by thermal and electrical forces only, and if there is a difference in concentration on either side of the membrane such that current flows across the membrane from the side where the concentration is at a high level, then one would expect the resistance of that membrane to fall, because it will fill up progressively with more and more of the ion. On the other hand, when the current flows from the side where the concentration is low, then the resistance of such a membrane will rise, because the ions carrying the current across the membrane are, so to speak, swept out of the membrane. Making appropriate assumptions about the membrane field, one can calculate a current-voltage relation for a system in which there is 2.5 mM potassium

outside, and something like 150 mM inside, which are
approximately the concentrations that one would find in a
striated muscle of the frog or in a cardiac muscle. There
would be zero current potential at the equilibrium potential
for potassium, around -100 or -90 mV. For outward current,
one would expect the current to rise more steeply than a
linear relation, and for inward current, coming from the low
concentration outside, one would expect to get less than the
linear current.

If the concentrations on each side of such a membrane
were precisely equal, then one would expect no rectification
at all; that is, the current-voltage relation would be a
straight line through zero potential, and the system would
behave in a linear manner. Now, it is very clear that in
cardiac and striated muscle that is not the case. What one
does in fact see, within a voltage range that does not excite
any of the potassium conductance for the action potential, is
an inward rectification. That is, one can pass very large
inward currents of potassium from the region which has a
small potassium concentration, and one can pass only very
small outward currents from the high internal potassium
concentration. This type of curve originally was called
anomalous rectification, because it was counter to the
expectation of a simple electrodiffusion model. Then, more
descriptively, it was called inward rectification, because
it is the inward current that is large and apparently
dominant.

There is no very obvious reason why a striated muscle
should have this kind of current-voltage relation in the
potassium channels which are open in the normal resting state.
In cardiac muscle, however, ingenious use is made of systems
which have this kind of current-voltage relation, in order
to pacemake in the Purkinje fiber. This process has been
described by Noble & Tsien (1968). The I_{K2} system inwardly
rectifies very strongly and in addition activates and
deactivates. One must imagine a series of channels which
can be turned on and off, each of which has a current-voltage
relation of this type. The I_{K2} channels are turned on during
the plateau, so that the activation variable of this system
(*s*) rises fairly rapidly and stays high during the plateau.
Nonetheless, because the potential is at the point where the
channels are carrying a very small current, little is
contributed to the current during the plateau. On the other
hand, when the cardiac action potential repolarizes to the
diastolic potential, when the channels are still turned on,
there is a lot of current, which maintains the diastolic
potential. The slow turning off of this system during the

interval between the action potentials accounts for the slow depolarization towards the next action potential. Thus this type of rectification curve does have a particular and rather neat function in the pacemaking of the cardiac action potential in the Purkinje fiber.

Although such a rectification curve does not appear to have any function in striated muscle, for various reasons it is a good deal easier to examine it experimentally there (Adrian & Freygang, 1962). In striated muscle, the inward rectifier does have an activation and an inactivation, as was shown by Almers (1972b). In fact, however, the voltage range over which this activation occurs is considerably more negative, as will be shown below.

As stated earlier, on any simple system, when the concentrations are equal on either side of the membrane, one would expect a linear current-voltage relation. One can test this expectation by putting a muscle fiber into a solution which has roughly 100 mM of potassium outside, which of course depolarizes the fiber very rapidly. The fiber goes through a period when the potassium conductance is high, because the action potential potassium conductance is turned on. In striated muscle, this system inactivates, and one is left with a membrane through which the only channels are a chloride channel and this inwardly rectifying potassium channel. If one removes the chloride, by using sulfate solutions externally, one is left with a system which passes only potassium ions. Fig. 1 shows the current-voltage relation that one obtains in a fiber which is put in an isotonic sulfate solution with 100 mM potassium. This kind of behavior was first shown by Katz in 1949. When one moves the potential negative in the hyperpolarizing direction, to drive an inward current, one finds very large currents. The current is so large, in fact, that it is rather difficult to hyperpolarize fibers by more than 40 or 50 mV. If the membrane potential is depolarized, the current rises initially and then actually falls, so that the slope conductance is negative in some region. Then for large positive internal potentials, the current begins to increase again, but rather slowly. And at no voltage within this range is the current very much more than a small fraction of what one can drive in the inward direction. This result can be seen if a more or less constant current is passed. A constant current that makes the internal potential positive behaves as if the resistance is very high and the time constant of the membrane is very long. For inward currents, the time constant of the membrane is very short because the membrane resistance is very low.

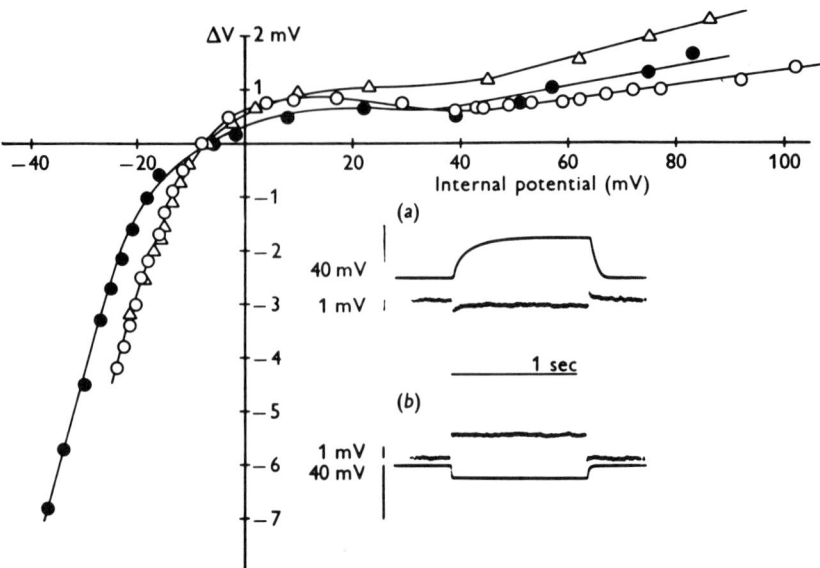

Fig. 1. The relation between membrane current (ΔV) and membrane potential for three sartorius muscle fibers in a sulfate solution containing 100 mM K. Outward current is plotted upwards. The photographic records show V and ΔV for outward current (a) and inward current (b). Reproduced from Adrian (1964).

Fig. 2 shows a striking property of this system, with respect to its selectivity. In many systems, potassium and rubidium ions behave in an almost identical manner. That is, an action potential that looks perfectly normal can be produced if potassium is replaced with rubidium both inside and outside the muscle fiber. The sodium-potassium pump in the muscle fiber handles rubidium in a way almost identical to potassium. The difference is that this inward rectifying system does not allow rubidium to go inwards. The figure shows the current-voltage relations as described previously for potassium, for a striated muscle fiber in 100 mM potassium sulfate. There are very large inward currents but only small outward currents. When the fiber is placed in 100 mM rubidium sulfate, the current-voltage relation is essentially linear, and only small currents can be passed in either direction. In addition, there is no hump of outward current; there is no region of negative slope

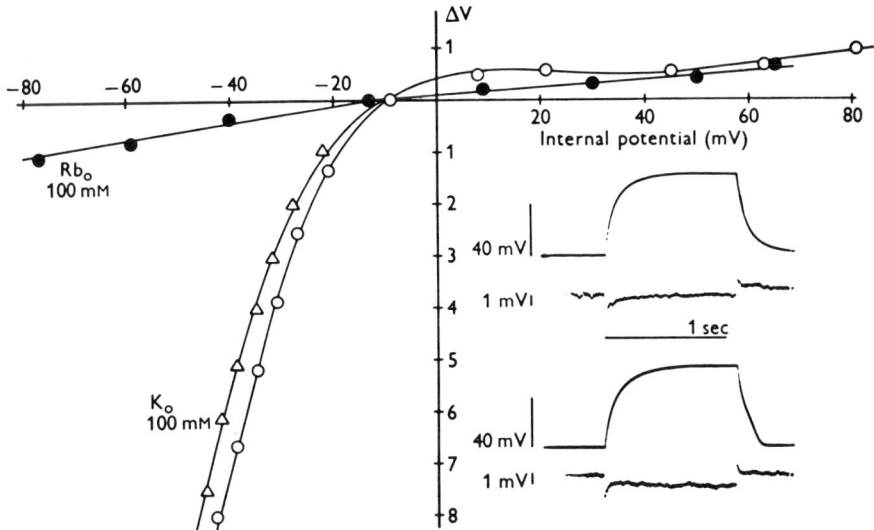

Fig. 2. The relation between membrane current (ΔV) and membrane potential for one fiber in sulfate solutions containing 100 mM K (open symbols before and after) and 100 mM Rb (filled circles). The photographic records are for outward current in Rb (above) and in K (below). Reproduced from Adrian (1964).

conductance. It is even more striking that, if 100 mM potassium is left, and only 10 mM rubidium is added, the inward current is halved. Thus rubidium will not move through the channel, and, in addition, it certainly strongly interferes with the movement of potassium.

Fig. 3 shows the kind of currents that result if one voltage-clamps the membrane of a striated muscle, holding it in this case at the resting potential of about -80, and applying a small depolarizing pulse in a low-potassium Ringer solution made up with sulfate replacing chloride. The currents are shown on a long time scale for a series of hyperpolarizing steps of larger and larger magnitude. Initially the currents are large; they then turn off and come to a steady level. This decline is the result of two processes, a depletion of potassium in the transverse tubules and a deactivation of the system through which the large inwardly rectifying currents move (Almers, 1972*a*, *b*).

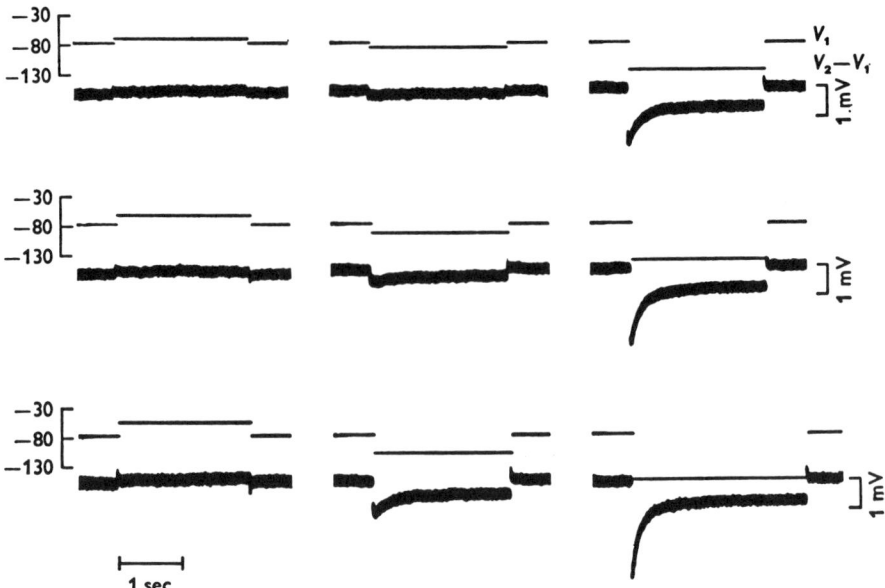

Fig. 3. Membrane currents for clamped membrane potentials between −54 and −154 mV. The fiber is in a sulfate solution with 5 mM K. Holding potential is −77 mV and temperature is 1.5° C. On current record 1 mV ≡ 6.7 μA/cm². Reproduced from Adrian, Chandler & Hodgkin (1970).

Fig. 4 is a plot of the initial current, illustrating very marked inward rectification, and the final current, shown by crosses. Although it is large at first, at -150 mV the current inactivates or is reduced. Thus in contrast to the same kind of system in cardiac muscle, the position of the activation in striated muscle is different. In striated muscle the activation is right out in the negative potential range; in cardiac muscle, it is in the region between the diastolic potential and the plateau potential.

Fig. 5 shows a very general model made to account for this inwardly rectifying system (Adrian, 1969). Rather than supposing that the ion in the membrane is subject only to electrical and thermal forces, we postulate that the passage of an ion across a membrane involves combination with some carrier molecule, here called *A*, and that the rate-limiting steps are not the transfer across the membrane, but rather, the combination of the carrier with some number of ions in the external solution. There is a reaction with ions outside

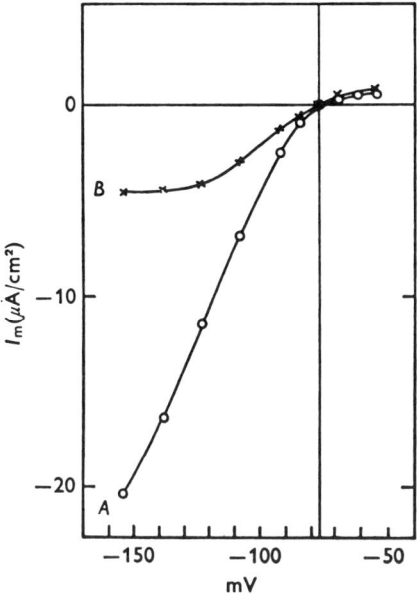

Fig. 4. *Current-voltage relation between -54 and -154 mV for fiber in 5 mM K sulfate solution. A, instantaneous currents; B, current at 2.1 sec after applying voltage. Reproduced from Adrian et al. (1970).*

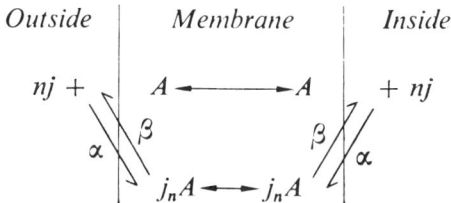

Fig. 5. *Carrier scheme for predicting current-voltage relations. A membrane carrier (A) and an ion-carrier complex (j_nA) are postulated to be very rapidly distributed across the membrane in accordance with their net charge and the membrane potential difference. The rate constants for the reaction of ion carrier are assumed to be much less than the rates of transfer of carrier and ion-carrier complex. The valence of the carrier and the number of combining ions (n) are variable parameters of the model. For details and equations see Adrian (1969).*

onto a carrier; the transfers are supposed to be sufficiently
rapid to be nonrate-limiting. The rate-limiting steps in
these systems are the steps onto and off the membrane
carrier. This is a very general representation of carrier
models. If one supposes that the quantity of the carrier is
constant, one can suggest that the carrier may have a
particular charge, and the combination of carrier and some
number of ions will therefore have another charge. The
bizarre current-voltage relations that one can get from
these systems depend essentially upon the fact that there is
a potential difference across this membrane, and the
carriers and carrier-ion complex are charged. Clearly, the
ratio of concentrations on the two sides inside the
membrane will depend on the net charge of each species.
This very general carrier model can lead to a number of
current-voltage relations.

   Fig. 6 shows current in arbitrary units on the ordinate,
and membrane potential on the abscissa. The curve marked
$z_A = 0,-1$ is calculated from a system of the kind shown in
Fig. 5, where one monovalent cation combines with one
carrier and the valence of the carrier is either 0 or -1.
Here the current-voltage relation saturates for large
voltages in either direction and is more or less linear
through the origin. The two curves marked $z_A = 1,-2$ show
that if the valence of the carrier is 1 or -2, the current
does more than saturate; it actually diminishes down to
zero in either direction. Of course, about the zero
potential it is still symmetrical when the concentration of
ions is the same on either side of the membrane. If one
takes these two valences for the carrier and calculates the
current-voltage relation for a 10:1 difference in
concentration on either side of the membrane, which is more
nearly the normal situation, a zero current potential is
produced at -58 mV. Then the current-voltage relation has
a very small inward current, but a largish hump of outward
current. Neither of these curves represents or resembles
the situation in striated muscle.

   These two models, however, can be modified in two ways
to produce current-voltage relations which at least resemble
those in the experimental situation. The first is the
assumption that Hodgkin, Huxley & Katz (1949) made in an
early qualitative explanation of their results on sodium
current. That is, one can assume a carrier which is
charged and can cross the membrane, and also a store form of
the carrier on one side of the membrane which is unable to
cross the membrane and which essentially buffers the quantity

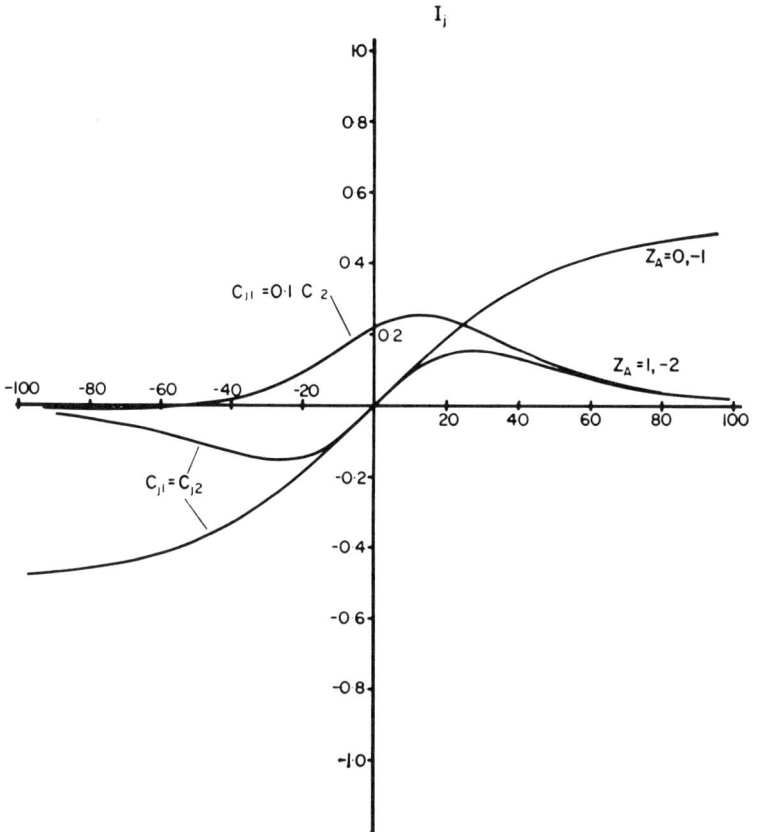

Fig. 6. Current-voltage relations for a membrane with a constant quantity of charged carrier and assuming the scheme in Fig. 5. The curves show the predictions of various values of the carrier charge (Z_A) and for two assumptions about the ionic concentrations on either side of the membrane: for equal concentrations and for a 10:1 concentration ratio. The current is in arbitrary units.

of carrier on one particular side of the membrane. With something of this kind, one can reconstruct current-voltage relations which are very similar to the experimental ones.

Fig. 7 shows a series of current-voltage relations for a model which buffers the concentration of carrier on one side of the membrane and makes the same assumptions: that the rate-limiting steps are the combination step of the ion and the carrier, and that the carrier and ion can cross the membrane. This produces a series of highly flexible current-voltage relations; they bend all over the place. For

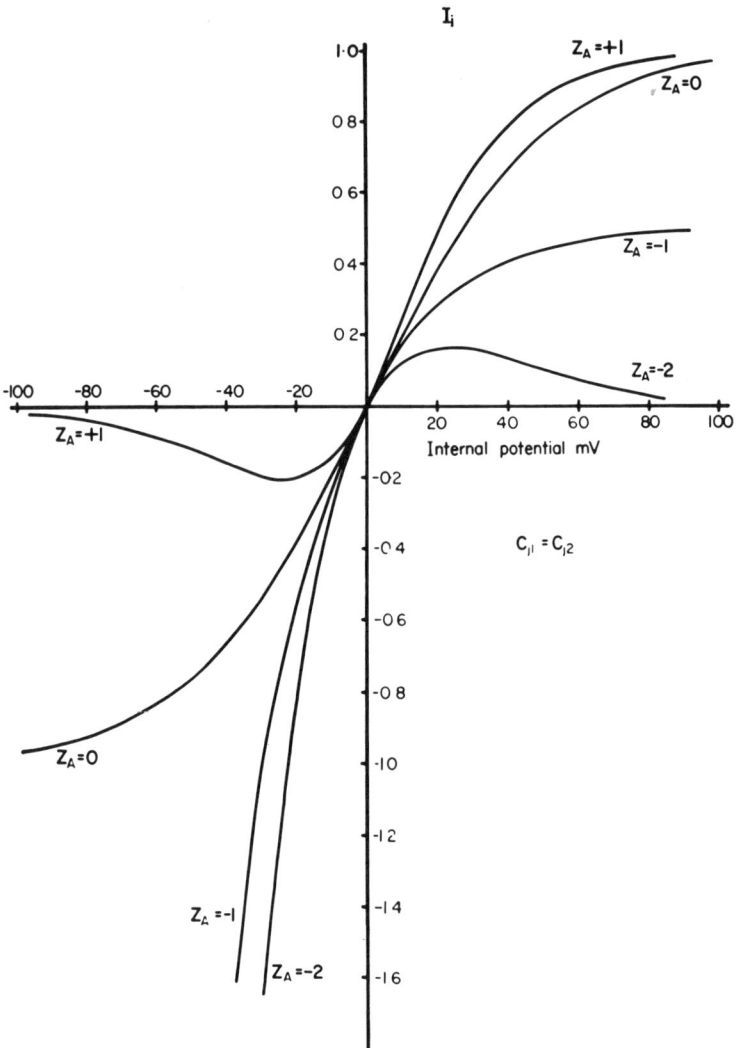

Fig. 7. As in Fig. 6, current-voltage relation for
various carrier charges and equal ionic concentrations. For
these curves the carrier concentration is assumed to be
constant at the inner boundary of the membrane--it is assumed
that there is a store of "noncarrying carrier" which can
buffer the concentration of "carrying carrier."

a doubly negative carrier molecule, the outward current at
positive internal potentials is small, rises to a maximum
as the internal potential is made less positive, and then
internal current increases indefinitely for negative internal
potentials. This current-voltage relation closely resembles
the actual experimental situation.

Alternately, one can assume that there is a large
surface potential at one face of the membrane; that is, that
there are a large number of bound charges on the membrane;
and this will substantially alter the profile of potential
in the membrane. The membrane will appear to be much more
asymmetrical than the simple systems that have been shown.
This method produces the next kind of current-voltage
relation. Fig. 8 shows part of a symmetrical current-voltage
relation which would result from assuming a constant quantity
of carrier, plotted for no potential difference, and no
surface membrane potential. If a 50 mV surface potential is
put on one or the other face of the membrane, the
current-voltage relation will actually diminish for very
large inward currents, but one can reproduce to some extent
the inwardly rectifying kind of current-voltage relationship.

In this system, perhaps, carriers are slightly favored
in that they provide an easier and possibly simpler
explanation for some of the phenomena, particularly the high
degree of selectivity between potassium and rubidium.
However, nothing that is known about this inward rectifying
system makes it possible to say with any degree of
certainty that this behavior could not be produced by
channels, if sufficient appropriate assumptions were made.
As in many other cases, it is still very much an open
question whether pores or carriers are involved. But with
both carriers or channels it requires a rather complex
mechanism to account for the very large inward currents from
the low concentration side, when one would normally expect
the system to rectify in the opposite direction.

DISCUSSION

Cleemann: I would like to know if you or anybody has
tested the temperature sensitivity of these currents.
Temperature is often a consideration when you try to
distinguish between carrier models and channels.
Adrian: Do you mean the temperature coefficient of the
conductance? I don't think that has been done. What has
been shown, by Almers (1972a, b) is that the temperature

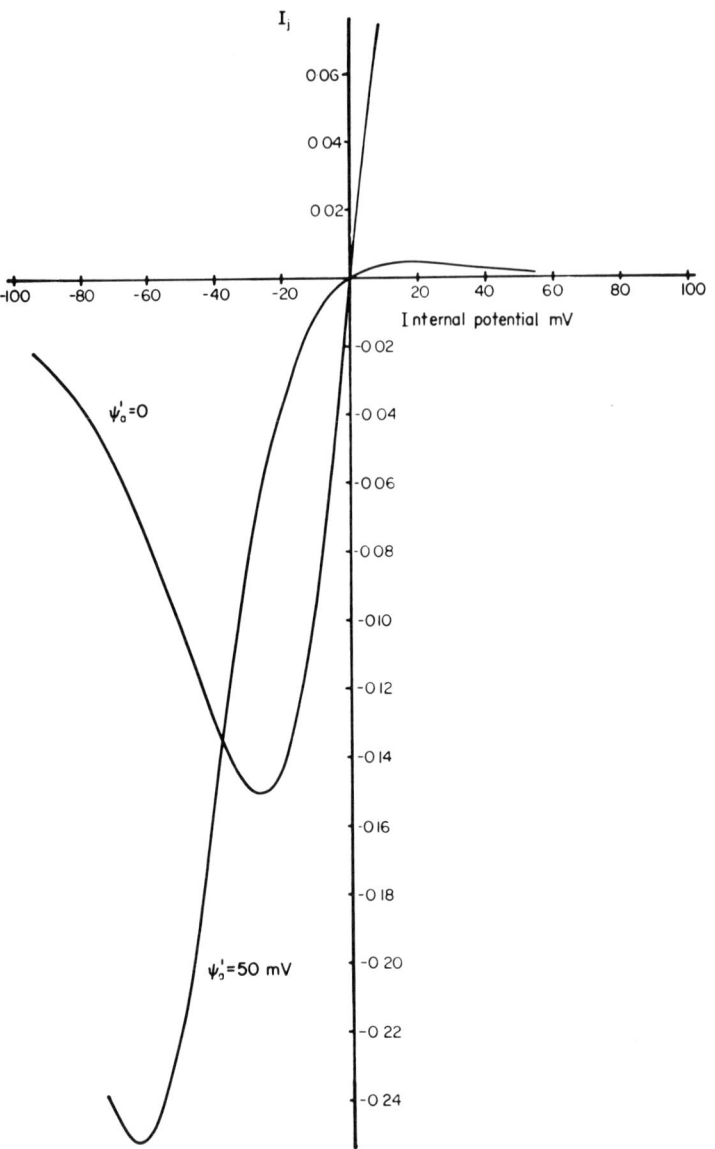

Fig. 8.   The effect of an internal surface charge on the
current-voltage curve of a system with a constant quantity
of carrier (Z_A = 1 or -2) carrying a cation across the
membrane.   The cation concentrations on either side of the
membrane are equal.

dependence of the process changing the conductance is high, or at least part of it is quite high, but I don't know of any work specifically on the temperature coefficient of the conductance, that is, on the conductance when all channels are open, although it looks to me as if Dr. Tsien has something to contribute.

Tsien: I believe that Cohen, Daut and Noble studied the temperature dependence of the pacemaker potassium current that you referred to on the board, and they found the Q_{10} for the maximum conductance when all the channels are open ($\bar{I}$). In other words, a very low temperature dependence of the type that is often found for the $\bar{I}$ functions of more typical types of excitable membrane channels. I guess the question that I'm uncertain about myself is how strong that evidence really is in deciding for one mechanism against another. I gathered that there hasn't been a similar experiment done in skeletal muscle for the rectification.

Cleemann: If you increase the extracellular potassium concentration, do you at some potentials get an increase in the outward membrane current? How does your carrier model account for that increase?

Adrian: It certainly doesn't.

Tsien: One of the interesting things about the inward rectifier in both systems you talked about is that it can be combined with Hodgkin-Huxley type gating. This is so reminiscent of the types of situation found in other excitable tissues, where "selectivity filters" are put in series with things that are called "gates." I wonder if that in itself is a bit of a tip off that we might be dealing with a mechanism which is more like a pore. I wondered whether anyone has ever described the situation for Hodgkin-Huxley type gating as combined with a carrier mechanism.

Adrian: Not so far as I am aware, and I think this is a perfectly reasonable comment. Our present thought about gating systems is that if we have a membrane, there is a hole through it, and in some way, there is a gate at some part, and a selectivity barrier at some other point. But I'm not sure that because we think of it this way, that it is necessary to suppose that there is a physical channel of the kind that is suggested by that sort of model. I feel that at the moment it is almost true to say that anything that a channel can do a carrier can do, and that anything that a carrier can do a pore can do, provided you make enough assumptions about the system.

Reuter: I wonder whether one couldn't distinguish between the possibility of a carrier and a pore, on the basis of a combination of impedance measurements and noise measurements, whether the carrier shouldn't behave in a very different way in this respect from a channel.

Adrian: I'm sure that's true, and I think that noise studies in this system might be exceedingly interesting. I don't know if anyone has done them or is doing them. But I'm sure that's something that should be done.

Hille: Just from a comparative biological point of view I would like to remind people that this happens not only in muscle, but also in eel electroplates, where Harry Grundfest did a fair amount of work showing the effects of rubidium and cesium 13 years ago. It also happens in the eggs of various kinds of echinoderms and tunicates, which may in the end turn out to be objects more amenable to experiments.

Adrian: Yes, this is true.

Armstrong: Wolf Almers and I at one point were thinking of ways of telling channels from carriers, and we decided that pores definitely would show electrical impedance at the same time that they have the single filing property which one sees with tracers. That is, the inward component of the current would be a linear function of the external current. The single filing, that sort of difference between the properties seen with tracers and with current, does not seem to be reproduced by the pore models, in which, if one has single filing, one has to postulate two ions that bind to the carrier.

Adrian: I think that's so, yes.

Armstrong: Consequently, both current and the tracer plots would show a second-order dependence in that model on the external potassium. Now, I don't know whether the experiment has ever been done.

Adrian: I think Horowitz has done that experiment.

Armstrong: Certainly they have single filing in muscle fibers, but whether it behaves as though it had electrical independence I don't know.

Adrian: Well it is certainly true, and this was the point that was made by Lars Cleemann, that there are regions of the current-voltage relation where it's clear that the systems are not independent. That is, if you take away the external potassium concentration, you decrease the outward current; you decrease the outward potassium movement. So the systems are not independent. And I think I'm right in

saying that Horowicz, Gage & Eisenberg* showed that the
fluxes behave in the same way as would be predicted from a
carrier model which took two potassiums across on each
carrier.

Armstrong: Yes, as I think Dr. Cleemann will discuss,
the decrease in the outward current could be attributed to a
decreased number of open pores, so I guess that wouldn't
properly be a demonstration of independence in the sense that
I mean, which would apply only, for example, if one had a
single open pore and could measure its current relation.

REFERENCES

1. Adrian, R.H. (1964). The rubidium and potassium
    permeability of frog muscle membrane. *J. Physiol. 175,*
    134-159.

2. Adrian, R.H. (1969). Rectification in muscle membrane.
    Prog. Biophys. biophys. Chem. 19, 339-369.

3. Adrian, R.H., Chandler, W.K. & Hodgkin, A.L. (1970).
    Slow changes in potassium permeability in skeletal muscle.
    J. Physiol. 208, 645-668.

4. Adrian, R.H. & Freygang, W.H. (1962). The potassium and
    chloride conductance of frog muscle membrane. *J. Physiol.
    163,* 61-103.

5. Almers, W. (1972a). Potassium conductance changes in
    skeletal muscle and the potassium concentration in the
    transverse tubules. *J. Physiol. 225,* 33-56.

6. Almers, W. (1972b). The decline of potassium
    permeability during extreme hyperpolarization in frog
    skeletal muscle. *J. Physiol. 225,* 57-83.

7. Hodgkin, A.L. & Huxley, A.F. (1952). A quantitative
    description of membrane current and its application to
    conduction and excitation in nerve. *J. Physiol. 117,*
    500-544.

8. Hodgkin, A.L., Huxley, A.F. & Katz, B. (1949). Ionic
    currents underlying activity in the giant axon of the
    squid. *Arch. Sci. Physiol. 3,* 129-150.

9. Katz, B. (1949). Les constants electriques de la
    membrane du muscle. *Arch. Sci. Physiol. 3,* 285-300.

10. Noble, D. & Tsien, R.W. (1968). The kinetics and
    rectifier properties of the slow potassium current in
    cardiac purkinje fibres. *J. Physiol. 195,* 185-214.

*Horowicz, P., Gage, P.W. & Eisenberg, R.S. (1968).
The role of the electrochemical gradient in determining
potassium fluxes in frog striated muscle. *J. Gen. Physiol.
51,* 193s-203s.

CALCIUM TRANSPORT IN EXCITABLE MEMBRANES

Peter A. McNaughton

Physiological Laboratory
Downing Street, Cambridge

Ionized calcium is normally about 10^5-fold more concentrated outside a cell than it is inside, a difference which has important functional implications for a number of excitable cells, such as muscles and the synaptic regions of nerve cells. This paper will summarize some recent work, performed at the Marine Biological Laboratory in Plymouth, on how one excitable cell--the squid giant axon--maintains this enormous Ca gradient across its membrane.

The first clues to the function of the Ca pump in squid axon emerged in the late 1960's. Baker, Blaustein, Hodgkin & Steinhardt (1969) showed that a Ca_o-dependent Na efflux could be observed when the Na gradient across the membrane was reversed by replacing the external Na with Li. The same authors also showed that the Ca influx was greatly increased by raising the internal sodium concentration. The sodium-calcium exchange they demonstrated was, of course, working in the wrong direction to expel calcium, but the same exchange also seemed capable of reversing at normal levels of extracellular sodium. Fig. 1, from Blaustein & Hodgkin (1969), shows that the Ca efflux from an unpoisoned axon is reduced about one-third by removal of Ca from the sea water bathing the axon, and is further reduced, to about one-half of its initial level, by removal of external Na. This experiment suggests that part of the Ca efflux might be coupled to an influx of Ca (Ca-Ca exchange) and part to an influx of Na (Na-Ca exchange). Effective Ca extrusion would be accomplished by the Na-Ca exchange, while the Ca-Ca exchange would be of less interest, as it performs no net Ca extrusion. Sodium-calcium exchange therefore seemed capable of working either to expel Ca in exchange for Na, or of reversing when Na$_O$ had been reduced to unphysiologically low levels.

Similar observations on the Ca efflux from cardiac muscle had also led Reuter & Seitz (1968) to propose that Na–Ca exchange mediated the Ca efflux in this tissue. Fig. 2, from their paper, shows that the Ca efflux from two cardiac preparations is reduced by removal of both external Na and external Ca, although, as in squid axon, there is a residual efflux not abolished by removal of both Na$_O$ and Ca$_O$.

Fig. 1 also shows the effect of depleting the internal ATP supply of the axon by poisoning with cyanide. Cyanide poisoning rapidly inhibits the ATP-dependent Na-K pump, but in the case of the Ca pump the effect is the opposite: the Ca efflux rises after a delay. The rise in efflux on poisoning can be explained by an increase in the concentration of ionized Ca inside the axon, resulting from a release of Ca from ATP-depleted mitochondria (Baker, Hodgkin & Ridgway, 1971). Even when this effect is taken into account, however, Fig. 1 demonstrates that poisoning does not totally inhibit the Ca pump.

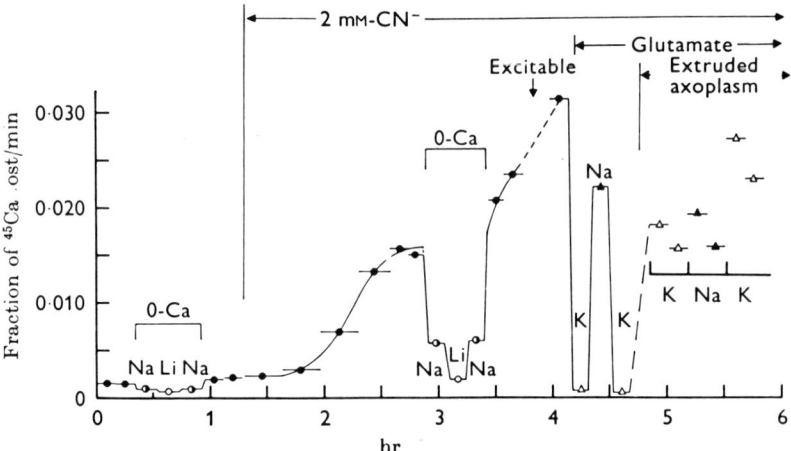

Fig. 1. Effect of removing first calcium and then sodium on Ca efflux in unpoisoned axon and in CN-poisoned axon. The second part of the figure shows the effect of K or Na glutamate and of extrusion. Abscissa: time. Ordinate: rate constant of calcium efflux. Solid circles: Na sea water. Half-filled circles: Ca replaced by Mg. Open circles: Ca replaced by Mg and Na replaced by Li. Open triangles: sea water replaced by K glutamate with 0-Ca and 0-Mg. Solid triangles: same but K replaced by Na. From Blaustein & Hodgkin (1969).

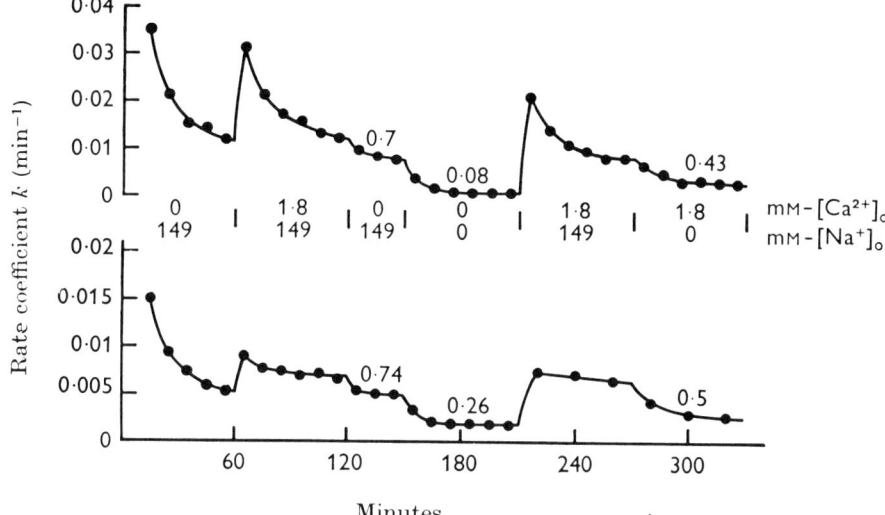

Fig. 2.   The effects of removing Ca_O and Na_O on the Ca efflux from a guinea-pig auricle (upper curve) and a ventricular trabeculum from a sheep (lower curve).   In Na-free solutions NaCl was replaced by sucrose.   From Reuter & Seitz (1968).

In the poisoned state the characteristics of the Ca efflux are quantitatively different from those in the unpoisoned state.   Almost all of the efflux is now sensitive to removal of Na_O and Ca_O, with only a tiny amount of "residual" efflux remaining.   The last part of Fig. 1 demonstrates that the sensitivity of Na_O resides in the axon membrane.   The Ca efflux from extruded axoplasm is found to be independent of Na_O.

These experiments and others (see Baker, 1972, for a review) suggested the model of an ATP-independent Na-Ca exchange shown in Fig. 3.   Calcium efflux is imagined to be driven by a coupled influx of Na running down its concentration gradient.   Depending on external and internal ionic concentrations, the pump is also capable of operating in a reversed Ca-Na exchange mode, a Na-Na mode, and a Ca-Ca mode.

This model has been extensively tested in the experiments described below.   While Na-Ca exchange still appears to be

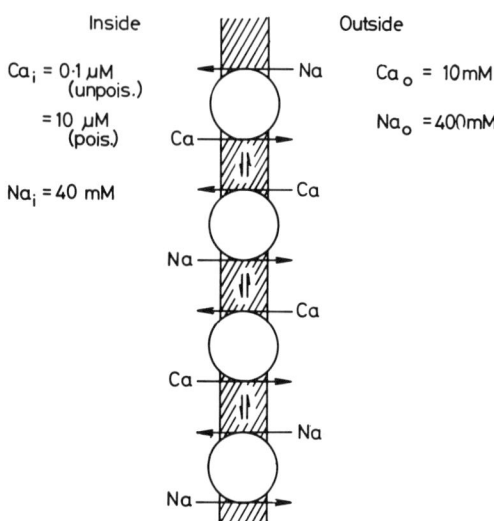

Fig. 3. An early model for sodium-calcium exchange

an important mechanism mediating Ca movements in nerve, the model of Fig. 3 has required major surgery in two respects:

1. Cellular metabolic energy, probably in the form of ATP, is now thought to interact with the pump.
2. An uncoupled Ca extrusion must be added to the exchange modes shown in Fig. 3.

ATP-DEPENDENCE OF THE Ca EFFLUX

In testing the ATP dependence of the Ca pump, it is important to distinguish changes in the activity of the pump itself from changes consequent on alterations in Ca_i. This has been accomplished in either one of the following ways:

1. By stabilizing Ca_i with Ca-EGTA buffers.
2. By preventing mitochondrial accumulation of Ca by blocking both respiration, using cyanide, and ATP-dependent Ca accumulation, using oligomycin.

When an axon pre-injected with a Ca-EGTA buffer is poisoned, the Ca efflux is observed to fall by about two-thirds (Baker & McNaughton, 1976a) or even more in axons with low Ca_i (DiPolo, 1976). Thus although the pump continues to operate when ATP has been reduced (although not totally removed) by poisoning, it does so with reduced efficiency. In Fig. 4, an intracellular injection of oligomycin has been used to stabilize Ca_i at a high level. The figure shows one reason for this decline in efficiency. Shortly after the time origin of Fig. 4A, cyanide poisoning was begun, and it

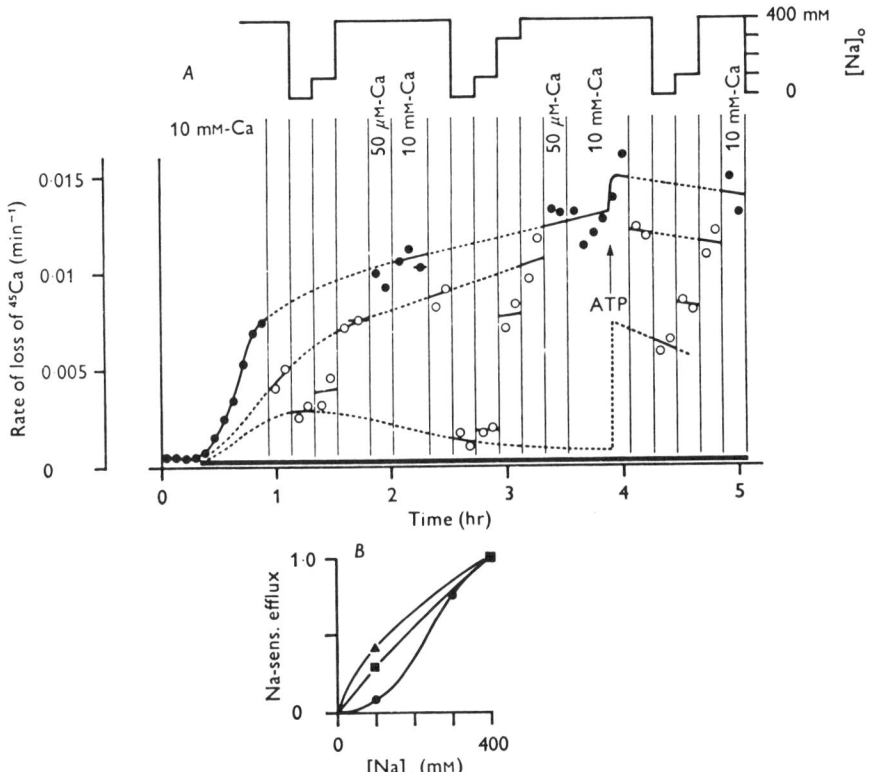

Fig. 4. *Alterations in affinity of the Ca efflux for Na_O during poisoning. Oligomycin was pre-injected into the axon, and was present in all external solutions. (A) Ca efflux during application of cyanide (black bar). Na concentration stepped from 2.5 to 400 mM as shown above the graph. Ca concentration either 50 µM or 10 mM as shown (solid circles) or nominally zero (open circles). (B) Activation of Ca efflux by Na, as determined in A, early after CN application (squares), later in poisoning (circles), and after injection of ATP (triangles). Curves drawn by eye. From Baker & McNaughton (1976a).*

produced a rapid rise in Ca efflux due to the discharge of Ca
from the mitochondria. The activation of Ca efflux by
external Na was then tested as the ATP level declined.
Shortly after poisoning was begun, the Ca efflux was found to
be activated by small amounts of Na_O, but after long poisoning
the affinity for Na_O was much lower, and the relationship
between Na_O-dependent Ca efflux became increasingly more
signoidal (see Fig. 4*B*). Injection of ATP resuscitated the
high Na affinity characteristic of the unpoisoned state.

The ATP-dependent transition between high- and
low-affinity activation of Ca efflux by Na_O was found to be
independent of Ca_i, as it could also be observed either when
Ca_i was maintained at a low level with Ca-EGTA buffers, or
when no special precautions were taken to control Ca_i during
poisoning. The change to low-affinity activation was
progressive, the affinity becoming lower and the activation
curve more signoidal as poisoning progressed (Baker &
McNaughton, 1976a). Both of these last-mentioned features
are in marked contrast to the behavior of the Na pump, where
the activation of Na efflux by K_O is simply reduced by
poisoning, with no change of affinity.

Jundt & Reuter (1977) have recently demonstrated
qualitatively similar affinity changes in cardiac muscle.
Fig. 5 shows that the affinity of the Na_O-dependent component
of the Ca efflux from guinea-pig auricle is also reduced by
poisoning, although to a smaller extent than in squid axon.
The smaller change may be due to a greater resistance of
cardiac muscle to poisoning.

The progressive changes in affinity for Na_O as ATP is
reduced can be accounted for by a model in which the pump is
assumed to bind ATP *before* Na can attach to an external site
and activate Ca extrusion. This reaction scheme is shown
below. The pump is denoted by *M*, and for simplicity the
binding of Ca_i is ignored, and only one Na_O is shown binding:

$$M \underset{K_M^{ATP}}{\overset{ATP}{\rightleftharpoons}} M \cdot ATP \underset{K_M^{Na_O}}{\overset{Na_O}{\rightleftharpoons}} M \cdot ATP \cdot Na_O \rightarrow Ca \text{ efflux}$$

Making the conventional assumption of equilibrium throughout
except at the rate-limiting ion translocation step, it can be
shown that

$$Na_O\text{-dependent Ca efflux} = \frac{\text{Maximum } Na_O\text{-dependent Ca efflux}}{1 + \dfrac{K_M^{Na_O}}{Na_O}\left(1 + \dfrac{K_M^{ATP}}{ATP}\right)}$$

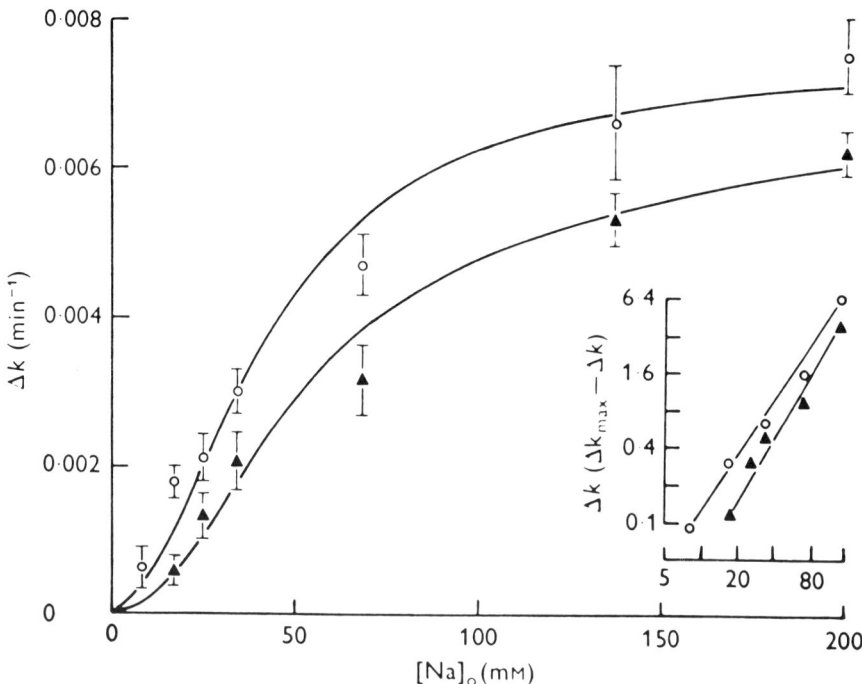

Fig. 5. Na_O-*dependent Ca efflux from guinea-pig auricles
without (circles) and with (triangles) metabolic inhibition.
Curves calculated from*

$$\frac{\Delta K}{\Delta K_{max}} = \frac{1}{1 + (\frac{K_{Na}}{[Na]_O})^2}$$

*where K_{Na} = 39 mM in normal and 53 mM in poisoned preparations.
Inset: Hill plot of the experimental data. The slopes of
the regression lines were not significantly different from 2.
From Jundt & Reuter (1977).*

This model predicts that at a constant level of ATP, the Ca
efflux will be activated by Na_O in a Michaelis fashion, but
that the affinity of activation will depend on the level of
ATP, becoming lower as ATP is reduced. The model does not
predict the increasing sigmoidicity of Na activation which
always accompanied poisoning, but this feature can be
accomodated if independent binding of several Na ions is
assumed to follow the binding of ATP. The activation of Ca

efflux by Na_O predicted by this model, for various levels of
ATP, is shown in Fig. 6.   This second model closely reproduces
the actual behavior of the Na_O-dependent Ca efflux during
poisoning.

Similar behavior in the case of the activation of Ca
efflux by Ca_i has recently been reported by DiPolo (1976),
who found that dialysing away most of the ATP reduces the
internal affinity of the pump for Ca.   This effect might also
be due to a requirement for ATP before ions are bound and
translocated.

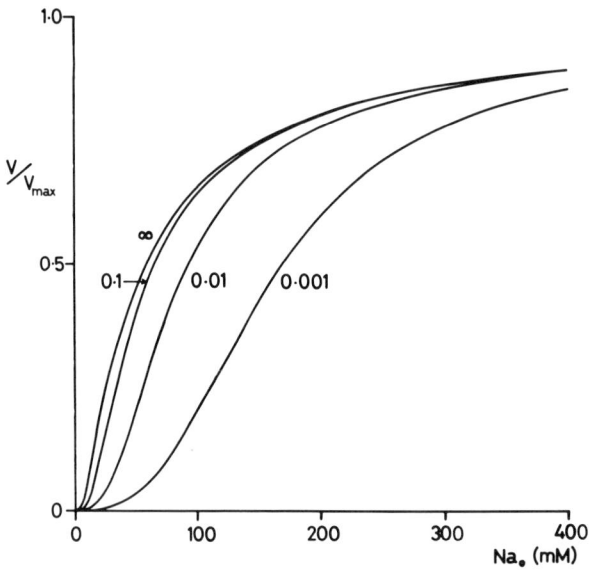

*Fig. 6.   The effect of reducing ATP on the sequential-
binding model described in the text.   The independent binding
of 3 Na^+ ions is assumed to follow the internal attachment
of ATP to the pump.   The curves are given by the equation
(Baker & McNaughton, 1976a):*

$$\frac{Na_O\text{-dependent Ca efflux}}{Max.\ Na_O\text{-dependent Ca efflux}} = \frac{1}{(1 + \frac{K_M^{Na_O}}{Na_O})^3 + \frac{K_M^{ATP}}{ATP}(\frac{K_M^{Na_O}}{Na_O})^3}$$

*The numbers near the curves give ATP/K_M^{ATP}.   The value of $K_M^{Na_O}$
is taken to be 15 mM, to reproduce the observed activation
of Ca efflux by Na_O in the unpoisoned state.*

EFFECT OF MEMBRANE POTENTIAL ON THE Na_o-DEPENDENT
COMPONENT OF THE Ca EFFLUX

The experiments described above suggest that ATP
interacts with the pump, and, according to the sequential-
binding model, the interaction is a prerequisite for Ca efflux
to occur. But these arguments provide no information as to
whether the ATP is actually metabolized. Clearly at least
part of the energy required for Na-Ca exchange could be
extracted from the Na gradient, so it is possible that ATP is
not consumed, but merely participates as a cofactor. On the
assumption that nNa^+ ions are exchanged for one Ca^{++}, the
equilibrium value of Ca_i that could be attained without using
metabolic energy is given (Blaustein & Hodgkin, 1969) by:

$$[Ca]_i = [Ca]_o \frac{[Na]_i^n}{[Na]_o^n} \exp \frac{(n-2)Fe_m}{RT}$$

where E_m is the membrane potential, and the other symbols
have their usual meanings.

Since $[Ca]_o/[Ca]_i$ is at least 10^5 (Baker *et al.* 1971;
DiPolo, Requena, Brinley, Mullins, Scarpa & Tiffert, 1976),
a value of $n = 4$ would be required to maintain the Ca gradient
using only the existing Na and potential gradients. Such
an electrogenic exchange should exhibit a substantial
potential-dependence; in fact it can be shown that the
product of the fractional increase in the forward rate (Na
in, Ca out) and the fractional depression in the back rate
(Na out, Ca in) caused by a 25-mV hyperpolarization should
be e^{2-n}. This prediction has been investigated in unpoisoned
axons in the experiments in Fig. 7, from Baker & McNaughton
(1976*b*). The potential-sensitivity of both Na_o-dependent
Ca efflux and Ca_o-dependent Na efflux are found to be much
too small to fit with a 4 to 1 exchange model, in the absence
of any additional energy input. The most likely way in
which the "passive-exchange" model outlined above could be
amended to account for the results of Fig. 7 would be to
include a direct contribution from cellular metabolic energy
to maintaining or modifying the Nao-dependent Ca efflux.

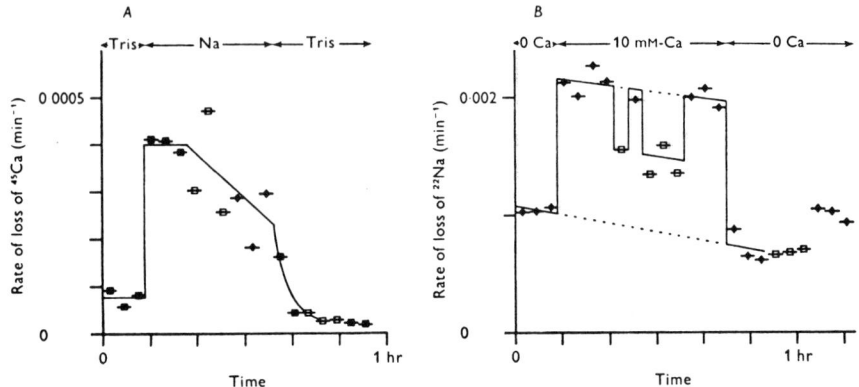

Fig. 7. Effect of altering the membrane potential, by passing a constant current, on Ca movements in unpoisoned axons. (A) Na_O-dependent Ca efflux. Membrane potential -40 mV (diamonds), -54 mV (solid squares), and -80 mV (open squares). Components other than the Na_O-dependent Ca efflux were diminished by including 300 μM La^{3+} in all sea waters. In other experiments a slight enhancement of the Ca efflux was observed during hyperpolarization. (B) Ca_O-dependent Na efflux. Membrane potential -40 mV (triangles), and -80 mV (squares). Ouabain (10^{-5} M) included in all sea waters. From Baker & McNaughton (1976b).

INHIBITION OF Na-Ca EXCHANGE BY A FALL IN INTRACELLULAR pH

Recently we noticed a property of the Ca influx into squid axons which may be of considerable importance in excitable tissues that derive a fraction of their releasable intracellular calcium from reversed Na-Ca exchange (i.e., Ca entry in exchange for Na). Fig. 8 shows that the Ca_O-dependent Na efflux, which Baker *et al*. (1969) had shown to be well correlated with the Ca influx, is reversibly inhibited by quite small reductions in pH_i, caused in this experiment by a raised pCO_2 in the sea water bathing the axon. By contrast, the Ca efflux (not shown) was much less affected, which implies that acidifying the intracellular contents would lead to a net loss of Ca from the cell. Such an effect might explain the negative inotropic effect on the heart of raised blood pCO_2, an action first noted by Jerusalem & Starling (1910).

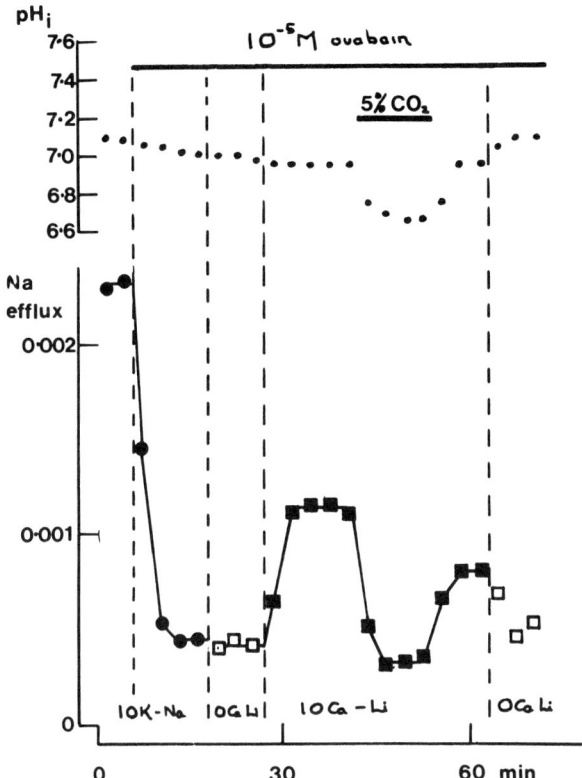

Fig. 8. Inhibition of Ca_O-dependent Na efflux by a fall
in intracellular pH. The pH_i was monitored throughout the
experiment with an intracellular pH electrode, and is shown
in the upper panel. Ouabain application is shown by the
upper black bar, and the use of sea waters containing 50 mM
HCO_3^- and bubbled back to pH 7.8 with 5% CO_2 is shown by the
lower black bar. From Baker & McNaughton (1977). Similar
experiments on the Ca efflux showed that the total Ca efflux
was much less affected by changes in pH_i, though the
Na-dependent component of Ca efflux was inhibited in a
similar manner to the Ca_O-dependent Na efflux.

CALCIUM-DEPENDENT CALCIUM EFFLUX, AND THE "RESIDUAL" EFFLUX

The discussion so far has been limited to the
Na_O-dependent Ca efflux, which probably represents Na-Ca
exchange. A much larger proportion (60%-90%) of the Ca efflux
from both unpoisoned squid axons and cardiac muscle is either

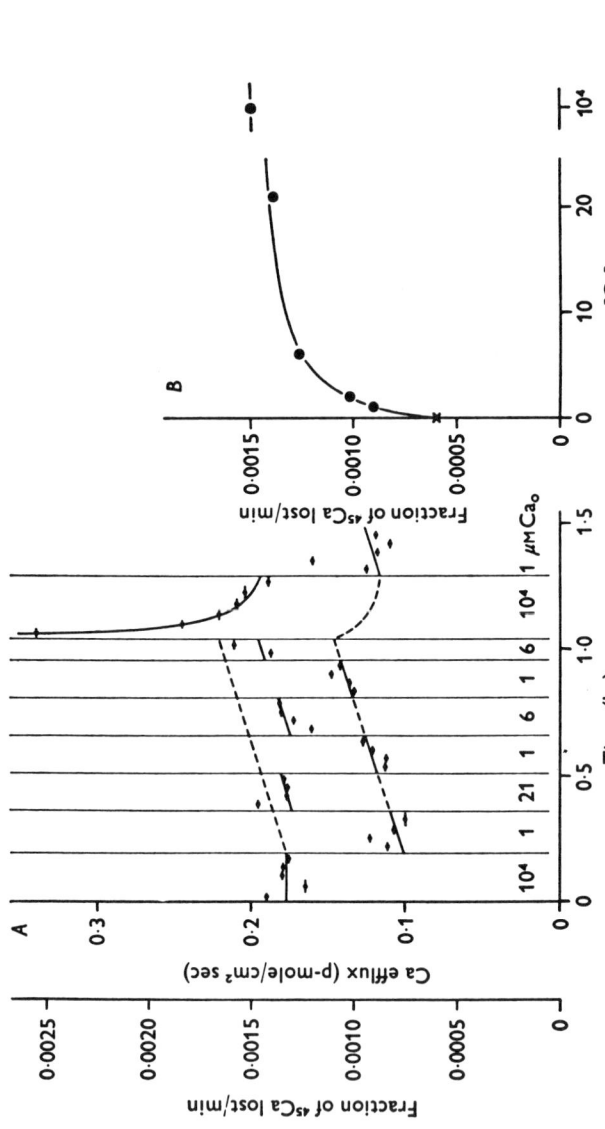

Fig. 9. (A) *Effect of alterations in Ca$_O$ (indicated at bottom) on apparent Ca efflux from an unpoisoned axon.* (B) *Ca efflux as a function of Ca$_O$, measured from the sloping baseline shown dotted in A. The curve near the points is a Michaelis relation with $K_M^{CaO} = 2$ μM. From Baker & McNaughton (1976a).*

dependent on Ca_O, or persists in the absence of both Ca_O and Na_O. The "residual" efflux, in particular, poses problems for the model outlined in Fig. 3, since according to this model practically no Ca efflux should be possible in the complete absence of both Ca_O and Na_O. The hypothesis that some other ion is exchanging for Ca seems improbable, since replacement of all extracellular cations by Li or choline (Baker & McNaughton, 1978) does not abolish the residual efflux. That this residual efflux does indeed play an important role, at least in cardiac muscle, is demonstrated by an experiment (Vassort, 1973) in which all the extracellular Na is replaced by Li. After a brief contracture, the muscle is found to relax and remains capable of generating twitches for some time. It is unlikely that all the Ca entering the cell in this time could be stored in intracellular vesicles; a more likely explanation is that the muscle has some means of extruding Ca in the absence of a Na gradient. The experiments described below suggest that the squid axon also has the ability to extrude Ca, uncoupled to the influx of any other ion.

An experiment to determine the properties of the Ca_O-dependent Ca efflux in an unpoisoned axon is shown in Fig. 9. In the presence of Na_O, removal of Ca produces a substantial drop in the Ca efflux, and the readmission of quite small concentrations of Ca fully reactivates this efflux. Similar results are obtained in the presence of Li, Tris, and K. The activation of Ca efflux by Ca_O can be represented by a Michaelis relation, with affinity for Ca_O of around 2 μM (Fig. 9B). However, there are considerable problems with this interpretation. The efflux drifts up towards its original level after some time in low Ca, and when the axon is returned to sea water containing normal Ca, there is a large transient increase in apparent Ca efflux (see Fig. 9A). Neither effect is consistent with Michaelis activation of the Ca efflux at the membrane, since under this hypothesis removal of Ca_O should cause a maintained drop in the Ca efflux, and replacement of Ca should generate a monotonic return to the original efflux level. A more serious objection to the idea that the Ca_O-dependence of the Ca efflux represents a true membrane Ca–Ca exchange is shown in Fig. 10, where the Ca influx, as a function of Ca_O, is compared with the Ca_O-dependent component of the Ca efflux. The large discrepancy between the two demonstrates that most of the Ca_O-dependence of the Ca efflux does not reflect a true membrane Ca–Ca exchange. Nor is Ca_O activating Ca efflux without exchanging for it, since reduction of Ca_O to very low

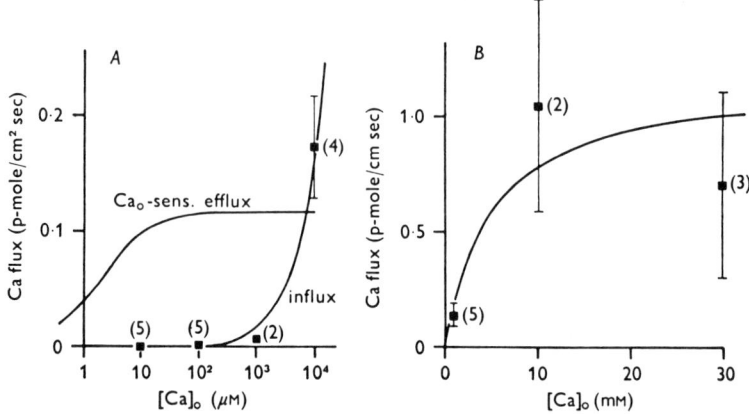

Fig. 10. *The Ca_O-dependent component of the Ca influx compared with the Ca influx in unpoisoned (A) and poisoned axons (B). The Ca efflux in A ha been redrawn from Fig. 9B with $K_M^{Cao} = 2$ μM; in B the Ca efflux (curve is typical of a poisoned axon ($K_M^{Cao} = 5$ mM). From Baker & McNaughton (1976a).*

levels using EGTA docs not abolish the Ca efflux, but rather causes a transient *rise* in efflux similar to that observed when an axon is transferred from low to high Ca.

These problems with the Ca-Ca exchange model led us to consider an alternative possibility, namely that the Ca-Ca exchange is occurring not at the membrane but at extracellular Ca binding sites. Thus, when Ca_O is removed, ^{45}Ca extruded from the axon would be absorbed by these binding sites, generating an apparent fall in efflux, and when Ca_O is replaced, the ^{45}Ca would be displaced from the binding sites, generating the observed transient rise in ^{45}Ca efflux. Clearly, the postulated sites would have to bind Ca with very high affinity in order to reproduce the observed activation of ^{45}Ca efflux by micromolar amounts of Ca_O (Fig. 9). That such high-affinity binding sites do indeed exist, presumably in the thin layer of Schwann cells and connective tissue adhering to a well-cleaned axon, is demonstrated in Fig. 11. Large amounts of Ca--about two Ca ions per square nanometer of measured axon surface area when Ca_O is 10 mM--are found to be rapidly and reversibly bound to the surface of the axon. The amount of bound Ca is affected neither by replacing the Na in the sea water with Tris nor by poisoning, but it is reduced by esterifying

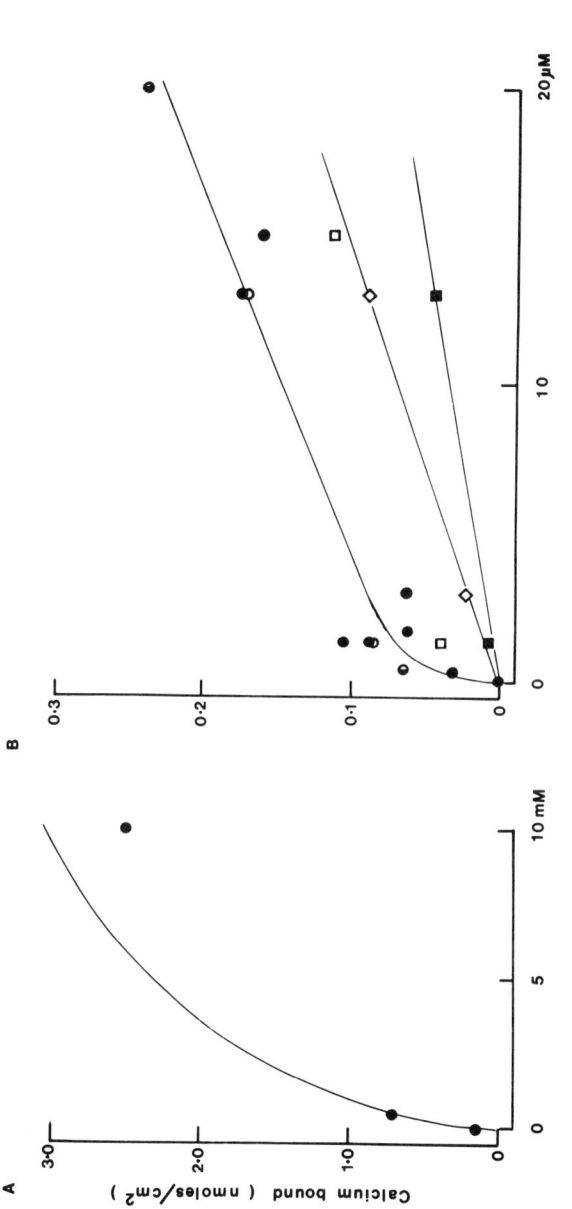

Fig. 11. Binding of calcium to highly cleaned axons under different conditions, as a function of Ca_O. (A) Axons in normal sea water. (B) Ca binding at low Ca to unpoisoned axons either in Na sea water (solid circles) or with all Na_O replaced with Tris (half-filled circles); to poisoned axons (open circles); to axons pretreated with Meerwein's reagent (open squares) or with pronase (2 mg/ml for 5 min) (solid squares); to axons in Na sea water with 300 µM added La^{3+} (diamonds). From Baker & McNaughton (1978).

carboxyl groups with Meerweins' reagent (Baker & Rubinson, 1977) by 300 μM lanthanum, or by brief digestion with pronase.

Treatment with pronase proved the most effective means of demonstrating that extracellular Ca binding is indeed responsible for the Ca_O-dependence of the Ca efflux, since pronase treatment sufficient to remove practically all Ca binding left other properties of the Ca pump, and the action potential, practically unaffected. In Fig. 12 a brief treatment with pronase was found to reduce the Ca_O-dependence of the Ca efflux, while a longer digestion (similar to that used in Fig. 11) abolished it almost entirely. In a pronase-treated axon, the Na_O-dependence of the Ca efflux closely resembled that in an untreated axon, demonstrating that this component of the efflux is not due to Na-Ca exchange at the extracellular binding sites, and therefore probably represents a true membrane exchange.

To examine more quantitatively whether the presence of a superficial Ca-binding matrix can modify the Ca efflux in the

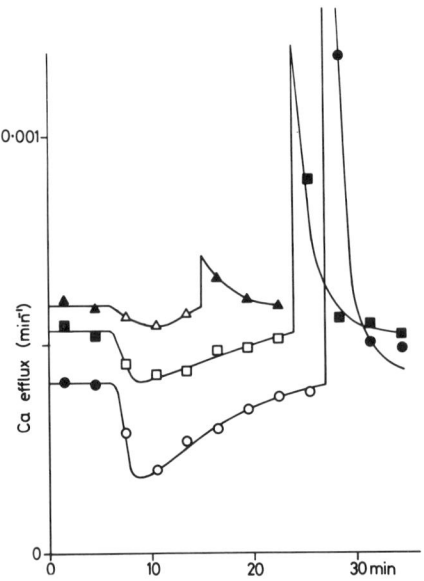

Fig. 12. The effect of pronase treatment on calcium efflux transients in an unpoisoned axon. Axon in Na sea water containing either 10 mM Ca_O (filled symbols) or 4.6 μM Ca_O (open symbols). Three separate experiments on the same axon have been superimposed: before treatment with pronase (solid and open circles); after treatment with 0.5 mg pronase/ml for 3 min (solid and open squares); and after treatment with 2 mg pronase/ml for 6 min (solid and open triangles). From Baker & McNaughton (1978).

observed manner, a mathematical model was set up, and its
behavior was compared with that of the axon. The model
assumes that the axon extrudes Ca at a constant rate into the
extracellular matrix of binding sites whose properties have
been determined in the experiment in Fig. 11. Fig. 13
compares the behavior of the axon and the model when Ca_O is
reduced from 10 mM to a variety of lower values. There are
small quantitative differences between the two, but the
agreement is sufficiently close to leave little doubt as to
the essential correctness of the assumptions underlying the
model.

The main conclusion of these experiments is that the
Ca_O-dependent Ca efflux and the residual efflux originate in
the same underlying membrane process--an uncoupled extrusion
of Ca across the axon membrane. Uncoupled extrusion therefore

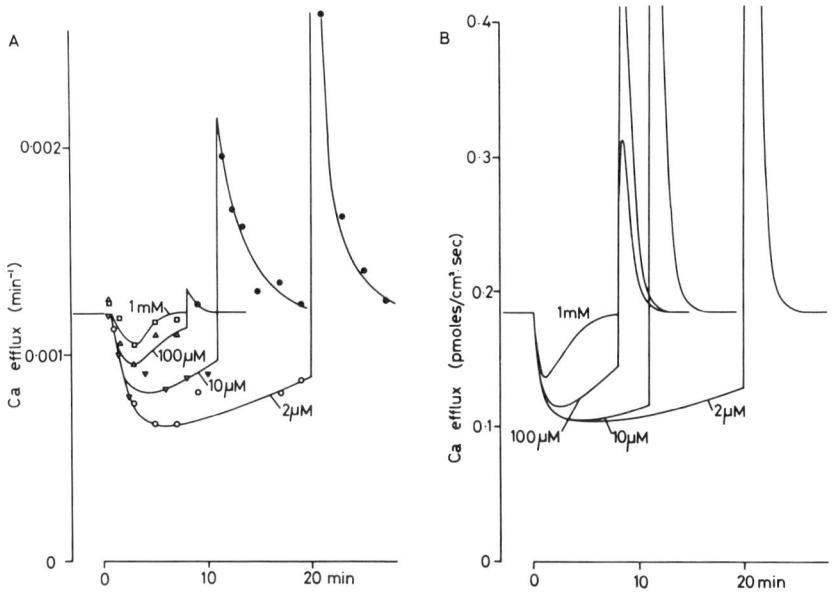

*Fig. 13. Transient changes in Ca efflux in an axon,
compared with the predictions of the model described in the
text. (A) Axon immersed in Na sea water containing 10 mM Ca
(solid circles); 1 mM Ca (open squares), 100 μM Ca (triangles),
10 μM Ca (inverted triangles), or 2 μM Ca (open circles).
The graph is a composite of four separate experiments on the
same axon. (B) Ca efflux transients predicted by the model.
The model was subjected to the same variations in Ca as the
axon. From Baker & McNaughton (1978).*

accounts for between 60% and 90% of the efflux from an
unpoisoned axon. The uncoupled extrusion is rapidly abolished
by poisoning the axon with cyanide and is resuscitated by
injecting ATP (Baker & McNaughton, 1976a), so it probably
metabolizes ATP in extruding Ca across the membrane. On
present information, it appears to resemble closely the Ca
extrusion from sarcoplasmic reticulum (Hasselbach, 1974) or
red blood cells (Schatzmann, 1966; Schatzmann & Vincenzi,
1969). In addition to the uncoupled extrusion, a very small
component can be detected in the presence of Li which
probably represents a true membrane Ca-Ca exchange (Baker &
McNaughton, 1978), but it seems unlikely that this component
contributes significantly to the Ca efflux in most
circumstances.

In the poisoned axon the picture is rather different.
It will be recalled from Fig. 10 that in the poisoned axon
the Ca influx and the Ca_O-dependent component of the efflux
are similar in size. The Ca_O-dependent component of Ca
efflux is maintained, in contrast to the transient nature of
the efflux changes of Fig. 9, and is not abolished by pronase
treatment sufficient to remove both binding sites and
Ca_O-dependent Ca efflux in an unpoisoned axon. It seems,
therefore, that there is a substantial amount of genuine
membrane Ca-Ca exchange in the poisoned axon. Transient
efflux changes similar to those observed in the unpoisoned
axon are also seen, but in the poisoned axon these are smaller
than the maintained changes which result from the true
Ca_O-dependence of the Ca efflux.

The picture of the Ca pump in the squid axon which
emerges from this work is summarized in the tentative model
shown in Fig. 14. It should be borne in mind that not all
features of the model have been firmly established; in
particular, it is not certain that the Na_O-dependence of the
Ca efflux actually reflects Na-Ca exchange. Nonetheless, the
model represents a reasonable set of working hypotheses.
A surprising feature of the model, perhaps, is that it
includes two apparently independent mechanisms for Ca
extrusion in the unpoisoned axon: uncoupled Ca efflux, and
Na-Ca exchange. These two modes resemble two of the major
partial reactions of the sodium pump, namely Na-K exchange
and uncoupled Na extrusion (see Glynn & Karlish, 1975, for a
review), and it is quite possible that the two major forms
of Ca efflux also represent different modes of the same
system. The experiment in Fig. 4 suggests that the two
modes may indeed be interchangeable, since in this experiment

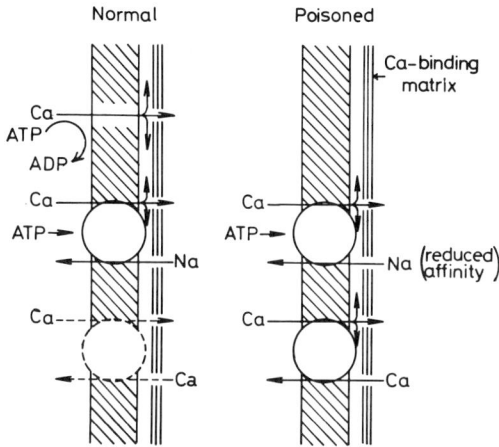

Fig. 14. A picture of the Ca efflux mechanisms in squid axon consistent with our current knowledge. Uncoupled extrusion in the unpoisoned axon membrane (left panel) metabolizes ATP. Coupled exchange modes, similar to those in Fig. 3, are also shown interacting with ATP, although it is not certain that ATP is actually metabolized. In the poisoned axon (right panel) uncoupled extrusion practically vanishes, but exchange modes persist with reduced affinity at the external face of the membrane. A Ca-binding matrix external to the membrane is unaffected by poisoning.

almost all the Ca efflux was Na_O-dependent before the injection of ATP, while after the injection the uncoupled extrusion had increased, at the expense of the Na_O-dependent component, to account for about two-thirds of the Ca efflux. Further investigation along these lines is clearly desirable.

DISCUSSION

Page: One small worry I have about experiments looking at the sodium sensitivity of the calcium efflux (in zero calcium solutions) is that one knows that calcium influx is very likely increased by going to zero sodium. Since there is no other calcium around except that which is coming off the efflux one might very well expect a reduction, an apparent reduction of the measured efflux, because the calcium is going back in again. I wonder how such considerations are affecting your measurements.

McNaughton: In squid axon, if you remove the extracellular sodium in the presence of high calcium, then there is a very substantial calcium influx. This of course reduces the specific activity of the calcium close to the membrane. If you go from sodium-calcium solution to zero sodium solution *with* calcium, then the efflux which was previously at a high level, goes down very sharply towards zero, and in fact when you go back to the normal solution, again with a sodium gradient, it really doesn't recover at all, presumably because of substantial activity change. We haven't used this protocol at all, we've always gone to zero calcium before we made the sodium change. Under those circumstances, there really doesn't appear to be a significant change in the specific activity of calcium adjacent to membrane. I don't think that the activity changes are important to this kind of experiment.

Page: I don't think I was suggesting that specific activity changes occur. I'm talking about total calcium which comes off which would be taken up again and the activity would not change.

McNaughton: You mean an actual increase in the total calcium concentration inside the cell?

Page: Yes.

McNaughton: Well, that shouldn't really have any effect, provided the pump depends linearly on calcium concentration. If you put some more cold calcium inside the cell, then the pump will go to a higher rate, but it will only be pumping out the same number of radioactive atoms.

Morad: In cardiac muscle one is also interested in the backward direction of your pump. Did I understand correctly that in the backward direction the reduction of K^+ reduces the calcium influx? This is rather problematic, if you would like to draw calcium in for contraction, because reducing potassium as we showed a number of years ago, under voltage clamp conditions, increases the developed tension.

McNaughton: Well, I think this is a most puzzling observation. It is, of course, well known that within limits, increasing the potassium concentration has a negative inotropic effect in cardiac muscle. In other words, apparently *depolarization* decreases the calcium influx, whereas the hyperpolarization experiment I showed you decreased the calcium influx. We haven't seen anything corresponding to that in squid axon. I can only assume

that there is some different process going on in the membrane
of cardiac muscle, or that the system is more complex.

Morad: Maybe the system works more in the forward
direction in cardiac muscle.

McNaughton: The effect that we described in fact works
the wrong way in both directions. The calcium influx is
enhanced by depolarization. In other words, if we go from
sodium containing solutions and add potassium, we observe
a small depression in Ca efflux. Whereas if you look at the
pump operating in the backwards direction, at the calcium
influx, when you depolarize the axon either with potassium or
with electrical polarization, you observe an increase. So
I'm afraid that the pump that we're looking at works in the
wrong way in both directions. I can only think that you're
looking at quite a different system in cardiac muscle.

REFERENCES

1. Baker, P.F. (1972). Transport and metabolism of calcium
   ions in nerve. *Progr. Biophys. molec. Biol. 24*, 177-223.
2. Baker, P.F., Blaustein, M.P., Hodgkin, A.L. & Steinhardt,
   R.A. (1969). The influence of calcium on sodium efflux
   in squid axons. *J. Physiol. 200*, 431-458.
3. Baker, P.F., Hodgkin, A.L. & Ridgway, E.B. (1971).
   Depolarization and calcium entry in squid giant axons.
   J. Physiol. 218, 709-755.
4. Baker, P.F. & McNaughton, P.A. (1976a). Kinetics and
   energetics of calcium efflux from intact squid giant
   axons. *J. Physiol. 259*, 103-144.
5. Baker, P.F. & McNaughton, P.A. (1976b). The effect of
   membrane potential on the calcium transport systems in
   squid axons. *J. Physiol. 260*, 24-25P.
6. Baker, P.F. & McNaughton, P.A. (1977). Selective
   inhibition of the Ca-dependent Na efflux from intact
   squid axons by a fall in intracellular pH. *J. Physiol.
   269*, 78-79P.
7. Baker, P.F. & McNaughton, P.A. (1978). The influence of
   extracellular calcium binding on the calcium efflux from
   squid axons. *J. Physiol. 276*, 127-150.
8. Baker, P.F. & Rubinson, K.A. (1977). TTX-resistant
   action potentials in crab nerve after treatment with
   Meerwein's reagent. *J. Physiol. 266*, 3P.
9. Blaustein, M.P. & Hodgkin, A.L. (1969). The effect of
   cyanide on the efflux of calcium from squid axons.
   J. Physiol. 200, 497-527.

10. DiPolo, R. (1976). The influence of nucleotides on
    calcium fluxes. *Fedn. Proc. 35*, 2579-2582.
11. DiPolo, R., Requena, J., Brinley, F.J. Jr., Mullins,
    L.J., Scarpa, A. & Tiffert, T. (1976). Ionized calcium
    concentrations in squid axons. *J. gen. Physiol. 67*,
    433-468.
12. Glynn, I.M. & Karlish, S. (1975). The sodium pump.
    A. Rev. Physiol. 37, 13-55.
13. Hasselbach, W. (1974). Sarcoplasmic membrane ATPases.
    In *The Enzymes*, vol. 10, 431-467. New York: Academic
    Press.
14. Jerusalem, E. & Starling, E.H. (1910). On the
    significance of carbon dioxide for the heartbeat.
    J. Physiol. 40, 279-294.
15. Jundt, H. & Reuter, H. (1977). Is sodium-dependent
    calcium efflux from mammalian cardiac muscle dependent
    on metabolic energy? *J. Physiol. 266*, 78P.
16. Reuter, H. & Seitz, N. (1968). The dependence of
    calcium efflux from cardiac muscle on temperature and
    external ionic composition. *J. Physiol. 195*, 451-470.
17. Schatzmann, H.J. (1966). ATP-dependent Ca^{2+} extrusion
    from human red cells. *Experientia 22*, 364-365.
18. Schatzmann, J.H. & Vincenzi, F.J. (1969). Calcium
    movements across the membrane of human red cells.
    J. Physiol. 201, 369-395.
19. Vassort, G. (1973). Influence of sodium ion on the
    regulation of frog myocardial contractility. *Pflügers
    Arch. ges. Physiol. 339*, 225-240.

VOLTAGE DEPENDENCE OF TETRODOTOXIN ACTION
IN MAMMALIAN CARDIAC MUSCLE

H. Reuter, M. Baer, P.M. Best*

Department of Pharmacology
University of Bern, Friedbuhlstrasse 49
3008 Bern, Switzerland

Tetrodotoxin (TTX) specifically blocks the early Na
current that underlies the excitability of many cell
membranes (Narahashi, 1974). Electrophysiological and
binding studies in nerves and skeletal muscle suggest a
first order binding reaction between TTX and Na channels.
The equilibrium dissociation constant of this reaction is on
the order of 5 nM. The most accurate kinetic data of TTX
action were obtained from measurements of the Na current
(I_{Na}) in voltage clamp experiments in squid axons and nodes
of Ranvier of myelinated nerve fibers. However, indirect
measurements of the Na current, like the upstroke velocities
of action potentials ($\dot{V}_{max}$), gave similar results in these
tissues (e.g. Ulbricht & Wagner, 1975). Moreover, in frog
skeletal muscle the binding isotherms of ^3H-labelled TTX
agree quite well with dose-response relationships obtained
from measurements of $\dot{V}_{max}$ (Almers & Levinson, 1975;
Jaimovich, Venosa, Shrager & Horowicz, 1976).
    In cardiac muscle much less is known about I_{Na} or Na
channels than in nerve or skeletal muscle fibers.
Unfortunately, none of the voltage clamp methods applied to
cardiac muscle measure I_{Na} accurately. On the other hand, a
better understanding of the function of Na channels in the
heart is clearly of practical relevance. Many drugs that
are used as antiarrhythmic agents seem to act on Na channels
(see Gettes & Reuter, 1974). Therefore, we have tried to

**Dr. Best's present address is Department of Medicine,
University of Chicago, Chicago, Ill. 60637 U.S.A.*

obtain more information about the similarities or
dissimilarities of cardiac Na channels with those of nerve
and skeletal muscle fibers by systematically investigating
the action of TTX. In these studies, as a measure for the
action of TTX, we used the upstroke velocity, $\dot{V}_{max}$, of the
cardiac action potential.

I. ELECTROPHYSIOLOGICAL METHODS

Intracellular action potentials were recorded, by
conventional microelectrode techniques, from short (<0.8 mm)
segments of isolated guinea-pig papillary muscles (diameter
$\sim$ 0.5 mm). In order to electrically isolate such a muscle
segment, the papillary muscles were mounted in a single
sucrose gap arrangement (Beeler & Reuter, 1970). The sucrose
gap permitted injection of current into the muscle end from
which action potentials were recorded. The constant
depolarizing or hyperpolarizing currents displaced the
membrane potential rather uniformly in this muscle segment.
Alternatively, the membrane potential could be changed by
the addition of KCl to the Tyrode solution superfusing this
muscle segment. Action potentials were differentiated
electronically. The differentiated maximal upstroke velocity
($\dot{V}_{max}$), the action potential, and the stimulus current were
recorded on an oscilloscope. The blocking action of TTX on
$\dot{V}_{max}$ reached a steady state within 10 min of exposure to the
toxin (Fig. 4) and its reversibility was at least 95%.
    To demonstrate that $\dot{V}_{max}$ of the action potential depends
on the inward Na current, in a series of experiments, we
altered the external Na concentration ($[Na]_o$). Fig. 1 shows
a plot of $\dot{V}_{max}$ vs. $[Na]_o$. The relation ($r = 0.98$) between
$\dot{V}_{max}$ and $[Na]_o$ strongly suggests that the membrane current
underlying $\dot{V}_{max}$ is an inward Na current, I_{Na}. However, the
relation between the Na conductance of the membrane and $\dot{V}_{max}$
is unknown and is probably rather complex and nonlinear
(Hunter, McNaughton & Noble, 1975). This could complicate
the interpretation of our results with TTX (Cohen &
Strichartz, 1977). On the other hand, in a kinetic study
of TTX action in myelinated nerve fibers of *rana esculenta*,
Ulbricht & Wagner (1975) found a good qualitative
correspondence of the toxin's effects on I_{Na} and $\dot{V}_{max}$.
A more extensive discussion of this problem, taking into
account the criticisms by Cohen & Strichartz, will be
presented elsewhere (Baer, Best, Reuter & Tsien, in

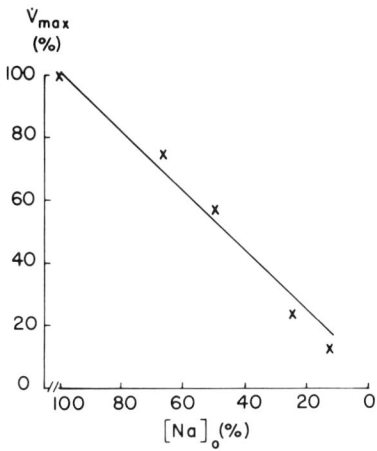

Fig. 1. Relationship between normalized upstroke velocities ($\dot{V}_{max}$, ordinate) and normalized Na concentration in the Tyrode solution ($[Na]_o$, abscissa). The experimental values (crosses) are the means of 1-10 experiments; the linear regression line is a least squares fit to the experimental data (correlation coefficient r = 0.98). 100% $\dot{V}_{max}$ = 162 ± 10 V/sec (mean ± S.E.); 100% $(Na)_o$ = 149 mM/l.

preparation; see also Baer, Best & Reuter, 1976). We believe that the very pronounced and self-consistent TTX effects we observed provide a qualitatively, though not quantitatively, correct picture of the "anomalous" action of this toxin in cardiac muscle.

II. POTENTIAL-DEPENDENT EFFECT OF TTX ON $\dot{V}_{max}$

Fig. 2 shows records of the action potential upstroke and $\dot{V}_{max}$ which arise from two different resting potentials in the absence (control records 1 and 2) and presence of 5 µM TTX (records 3 and 4). From the same records, $\dot{V}_{max}$ values are also plotted as functions of membrane potentials during the upstrokes of the respective action potentials (phase plane plots; lower part of Fig. 2). Fig. 2 illustrates the following results:

1. The reduction of $\dot{V}_{max}$ by TTX is much greater at the lower resting potential (-73 mV; records 2 and 4) than at the higher one (-88 mV; records 1 and 3). This implies an apparent voltage dependence of the TTX action.

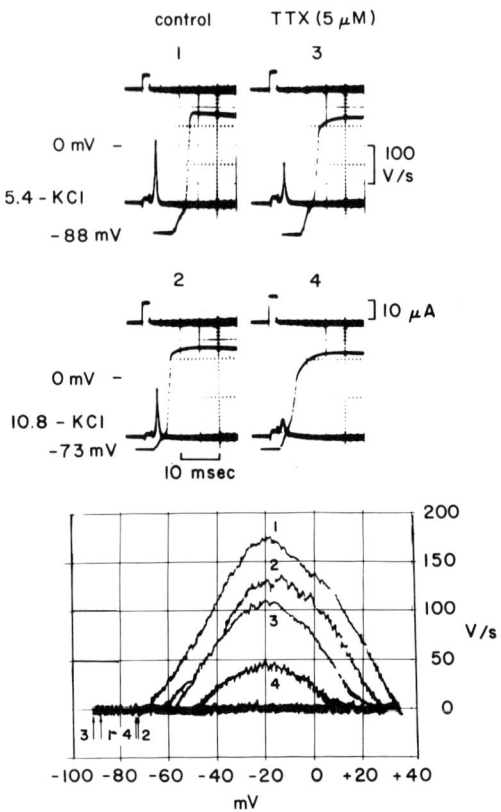

Fig. 2. Top: *Experimental records of upstrokes of action potentials (mV), their first derivatives ($\dot{V}_{max}$, V/sec), and stimulus currents (μA). For clarity, the action potential traces are delayed. All records were obtained from the same microelectrode impalement: records 1 and 3, resting potentials -88 and -91 mV (KCl in Tyrode solution 5.4 mM/l); records 2 and 4, resting potentials -73 and -74 mV (KCl 10.8 mM/l). Records 1 and 2 without (controls) and 3 and 4 with TTX (5 µM/l). Bottom: Phase plane (x-y) plots of upper records: $\dot{V}_{max}$ (V/sec) ordinate is plotted against membrane potentials during the upstroke (mV, abscissa). The record numbers correspond to those in the upper part of the figure; the arrows and numbers on the left-hand side correspond to the respective resting potentials in the upper records.*

2. The fact that under all conditions in Fig. 2 (phase plane plot) $\dot{V}_{max}$ peaks at the same membrane potential (-20 mV) suggests that the kinetic parameters of I_{Na}, (m^3h in the Hodgkin & Huxley (1952) formulation) are not greatly affected by the different experimental conditions in records 1-4.

3. Despite the large decrease of $\dot{V}_{max}$ in the presence of TTX, the total amplitude of the action potential is reduced very little. This indicates that the much smaller and slower secondary inward current, I_{si}, which is primarily carried by Ca ions (Reuter, 1973) and which is responsible for the plateau phase of the action potential, is not much affected by TTX. In voltage clamp experiments in cat papillary muscles and cow ventricular trabeculae, we have confirmed the lack of any effect of TTX (up to 2×10^{-5}M) on I_{si}.

The inhibitory effect of TTX on $\dot{V}_{max}$ at two different membrane potentials is plotted for various TTX concentrations in Fig. 3. The concentration-response curves were obtained at resting potentials of -89 ± 3 mV (crosses) or -71 ± 3 mV (circles). There was a tenfold increase in the sensitivity of $\dot{V}_{max}$ to TTX when the muscles were depolarized.

The results were independent of the method of depolarization; i.e. K^+-depolarization or constant current depolarization. This implies that the membrane potential dependence of the TTX effects is not the result of the increased K^+ concentration *per se*. Moreover, since $\dot{V}_{max}$ decreases during depolarization (Figs. 2, 3A), we thought it might be possible that the apparent increase in sensitivity to TTX during depolarization is due to the reduction of $\dot{V}_{max}$, rather than implying a voltage-dependent increase in the apparent affinity of TTX to its receptor (see also Cohen & Strichartz, 1977). However, a decrease of $\dot{V}_{max}$ by 50% reduction of $[Na]_o$, without a depolarization, caused only a slight (1.4 fold, $P > 0.1$) shift of the TTX concentration-response curve. Reduction of $\dot{V}_{max}$ by other means (high $[Ca]_o$, lidocaine) also did not increase the sensitivity of $\dot{V}_{max}$ to TTX. As a matter of fact, high $[Ca]_o$ reduced the sensitivity of $\dot{V}_{max}$ to TTX, an effect opposite to that seen with depolarization. These results indicate that the increased sensitivity of $\dot{V}_{max}$ to TTX during depolarization is directly related to the change in membrane potential and not to the potential dependent decrease in $\dot{V}_{max}$.

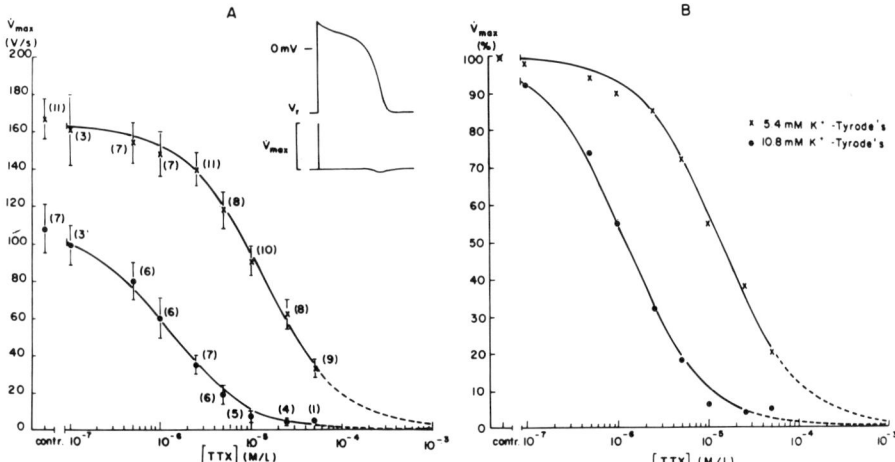

Fig. 3. (A) Concentration-response curves showing the
effects of TTX (abscissa; log scale) on maximum upstroke
velocity ($\dot{V}_{max}$; ordinate; mean ± S.E.) of the action
potential in 5.4 (crosses) and 10.8 (circles) mM K^+-Tyrode's.
The inset is a schematic drawing of resting potential (V_r)
and action potential (upper trace); the latter was
differentiated electronically (lower trace), and the maximal
upstroke velocity ($\dot{V}_{max}$) was measured. (B) Normalized mean
values of panel A. Solid lines are concentration-response
curves calculated from the equation

$$\dot{V}_{max}\ (TTX) = \dot{V}_{max}\ (control)\ (1-[1/\{1+(K_m/[TTX])\}]);$$

$\dot{V}_{max}$ (control) and K_m were 167 V/sec and 1.4×10^{-5} M/l in
5.4 K^+-Tyrode's, and 109 V/sec and 1.2×10^{-6} M/l in 10.8
K^+-Tyrode's. Numbers of determinations per mean value given
in parentheses in (A). (From Baer et al. 1976.)

This conclusion is supported by our finding of a pronounced shift of the relation between $\dot{V}_{max}$ and membrane potential (Weidmann, 1955) towards more negative potentials in the presence of 5 µM TTX. A similar effect also has been described by Hille (this volume) for the action of local anesthetics on the "inactivation" of the Na conductance in nerve and skeletal muscle fibers. The interpretation of these results is that TTX binds more strongly to "inactivated" Na channels, i.e. when the membrane is depolarized, than to "available" Na channels when the membrane potential is high. Thus, with increasing depolarization more and more Na channels will be occupied by TTX, even if the concentration in the bulk solution is constant.

III.   FREQUENCY-DEPENDENT EFFECT OF TTX ON $\dot{V}_{max}$

Another surprising effect of TTX in cardiac muscle is its dependence on the rate of stimulation. In the experiment illustrated in Fig. 4, $\dot{V}_{max}$ was continuously recorded at two rates of stimulation, 0.1 Hz and 2 Hz. In the control run without TTX (crosses), a sudden change in the frequency had little, if any, effect on $\dot{V}_{max}$. When TTX (5×10^{-6} M; circles) was applied at a stimulation rate of 0.1 Hz, $\dot{V}_{max}$ was reduced from 192 to 130 V/sec. The steady state of the TTX action was reached in about 5 min. After 7 min incubation with TTX, the stimulation rate was suddenly changed from 0.1 to 2 Hz. Within 15 sec, this caused a further reduction of $\dot{V}_{max}$ to 110 V/sec. When the rate of stimulation was reduced to 0.1 Hz, within 2 sec $\dot{V}_{max}$ recovered to 130 V/sec again. The experiment shows that frequency of stimulation is a modulating factor of the TTX action in cardiac muscle. Comparable experiments in myelinated nerve fibers did not show an increase of the TTX effect on $\dot{V}_{max}$ at high rates of stimulation (Schwarz, Ulbricht & Wagner, 1973). However, a similar frequency dependence of the inhibition of I_{Na} by local anesthetics has been observed both in nerve fibers (see Hille, this volume) and by measuring $\dot{V}_{max}$ in cardiac muscle (Chen & Gettes, 1976).

The explanation of our result (Fig. 4) is similar to that given for local anesthetics by Hille and his coworkers (this volume). During stimulation of the preparation at high frequencies, more TTX is bound to the Na channels than at low frequencies. This could be due either to an increased binding affinity of TTX to the Na channels or, equivalently, to a reduced dissociation of TTX from the Na channels during

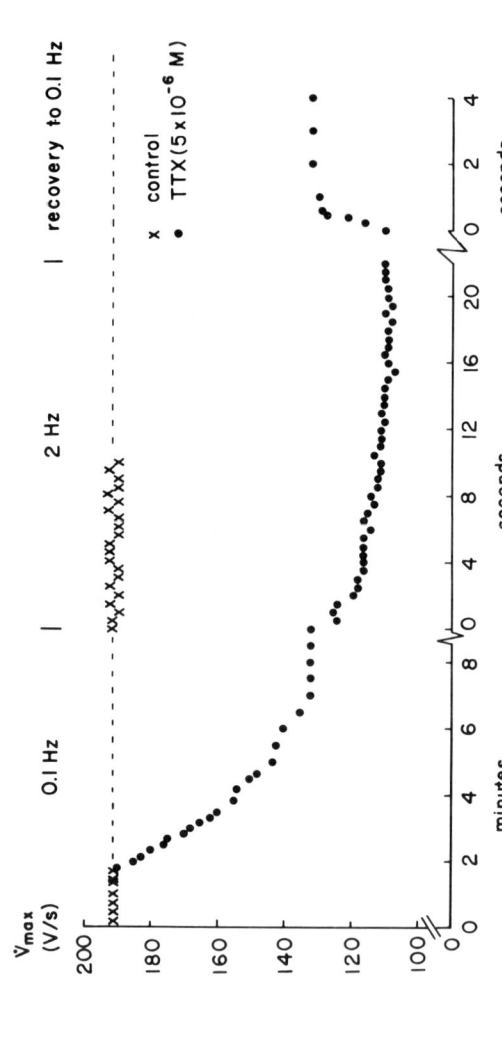

Fig. 4. Dependence of TTX action on stimulation frequency. $\dot{V}_{max}$ (V/sec; ordinate) of action potentials recorded during a continuous microelectrode impalement. Crosses indicate control records without TTX; circles are records with TTX (5 μM/1). TTX was first applied during a period of stimulation at 0.1 Hz. After a steady state of drug action was reached, stimulus frequency was changed to 2 Hz and back to 0.1 Hz. Note changes in time scale on the abscissa.

depolarization. If this is true, TTX should reduce the rate of recovery of the Na channels from "inactivation" occurring during depolarization of the membrane. Without TTX the recovery is very rapid at the normal resting potential. We have investigated this question by progressively increasing the interval between two action potentials. The corresponding increase in $\dot{V}_{max}$ of the second action potential was used as index of the recovery of I_{Na} from inactivation occurring during the first action potential (Gettes & Reuter, 1974). In three experiments TTX (5×10^{-6} M) prolonged the recovery of $\dot{V}_{max}$ 10-20 fold.

IV. VOLTAGE-DEPENDENT DRUG ACTIONS

Several drugs are known to react rather specifically with either K channels or Na channels of excitable membranes. The drugs listed in Table I are remarkable, since their action on ionic channels is not a simple function of drug concentration. The potencies of these drugs have been shown to depend on the membrane potential of some excitable cells. There may be either an increase (TEA, local anesthetics, phenytoin, quinidine, batrachotoxin) or a decrease (4-aminopyridine, scorpion toxin) of the apparent affinity when the membrane potential is depolarized. However, if we take the examples of TEA and local anesthetics, the hypotheses for the modes of action by which the potencies of these drugs are affected by the membrane potential are quite different. Both drugs act primarily from the inside surface of the membrane. During depolarization, the positively charged TEA molecules seem to be pulled into the K channels which *open* during depolarization; i.e. the TEA blocking effect is stronger when the K channels are open. Potassium ions that enter the channels from the outside during hyperpolarization clear TEA ions from the channels (Armstrong, 1971). In contrast, the potency of local anesthetics on Na channels is increased when the channels first open and then close (inactivate) during depolarization, as if the local anesthetics are pulled into the open channels and thereafter are trapped in the *inactivated* channels. Hyperpolarization of the membrane potential removes inactivation, whereby the affinity of the Na channels for local anesthetics is reduced. This helps clearing of the drug molecules from the channels (see Hille, this volume).

TTX in cardiac muscle shows a remarkable similarity to
the action of local anesthetics on Na channels, although TTX
acts from the outside of the membrane (Narahashi, 1974).
Moreover, with the possible exception of the TTX action on
myelinated nerve fibers at low pH (Ulbricht & Wagner, 1975),
which is explained by a competition between protons and TTX
in Na channels, membrane potential does not seem to have any
effect on TTX action or binding in other excitable tissues
(e.g. Almers & Levinson, 1975).

In 1976 R.W. Tsien and H. Reuter considered a reaction
scheme (unpublished) which would, at least qualitatively,
account for our experimental observations. If we suppose
that TTX reacts with a receptor R in the Na channel, then

$$\text{TTX} + R \underset{k_2}{\overset{k_1}{\longleftrightarrow}} \text{TTX} \cdot R$$

where $k_2/k_1 = K_D$. If the receptors of "available" channels
R_a, at high membrane potentials dissociate TTX molecules
more rapidly than receptors of inactivated channels, R_i, at
lower (depolarized) membrane potentials, the following
reaction scheme may apply.

$$\text{TTX} + R_a \underset{\beta_h}{\overset{\alpha_h}{\longleftrightarrow}} \text{TTX} + R_i$$
$$k_{a2} \Big\updownarrow k_{a1} \qquad k_{i2} \Big\updownarrow k_{i1}$$
$$\text{TTX} \cdot R_a \qquad\qquad \text{TTX} \cdot R_i$$

where $k_{a2}/k_{a1} = K_{Da}$ and $k_{i2}/k_{i1} = K_{Di}$, and $K_{Di} \ll K_{Da}$. α_h
and β_h are the voltage-dependent rate constants of the
inactivation process of the Na channels (Hodgkin & Huxley,
1952) which shift the reaction towards TTX $\cdot$ R_i during
depolarization and towards TTX $\cdot$ R_a during hyperpolarization.
This scheme is very similar to the one presented by Hille
(this volume) for the action of local anesthetics.

V.   CONCLUSIONS

The results on TTX action in cardiac muscle show
important differences when compared with the action of the
toxin in nerve and muscle membranes:

1.   When the effect of TTX on $\dot{V}_{max}$, an indirect measure
of I_{Na}, is compared in nerve, skeletal muscle and cardiac
muscle, the potency of TTX in cardiac muscle is 3-4 orders

of magnitude smaller than in the other tissues. However, in skeletal muscle the sensitivity to TTX can be reduced considerably by denervation (Redfern & Thesleff, 1971), and in nerve by the application of chemicals like carbodiimide or Meerwein's reagent (Baker & Rubinson, 1975, 1977).

2. Our results suggest a voltage dependence of the TTX action in cardiac muscle. This, again, is different from the action of the toxin in nerve and skeletal muscle. However, the voltage dependence is similar to that observed with local anesthetics in nerve, skeletal and cardiac muscle: Depolarization and rapid stimulation increase the sensitivity to TTX, while hyperpolarization and high $[Ca]_o$ reduce it.

DISCUSSION

Tsien: Last summer, when we discussed a model of the type that Hille presented yesterday, we talked about a model where the explanation of your results had to do with the relationship between inactivation and binding of TTX to the inactivated form. I was very excited about the model at the time, because it seemed to fit all of your results, but you convinced me later that the model didn't work. What has caused you to change your mind? Is it the location of the inactivation line?

Reuter: No, I didn't really convince you at the time that the model didn't work. There was one puzzling result, namely, that the steepness of the apparent inactivation curve was changed by TTX in one experiment. We have done more experiments since, and in most we see a parallel shift. The way we discussed it a year ago, the model would not very easily accomodate such a change in the slope of the inactivation of the sodium channels. Bert Hille gave another explanation yesterday that would account for such a change in the slope.

Tsien: When you do the TTX experiment in nerve and plot the inactivation curve, you don't find any change in either the location, in terms of voltage, or in the steepness of the inactivation curve. Is that correct?

Reuter: That's right.

Morad: Suppose I were to argue that increased extracellular K^+ concentration somehow enhances the effect of TTX independent of the effect of K^+ on membrane potential. Depolarization by constant current or ten fold increase in frequency will also accumulate K^+ and have the same effect on TTX binding as would the direct application of K^+.

Reuter: I don't think that the passage of current really changes the extracellular potassium concentration very much, because the V_m change is rather small. We are not very far in the region of the inward rectification; it's 10-15 mV depolarization from the resting potential.

Morad: Dr. Cleemann shows (see Cleemann & Morad, this volume) that's the range in which you get a tremendous amount of accumulation very rapidly.

Reuter: Also, in mammalian muscle?

Morad: Our results are in frog ventricle, but I should think the K^+ accumulation profile to be quite similar to frog, based on the steady state I-V relations.

Reuter: I think the important difference is that in the mammalian muscle, the extracellular space seems to be wider, while in the frog heart, you have rather narrow intercellular clefts.

Morad: But the K^+ accumulation would be in the paracellular space, in a few angstrom space, and that's the space that we are interested in. I imagine that's where TTX concentration is important

Reuter: I'm interested in the accumulation space which would be equivalent to the extracellular space in this particular case. I'm not aware of any studies in this matter, Martin, and I would be grateful, definitely, to see the magnitude of K^+ accumulation in mammalian cardiac muscle under experimental conditions.

TABLE I

Drugs Acting in a Voltage-Dependent Manner on the K and Na Channels of Some Excitable Tissues

Drug	Reference
K Channel	
Tetraethylammonium (TEA)	*Armstrong, 1971*
Aminopyridines	*Yeh et al. 1976*
	Meves & Pichon, 1977
Quinidine	*Yeh & Narahashi, 1976*
Na Channel	
Local anesthetics	*Hille, this volume*
Phenytoin	*Schwarz & Vogel, 1977*
Batrachotoxin	*Bartels-Bernal et al. 1977*
Scorpion toxin	*Catterall et al. 1976*
Tetrodotoxin	*Baer et al. 1976; this paper*

REFERENCES

1. Almers, W. & Levinson, S.R. (1975). Tetrodotoxin binding to normal and depolarized frog muscle and the conductance of a single sodium channel. *J. Physiol.* 247, 483-509.
2. Armstrong, C.M. (1971). Interaction of tetraethylammonium ion derivatives with the potassium channels of giant axons. *J. gen. Physiol.* 58, 413-437.
3. Baer, M., Best, P.M. & Reuter, H. (1976). Voltage-dependent action of tetrodotoxin in mammalian cardiac muscle. *Nature, Lond.* 263, 344-345.
4. Baker, P.F. & Rubinson, K.A. (1975). Chemical modification of crab nerves can make them insensitive to the local anesthetics tetrodotoxin and saxotoxin. *Nature, Lond.* 257, 412-414.
5. Baker, P.F. & Rubinson, K.A. (1977). TTX-resistant action potentials in crab nerve after treatment with Meerwein's Reagent. *J. Physiol.* 266, 3-4P.
6. Bartels-Bernal, E., Rosenberry, T.L. & Daly, J.W. (1977). Effect of batrachotoxin on the electroplax of electric eel: Evidence for voltage-dependent interaction with sodium channels. *Proc. natn. Acad. Sci. U.S.A.* 74, 951-955.
7. Beeler, G.W., Jr. & Reuter, H. (1970). Voltage clamp experiments on ventricular myocardium fibres. *J. Physiol.* 207, 165-190.
8. Catterall, W.A., Ray, R. & Morrow, C.S. (1976). Membrane potential dependent binding of scorpion toxin to action potential Na^+ ionophore. *Proc. natn. Acad. Sci. U.S.A.* 73, 2682-2686.
9. Chen, C.M. & Gettes, L. (1976). Combined effects of rate, membrane potential, and drugs on maximum rate of rise ($\dot{V}_{max}$) of action potential upstroke of guinea pig papillary muscle. *Circulation Res.* 38, 464-469.
10. Cohen, I.S. & Strichartz, G.R. (1977). On the voltage-dependent action of tetrodotoxin. *Biophys. J.* 17, 275-279.
11. Gettes, L.S. & Reuter, H. (1974). Slow recovery from inactivation of inward currents in mammalian myocardial fibres. *J. Physiol.* 240, 703-724.

12. Hille, B. Local anesthetic action on inactivation of the Na channel in nerve and skeletal muscle: Possible mechanisms for antiarrhythmic agents. This volume.

13. Hodgkin, A.L. & Huxley, A.F. (1952). A quantitative description of membrane current and its application to conduction and excitation in nerve. *J. Physiol. 117*, 500-544.

14. Hunter, P.J., McNaughton, P.A. & Noble, D. (1975). Analytical models of propagation in excitable cells. *Progr. Biophys. mol. Biol. 30*, 99-144.

15. Jaimovich, E., Venosa, R.A., Shrager, P. & Horowicz, P. (1976). Density and distribution of tetrodotoxin receptors in normal and detubulated frog sartorius muscle. *J. gen. Physiol. 67*, 399-416.

16. Meves, H. & Pichon, Y. (1977). The effect of internal and external 4-aminopyridine on the potassium currents in intracellularly perfused squid giant axons. *J. Physiol. 268*, 511-532.

17. Narahashi, T. (1974). Chemicals as tools in the study of excitable membranes. *Physiol. Rev. 54*, 813-889.

18. Redfern, P. & Thesleff, S. (1971). Action potential generation in denervated rat skeletal muscle. II. The action of tetrodotoxin. *Acta physiol. scand. 82*, 70-78.

19. Reuter, H. (1973). Divalent cations as charge carriers in excitable membranes. *Progr. Biophys. mol. Biol. 26*, 1-43.

20. Schwarz, J.R., Ulbricht, W. & Wagner, H.H. (1973). The rate of action of tetrodotoxin on myelinated nerve fibers of *Xenopus Laevis* and *Rana Esculenta*. *J. Physiol. 233*, 167-194.

21. Schwarz, J.R. & Vogel, W. (1977). Diphenylhydantoin: excitability reducing action in single myelinated nerve fibres. *Eur. J. Pharmac. 44*, 241-249.

22. Ulbricht, W. & Wagner, H.H. (1975). The influence of pH on equilibrium effects of tetrodotoxin on myelinated nerve fibres of *Rana esculenta*. *J. Physiol. 252*, 159-184.

23. Weidmann, S. (1955). Effects of calcium ions and local anaesthetics on electrical properties of Purkinje fibres. *J. Physiol. 129*, 568-582.

24. Yeh, J.Z. & Narahashi, T. (1976). Mechanism of action of quinidine on squid axon membranes. *J. Pharmac. exp. Ther. 196*, 62-70.

25. Yeh, J.Z., Oxford, G.S., Wu, C.H. & Narahashi, T. (1976). Dynamics of aminopyridine block of potassium channels in squid axon membrane. *J. gen. Physiol. 68*, 519-535.

THE EFFECT OF CESIUM IONS ON ^{42}K EFFLUX
IN CARDIAC PURKINJE FIBERS

E. Carmeliet

Laboratory of Physiology
Campus Gasthuisberg
3000 Leuven, Belgium

I. INTRODUCTION

In cardiac Purkinje fibers, exposure to a Tyrode solution containing 20 mM Cs blocks the pacemaker current i_{K2}, shifts the steady state current-voltage relation in the inward direction, and abolishes inward-going rectification (Isenberg, 1976). Since the Cs-sensitive current shows a reversal potential that follows changes in external K concentration with a slope of 60 mV per decade, Isenberg concluded that Cs ions can be used as a tool to separate inward rectifying K current from the net membrane current of the cardiac Purkinje fiber. The question remains, however, whether total K movement is blocked by Cs ions, in other words, whether all K movement occurs through the channel showing inward-going rectification. In order to solve this problem, we measured ^{42}K efflux in Purkinje fibers and its modification by different concentrations of Cs. The effect of Cs was studied further at different external K concentrations to see whether the characteristic effect of K_o on ^{42}K efflux is abolished in the presence of Cs.

143

II.   METHODS

   The flux experiments were performed on nonstimulated cow
Purkinje bundles.   Details of the technique used to measure
^{42}K efflux can be found in previous publications (Carmeliet &
Verdonck, 1977).   The data are expressed as changes in rate
coefficient.   In separate experiments the change in resting
potential was measured by conventional microelectrodes.   As a
rule, the preparations were at rest.   In some experimental
conditions, however, spontaneous activity was recorded.
Occasionally the preparations were stimulated to study
changes in action potential configuration.

III.   RESULTS

   When Cs ions are added to the bathing solution, the rate
coefficient for ^{42}K efflux of nonstimulated cow Purkinje
fibers drops immediately (Fig. 1, mean of 10 experiments).
The effect is not progressive with time and is quickly

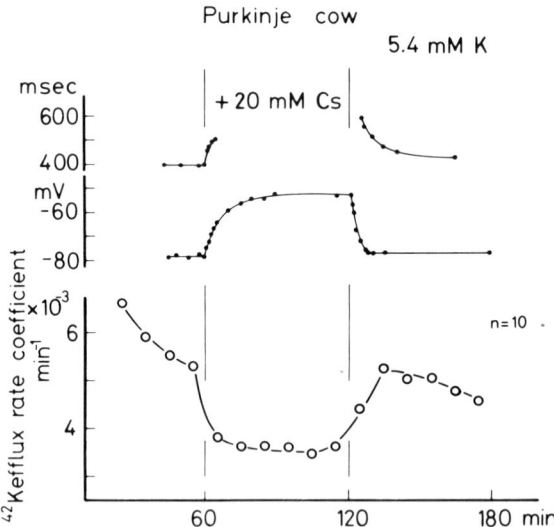

   Fig. 1.   Effect of 20 mM Cs on action potential duration,
maximum diastolic potential, and rate coefficient of ^{42}K
efflux in cow Purkinje fibers.   The upper two curves are
typical results for one preparation.   The lower curver for ^{42}K
efflux represents the mean of 10 preparations.

reversible upon washing. The change in rate coefficient cannot be explained by the concomitant change in membrane potential. Measurements of the maximum diastolic potential and the duration of the action potential in a separate series of Purkinje fibers show that the membrane depolarizes from -80 to -60 mV with a half time of about 3 min; the action potential is slightly prolonged, and the fiber becomes irresponsive to short stimuli after about 2-4 min. Since depolarization would rather increase K efflux, even when inward-going rectification is taken into account (Haas & Kern, 1966), the fall in K efflux indicates a fall in the permeability of the membrane to K ions.

The changes in ^{42}K efflux is dependent on the external Cs and K concentration. In 5.4 mM K-Tyrode, the fall in ^{42}K efflux is 14% in 1 mM Cs, 25% in 5.4 mM Cs, and 37% in 20 mM Cs (Fig. 2). When the same concentrations of Cs ions are tested in 16.2 mM K-Tyrode, the change in rate coefficient in absolute terms is more pronounced (Fig. 3). It should be noted that ^{42}K efflux in control conditions is also larger in the elevated K-Tyrode. In relative terms, however, the fall in K efflux by a certain Cs concentration is the same in 5.4 and 16.2 mM K, which suggests that Cs ions block a certain proportion of the K channels, or reduce the K conductance per channel by a fixed amount, irrespective of the total K channels or the initial conductance per channel.

Another important conclusion from Figs. 2 and 3 is that the block of ^{42}K efflux by 20 mM Cs is far from complete. In 5.4 or 16.2 mM K-Tyrode, slightly less than 2/3 of the initial K efflux persists. The fraction remaining in 20 mM Cs furthermore is sensitive to K_O. In the experiment, summarized in Fig. 4 (mean of 5 experiments) the preparations were first bathed in a K-free solution. Upon addition of 20 mM Cs the ^{42}K efflux reduced to about 0.7 of the initial value. When K_O was increased to 5.4 and 16.2 mM successively, the rate coefficient increased concomitantly. In relative terms, the change in rate coefficients by K_O is the same as in the absence of Cs ions (Carmeliet & Verdonck, 1977).

It could be argued, however, that the ^{42}K efflux remaining in 20 mM Cs is not electrogenic; the increase in ^{42}K efflux by higher K_O would thus be simply a manifestation of an increased K-K exchange diffusion. This question was answered by measuring membrane potential and excitability in the presence of a fixed amount of Cs ions (20 mM) and variable concentrations of K ions (K-free, 5.4, and

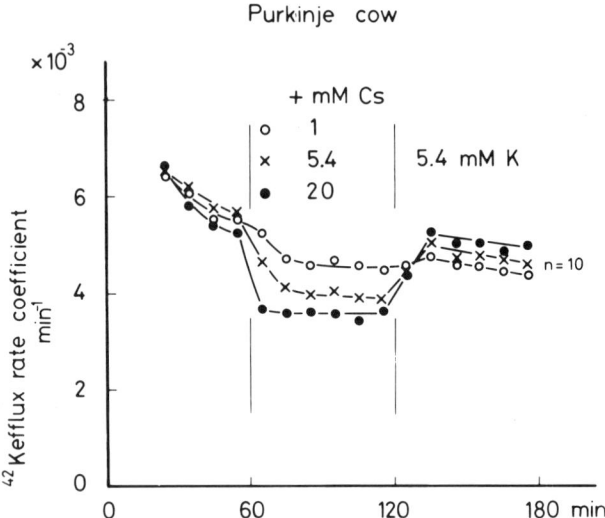

Fig. 2. Effect of 1, 5, and 20 mM Cs on the rate
coefficient of ^{42}K efflux in cow Purkinje fibers (mean of 10
preparations). Extracellular K concentration was 5.4 mM.

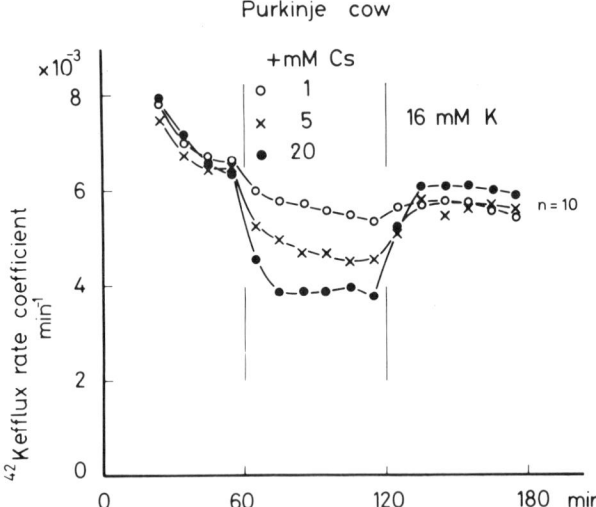

Fig. 3. Effect of 1, 5, and 20 mM Cs on the rate
coefficient of ^{42}K efflux in cow Purkinje fibers (mean of 10
preparations). Extracellular K concentration was 16 mM.

Purkinje cow

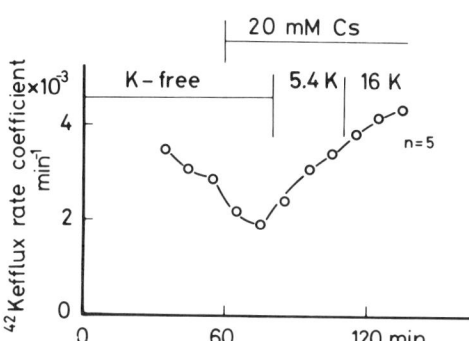

Fig. 4. Effect of changes in the extracellular K concentration between 0, 5.4, and 16 mM on ^{42}K efflux in the presence of 20 mM Cs. Mean of 5 cow Purkinje fibers.

16.2 mM K) (Fig. 5). The experiment was started in 5.4 mM K, 20 mM Cs; as already noted, the fiber depolarizes to about −50 mV under these conditions. Omission of external K resulted in a further depolarization and induced spontaneous activity. Return to 5.4 mM K caused the membrane to hyperpolarize and blocked the pacemaker activity; the fiber could still be stimulated by strong electric stimuli and showed prolonged action potentials. Finally K_O was increased to 16.2 mM: the membrane slightly depolarized, excitability markedly decreased, and action potentials of short duration could only be evoked by using strong stimuli. This experiment clearly demonstrates that the presence of Cs ions does not eliminate responses of the resting potential, action potential duration, and excitability to changes in external K_O. The described effect of K_O on ^{42}K efflux thus cannot be seen as a simple enhancement of K–K exchange, but, at least in part, as an effect on electrogenic K movement.

IV. SUMMARY AND CONCLUSIONS

Although Cs ions enter the cardiac cell (personal unpublished observations on guinea-pig auricles) the described decrease in ^{42}K efflux is probably not related to the internal Cs concentration. The fall in ^{42}K efflux was

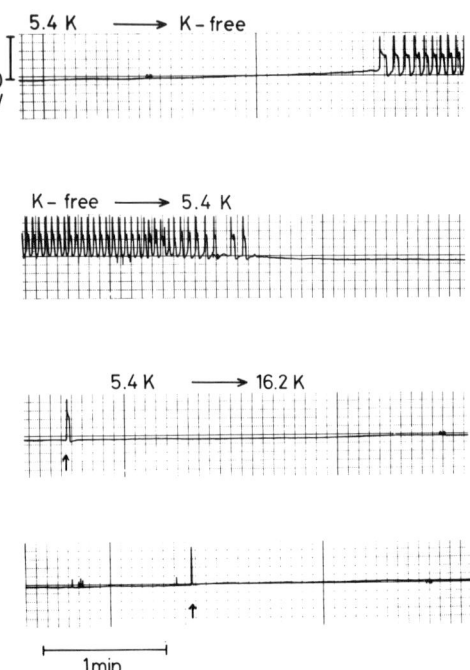

Fig. 5. Effect of changes in the extracellular K concentration on the resting potential, spontaneous activity, and action potential in a cow Purkinje fiber. Extracellular Cs concentration was 20 mM. Action potentials in 5.4 and 16.2 mM K were elicited by electrical stimulation (vertical arrows).

immediate, was not progressive with time, and was quickly reversible, thus suggesting that Cs ions act at the external surface of the membrane.

In this respect cardiac cells resemble skeletal muscle, where resting K conductance is largely decreased by the addition of Cs ions to the extracellular medium (Beaugé *et al.* 1973). They differ, however, quite appreciably from squid giant axon. In the squid giant axon Cs ions do not affect the resting conductance of the membrane, but only the K conductance activated during depolarization: ^{42}K efflux is not changed by Cs in the resting preparation, but drops to 7-20% of the control value in stimulated preparations

(Sjodin, 1966). These results are consistent with data
obtained in voltage-clamp experiments: the delayed K
current is blocked when Cs is applied from the inside;
external Cs only blocks K current when the electrochemical
gradient is such that K ions are forced inwards (Chandler &
Meves, 1965).

The blocking effect of Cs ions was not complete. About
2/3 of the resting K efflux remained in the presence of
20 mM Cs. This remaining K efflux was of electrogenic nature
and was sensitive to external K. Under similar conditions,
Isenberg (1976) found a disappearance of the inward-going
rectification for the steady-state current voltage relation.
We may conclude from these observations that the total
background K current must be subdivided in at least two
components: first, one which is Cs-sensitive and responsible
for the inward-going rectification, and second, one which
does not rectify but is still sensitive to K_O. In addition,
more recent observations indicate that the Cs-sensitive
component cannot be identified with the component that is
eliminated in Cl-free solutions (Carmeliet & Verdonck, 1977).
The ^{42}K efflux reduced by Cl-free solutions can be further
reduced by addition of Cs ions (unpublished observations).

DISCUSSION

Adrian: Would you compare Cs and Rb effects on heart
muscle?

Carmeliet: In sheep Purkinje fibers, Paul Müller* found
that rubidium had a potassiumlike effect, but that it was
quantitatively less pronounced. Thus when rubidium is added
to the perfusion solution, potassium efflux increases, but to
a lesser extent than after adding the same amount of
potassium. Cesium ions, on the other hand, always decrease
potassium efflux.

Morad: In frog ventricular muscle, using about 6 mM
potassium, I didn't find much change in time-dependent
currents. Now, was I to understand from your quoting
Isenberg that the Cs effect was only on the inward rectifier?

*J. Physiol. 177, 453, 1965

Carmeliet: Yes, in 20 mM cesium the inward-going rectification completely disappears. At lower concentrations, e.g., 1 mM Cs, only the pacemaker current is abolished. In our own experiments we also observed a complete disappearance of the pacemaker depolarization in the presence of 1 mM cesium.

Tsien: I think your results really fit in very nicely with earlier work of Peter Stanfield in skeletal muscle, where he used varying amounts of TEA and showed that TEA abolished the steeply rectifying part--leaving a linear component. You can see that kind of linear subtraction in earlier papers by Adrian and Freygang, and it certainly was in the discussion of one of your papers in 1965. The other thing is that the resting potential doesn't go to a very positive value; it stays at -60 at least, so all the evidence you presented really seems to fit in very nicely with the idea that part of the potassium movement is a rectifier, that the rectifier's real property is to shut off completely at strong depolarizations. I think that's an important point in making models for the types of mechanisms that can produce inward rectification. The model we really should be shooting for is not one which has an N-shape, but one which has a sharply turning-off characteristic of the kind that Lars Cleemann mentioned.

Hille: I think people who deal with the inward rectifier almost never remember a paper by Nakajima, Nakamura, and Grundfest on the eleolectroplates, in which they showed inward rectification similar to the steady-state current voltage relationships you showed; that is, very steep in the negative direction, and then coming to a peak, and then coming down, and then rising up slowly. They also showed that adding either cesium or rubidium made them perfectly linear. I agree with Drs. Tsien and Cleemann that to model such a system, you would have to model two components, one which was a linear characteristic, and superimposed upon that, an inward going rectifier which shut off 100% in regions like +20 from E_K.

McNaughton: I noticed that there seemed to be a bounce-back phenomenon when you removed cesium, which was cesium-dependent. In other words, the higher the concentration of cesium during the application, the more the K efflux increased after the cesium was taken off. This seemed to be fairly consistent in all the slides you showed. Do you have any idea what might be the cause of that?

Carmeliet: Your comment is correct, and the increase in rate coefficient above control levels was most pronounced in the 5.4 mM K series. Unfortunately, I have no explanation for this phenomenon.

REFERENCES

1. Beaugé, L.A., Medici, A. & Sjodin, R.A. (1973). The influence of external caesium ions on potassium efflux in frog skeletal muscle. *J. Physiol. 228*, 1-11.
2. Carmeliet, E. & Verdonck, F. (1977). Reduction of K permeability by Cl substitution in cardiac cells. *J. Physiol. 265*, 193-206.
3. Chandler, W.K. & Meves, H. (1965). Voltage clamp experiments on internally perfused giant axons. *J. Physiol. 180*, 788-820.
4. Haas, H.G. & Kern, R. (1966). Potassium fluxes in voltage clamped Purkinje fibres. *Pflügers Arch. ges. Physiol. 291*, 69-84.
5. Isenberg, G. (1976). Cardiac Purkinje fibers: Cesium as a tool to block inward rectifying potassium currents. *Pflügers Arch. ges. Physiol. 365*, 99-106.
6. Sjodin, R.A. (1966). Long duration responses in squid giant axons injected with [134]Cesium sulfate solutions. *J. gen. Physiol. 50*, 269-278.

POTASSIUM CURRENTS IN VENTRICULAR MUSCLE

Lars Cleemann*
Martin Morad

Department of Physiology
University of Pennsylvania
Philadelphia, Pennsylvania

In cardiac muscle, membrane current measured in voltage clamp experiments has been analyzed in terms of several time- and voltage-dependent current components (Trautwein, 1973; Noble, 1975). Noble & Tsien (1968, 1969a,b) described the K current in sheep Purkinje fibers by four separate components, I_{K1}, I_{K2}, I_{X1}, and I_{X2}. The description is based on the Hodgkin-Huxley formalism, but some additional complexity has been introduced by an apparent lack of ionic specificity and by nonlinear current-voltage relations. Brown & Noble (1969a,b) used the double sucrose gap voltage clamp technique to investigate the membrane current in strips of frog atrial muscle. A pacemaker current of the type labelled I_{K2} is not found in this normally quiescent preparation, but the results are otherwise analyzed using the same terminology.

It is also suggested, however, that part of the slow time-dependent current changes reflect extracellular K accumulation and not truly time-dependent permeability changes. Extracellular K accumulation not only may reduce the K equilibrium potential by increasing the unidirectional K influx; it also may increase the net outward K current. McGuigan (1974) used the double sucrose gap voltage clamp technique to investigate extracellular K accumulation in sheep and calf ventricular fibers. K accumulation was

*Present address: Department of Physiology, University of Wisconsin, Madison, Wisconsin

153

estimated from temporary changes in the reversal potential. The experiments suggested that extracellular K accumulation could explain only part of the slow time-dependent current changes.

On the basis of these observations, it may be asked: Which current changes are caused by extracellular K accumulation, and which are caused by truly time-dependent permeability changes? The experiments to be reported here were designed to answer this question. A second goal was to determine to what extent the membrane current is carried by K. This question had been investigated by Haas & Kern (1966), who measured unidirectional K efflux from voltage-clamped sheep Purkinje fibers. Their results confirmed inward-going rectification, but the time dependency of the K fluxes was not tested. The alternative approach, used in this investigation, is to determine the net K flux across the cell membrane from the rate of extracellular K accumulation. This approach requires detailed knowledge about the kinetics of the accumulation process.

In our experiment, the single sucrose gap voltage clamp technique (Morad & Orkand, 1971) was used to control the membrane potential and to measure the membrane current. The isometric contractile force was simultaneously monitored. Frog ventricular muscle, with diameters of 0.3 to 0.5 mm, and length of 0.5 mm were used.

K^+ ACCUMULATION IN VENTRICULAR MUSCLE

Fig. 1 demonstrates the three methods used to estimate extracellular K accumulation. The traces in the lower panel are recorded with a slow time base. The upper panel shows, with expanded time base, the action potentials from the last part of the lower panel. The traces are recorded during two sweeps. The preparation is stimulated normally during the first sweep, and action potentials are seen at the beginning and at the end of the sweep. A K-sensitive microelectrode (Walker, 1971) is inserted into the preparation and records in an extracellular position only a small potential artifact during the action potential. The twitches associated with the action potentials are recorded in the bottom trace. During the second sweep, the membrane potential is clamped to -40 mV for 5 sec. Outward current is passed during the clamp, and the K electrode records the sum of the K-induced potential and the potential

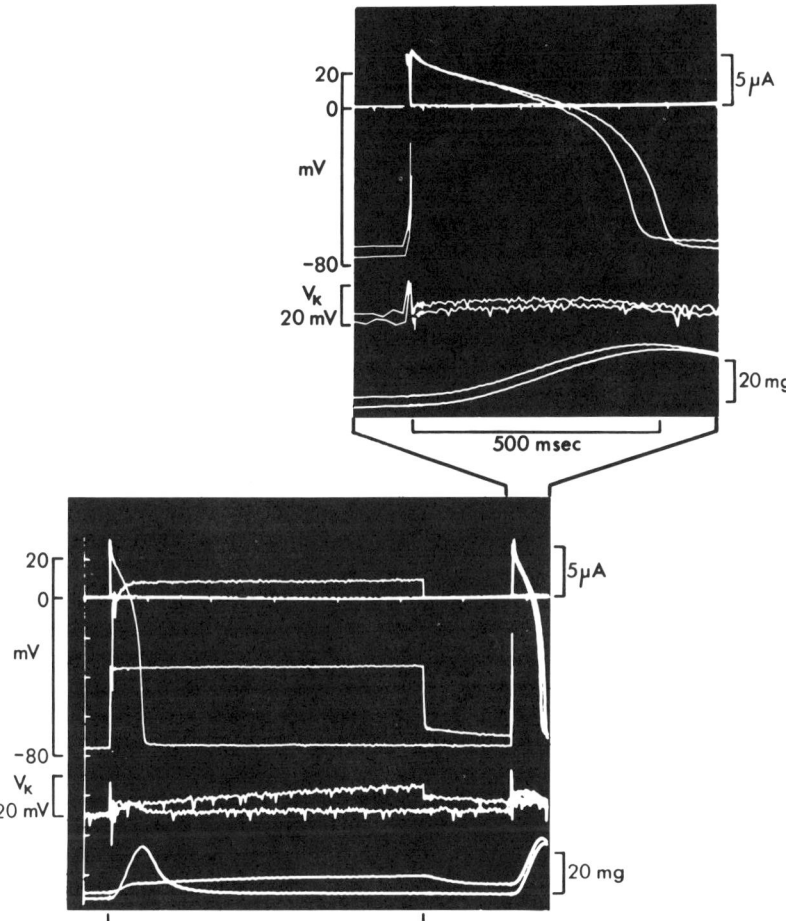

Fig. 1. After the release of a 5 sec clamp pulse
extracellular K accumulation is indicated by the afterpotential, the K electrode response, and the shortening of
the action potential. The lower panel shows the records
obtained during and after the clamp pulse superimposed on
reference records obtained during repeated stimulation of
normal action potentials. The shortening of the action
potential is seen most clearly in the upper panel where the
last part of the lower panel is shown with expanded time base.
From top to bottom each pair of traces is: the membrane
current, the membrane potential (overshooting during the
plateau of the action potentials), the K electrode signal
(V_K) and the isometric tension.

in the extracellular space. A small tension develops during
the early part of the clamp and is well-maintained throughout
the clamp. The clamp is released after 5 sec, the current
is again zero, and the membrane potential falls rapidly
towards the resting potential, but the steep fall is broken
before the normal resting potential is reached. Instead,
a slowly decaying afterpotential is observed. The magnitude
of this afterpotential is the key measurement in the majority
of experiments and is interpreted as resulting from K
accumulation outside the K-selective membrane. The K
electrode records a similar trace, which mainly represents
a change in the K activity. The third indicator for
extracellular K accumulation is the change in the action
potential duration observed when the preparation is
stimulated 1.6 sec after the release of the clamp. These
three signals--the afterpotential, the K electrode signal,
and the action potential shortening--all indicate
extracellular K accumulation. The following experiments
will relate this K accumulation to the measured membrane
current.

In the experiment shown in Fig. 2, the afterpotential
and the K electrode were used to study the voltage
dependency of the K efflux. Sample records are shown in
the top panels. The upper recordings show membrane current,
membrane potential, and contraction. The lower recordings
show with expanded ordinate the bottom part of the trace
from the intracellular microelectrode and also the K
electrode trace. The duration of the applied clamp pulses
is 8 sec, and the membrane current is measured at the end
of the clamps. The afterpotential and the K electrode
response are measured from the lower recordings shortly
after release of the clamp pulses. The membrane current at
the end of the clamp pulse, the afterpotential, and the K
electrode response are plotted versus the clamped membrane
current in lower graphs. It is observed that the
afterpotential and the K electrode both give N-shaped
relations similar to the current-voltage relation, thus
suggesting that the shape of the current-voltage relation
mainly reflects the behavior of the K current. It is clear,
however, that part of the measured membrane current is not
related to K accumulation. For instance, there is net
outward membrane current in the minimum region around -20 mV,
and yet, both the afterpotential and the K electrode response
suggest that the K concentration in the extracellular space
has been depleted. This may indicate that the active
reabsorption of K in this potential region occurs at a
faster rate than the passive K loss.

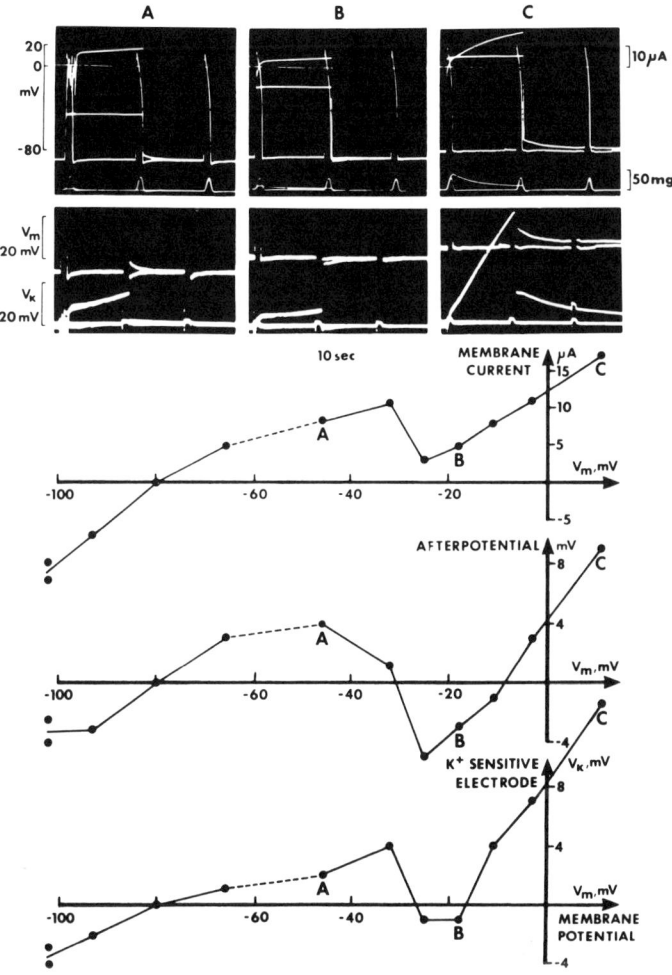

Fig. 2. The membrane current, the afterpotential, and the K electrode response are measured at the end of 8 sec clamp pulses, are plotted versus the clamped membrane potential, and yield similar N-shaped relations. The sample records (A,B, and C) are obtained at 3 different clamp potentials and show, in the upper panels membrane current, membrane potential, and isometric tension and, in the lower panels, the expanded lower part of the membrane potential (V_m) and the K electrode recordings (V_K). The clamp pulses to potentials above and below -60 mV are obtained with different intracellular punctures.

K+ ACCUMULATION AND MEMBRANE CURRENT

Figures 3, 4, and 5 illustrate in some detail the
correlation between the membrane current and the K
accumulation. The analysis is based on a model where K
accumulates in a single well-stirred space from which it is
redistributed at a rate proportional to change in
concentration. The redistribution of K is investigated in
the experiment shown in Fig. 3. The panel labelled ΔV shows

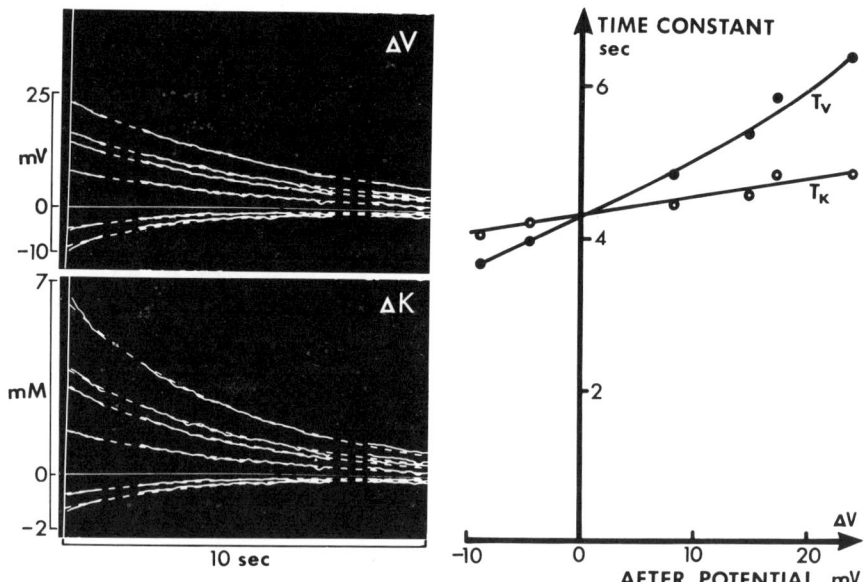

*Fig. 3. The time constant for the decay of the
afterpotential, T_V, is compared to the time constant for the
decay of the accumulated potassium, T_{Ke}. Both are plotted
versus the initial value of the afterpotential. The
accumulated K, ΔK_e, is calculated from the afterpotential,
ΔV, using the equation: $\Delta K_e = 4mM \cdot (exp(\Delta VF/RT)-1)$. The
insets show the measured afterpotentials (ΔV) and the derived
curves for the K accumulation (ΔK_e). Each curve is
approximated by an exponentially decaying curve in a dashed
line. The accumulated K is redistributed with a fairly well
defined time constant.*

the decay of afterpotentials following clamp pulses to
potentials in the range from -100 to +20 mV. The recordings
were stored on magnetic tape and were analyzed using a
computer. The afterpotentials have been approximated by
exponentially decaying curves drawn in a dashed line. The
exponentials can hardly be distinguished from the measured
curves and are clearly visible only in the gaps left open on
removal of action potentials which were stimulated 1 and 7.5
sec after the release of the clamp pulses. The time constant
and amplitude of each exponential were adjusted using a
least-squares method. The time constant (T_V) is plotted in
the graph as function of the amplitude of the
afterpotentials, and varies between 3.5 and 6.5 sec.

In the lower panel, the afterpotentials have been
converted point by point to the equivalent change in the
extracellular K concentration. This conversion is based on
a logarithmic relationship of the type suggested by the
Nernst equation or by the Goldman equation. The traces are
again well approximated by a single exponential, and the
time constant is plotted as function of the amplitude of
the afterpotential. The time constant now changes much less:
only from 3.8 to 4.6 sec. This result suggests that the
accumulation of K, with some degree of approximation, can be
described as occurring in a single compartment from which it
is redistributed with a single time constant.

The logarithmic relationship used to convert the
afterpotential to changes in the extracellular K concentration
is also tested in the experiment shown in Fig. 4, where the
membrane was clamped to various potentials in a series of
2-sec clamp pulses. The afterpotential is measured as
usual, shortly after release of the clamp pulse. A weighted
integral of the membrane current is used to estimate the
equivalent charge of the accumulated K. It is assumed that
K accumulated during the clamp pulse is redistributed with
the same time constant as determines the decay of the
afterpotential after the release of the clamp. The weighted
integral is therefore a convolution of the membrane current
with an exponential with this specific time constant. It is
also assumed that the majority of the membrane current is
carried by K. The estimated charge of the accumulated K is
plotted versus the afterpotential, and the experimental
results are approximated by a logarithmic curve with the
curvature predicted by the Nernst equation or the Goldman
equation. Experiments with altered extracellular K
concentrations have shown that the resting potential of frog
ventricular muscle can be approximated by the Goldman

equation, if it is assumed that ions other than K carry a
unidirectional inward current equal to that carried by
approximately 1 mM of extracellular K. The general agreement
between measurements and theory is quite satisfactory. Some
systematic differences can be observed, however. The
measured points are connected, corresponsing to a sequence of
steadily increasing clamp potentials, and the afterpotential

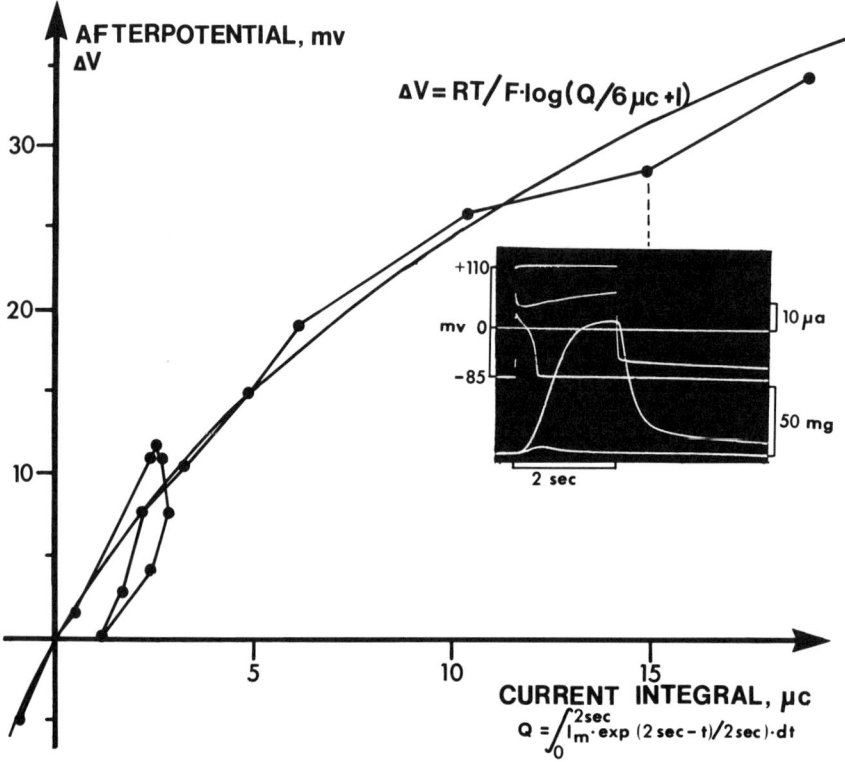

$$\Delta V = RT/F \cdot \log(Q/6\,\mu c + l)$$

$$Q = \int_{0}^{2sec} I_m \cdot \exp\,(2\,sec - t)/2\,sec) \cdot dt$$

*Fig. 4. The afterpotential, ΔV, is plotted versus the
current integral, Q. The current integral is an estimate of
the charge of the accumulated K assuming that the membrane
current is carried entirely by K and that the K during the
2 sec clamp pulses is redistributed with the same time
constant as after the clamp pulse (2 sec). The measured
points are approximated by a logarithmic curve of the shape
predicted from the Nernst equation or the Goldman equation.
This illustration is based on an experiment by T. Klitzner.*

is small not only when the clamp potential is close to the resting potential, but also in a potential region around −20 mV, where the current integral indicates significant outward current.

The correlation between the membrane current and the extracellular K accumulation is tested in greater details in the experiment shown in Fig. 5. The top panels show original records, and the bottom panels show the result of a computer simulation. The membrane potential was clamped to four different potentials, and at each membrane potential, clamps of different duration were applied to study the development of the afterpotential. The measured membrane current is approximated in the lower panels by a few straight lines, and the measured afterpotentials are indicated as circles. The afterpotentials are approximated by continuous curves calculated from the measured current. The three parameters in the left column are adjusted to give a best fit. The time constant for redistribution of K was assumed to have the same value both during and after the clamp pulses; the estimated value is 5 sec. The size of the extracellular space was determined as 0.013 mm^3 if the cell membrane is perfectly K-selective, or approximately one-tenth of a cubic mm with a more realistic K selectivity. This corresponds to approximately 12% of the total volume of the preparation estimated from its dimensions. Comparing this with the studies of Page & Niedergerke (1972), it may be suggested that K primarily accumulated in the subendothelial fraction of the extracellular space, and that diffusion through the narrow clefts between the endothelial cells is of importance for redistribution of accumulated K.

The fraction of the membrane current which appears to be unrelated to K accumulation is indicated in the current diagrams by straight horizontal lines. This current is called the residual current. It is independent of time and is proportional to the difference between the resting potential and the clamp potential, and it can consequently be characterized by a single number. The technique is not sufficiently sensitive to exclude completely time-dependent variations in the residual current. It may be noticed, however, that the time-dependent changes in the measured membrane current both at −58 and at +28 mV are much larger than the estimated magnitude of the residual current. It may therefore be concluded that the time-dependent current changes at these potentials mainly reflect time-dependent changes in the K current. The residual membrane current may

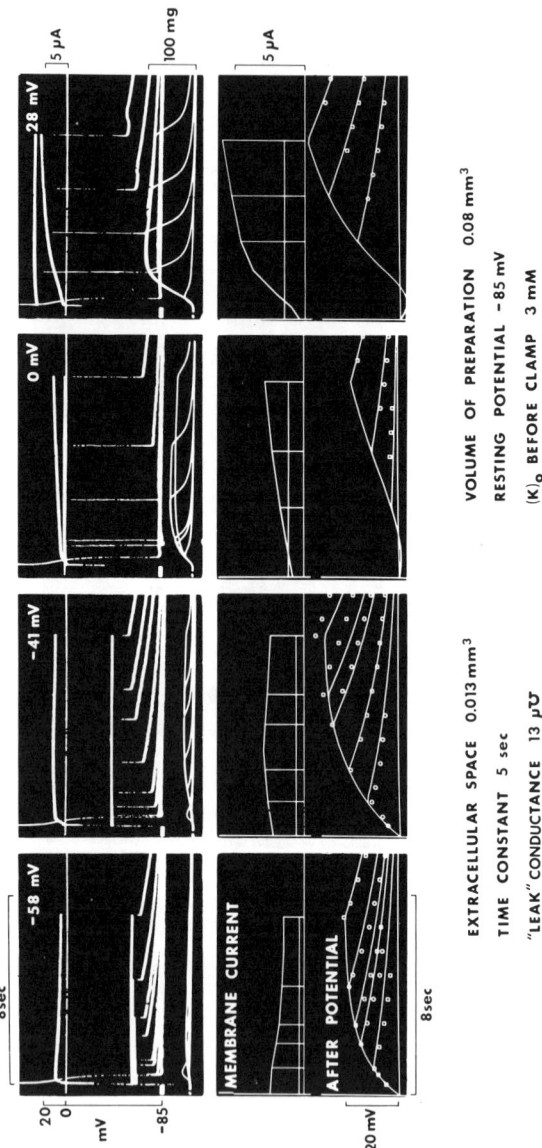

EXTRACELLULAR SPACE 0.013 mm³

TIME CONSTANT 5 sec

"LEAK" CONDUCTANCE 13 μ℧

VOLUME OF PREPARATION 0.08 mm³

RESTING POTENTIAL –85 mV

$(K)_o$ BEFORE CLAMP 3 mM

Fig. 5. At 4 different clamp potentials the development and decay of the afterpotential is studied using clamp pulses of various durations. The upper panels show the original records and the lower panels show the results from a computer simulations where the afterpotential was calculated from the measured membrane current. A good agreement between the measured afterpotentials (circles) and the calculated afterpotentials (continuous curves) is obtained by adjusting the volume of the extracellular accumulation space, the time constant for redistribution of K, and the residual current, which is the fraction of the measured membrane current not carried by K.

represent chloride current and leakage current across the
sucrose gap, and also possibly some passive K current which is
masked by simultaneous active K reabsorption.

INWARDLY RECTIFYING MEMBRANE CURRENTS

There is a marked difference between the records obtained
with clamp potentials below and above -20 mV. At -58 and
-41 mV, there is significant initial membrane current and
the afterpotentials develop at a fast rate. The estimated K
current develops more slowly at 0 and 28 mV, and the
afterpotentials appear to develop with some delay. The next
figures illustrate the qualitative differences between the
K currents below and above -20 mV.

Fig. 6 shows the difference between a current-voltage
relation measured early during clamps and at the end of the
3.5-sec clamp pulses. The final membrane current gives the
familiar N-shaped relation when plotted versus the clamped
membrane potential. The initial membrane current is not
measured using fast voltage clamp methods; it is simply the
slowly changing section of the current trace extrapolated to
the beginning of the clamp pulse. On depolarization, the
current first reaches a maximum, then falls with marked
negative slope conductance, and finally seems to approach
zero asymptotically. It has been shown that K accumulates
at an initial fast rate in the potential region where this
current is large, and it may therefore be suggested that this
current-voltage relation approximates the initial passive K
current. Current voltage relations of this approximate shape
have been described in various preparations and are commonly
called "inward-rectifying."

The mathematical equations used to give a quantitative
description of this current component are based on the model
shown in Fig. 7. The model is shown schematically in panel A.
A pore allows free passage of K through the major part of the
cell membrane, but a narrow section near the inner surface of
the membrane is normally completely blocked by an agent drawn
as a ball on a chain. The pore is only opened when a K ion
occasionally enters from the outside and pushes the blocking
agent out of the narrow section. The K ion may stick in the
opening. Conduction might then involve replacement of this
ion by other K ions from the inside or the outside until the
blocking agent finds its way back into the narrow section of
the pore. The number of K ions right inside and outside the

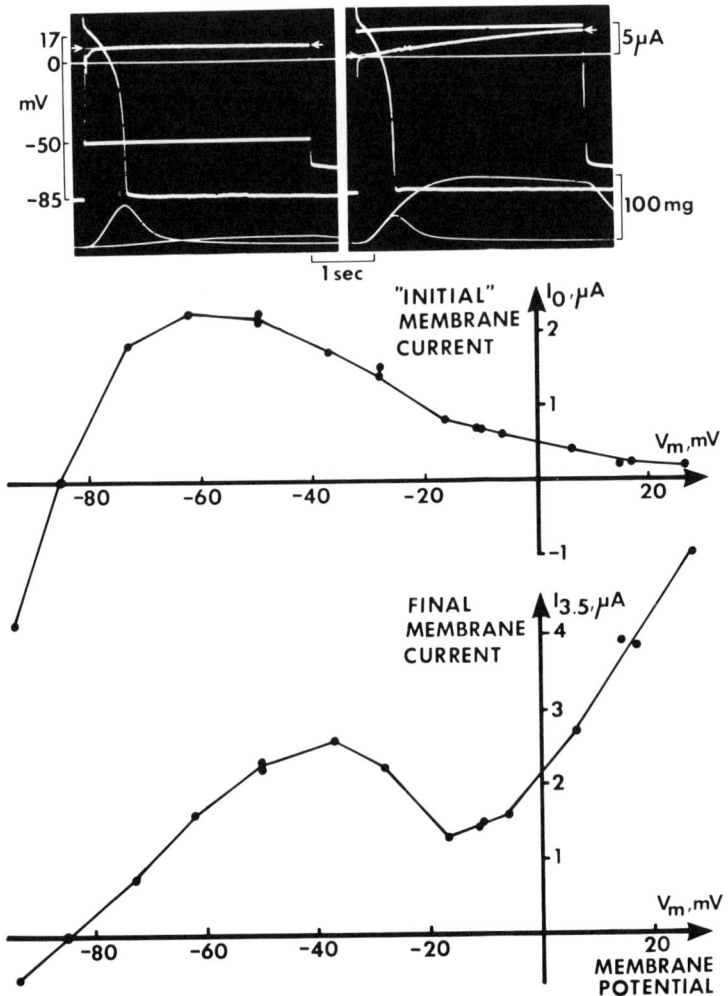

Fig. 6. The current-voltage relation for the "initial" membrane current is compared to the current-voltage relation for the final membrane current measured at the end of 3.5 sec clamp pulses. The "initial" membrane current is obtained by extrapolating the slowly changing part of the current trace back to the beginning of the clamp pulse.

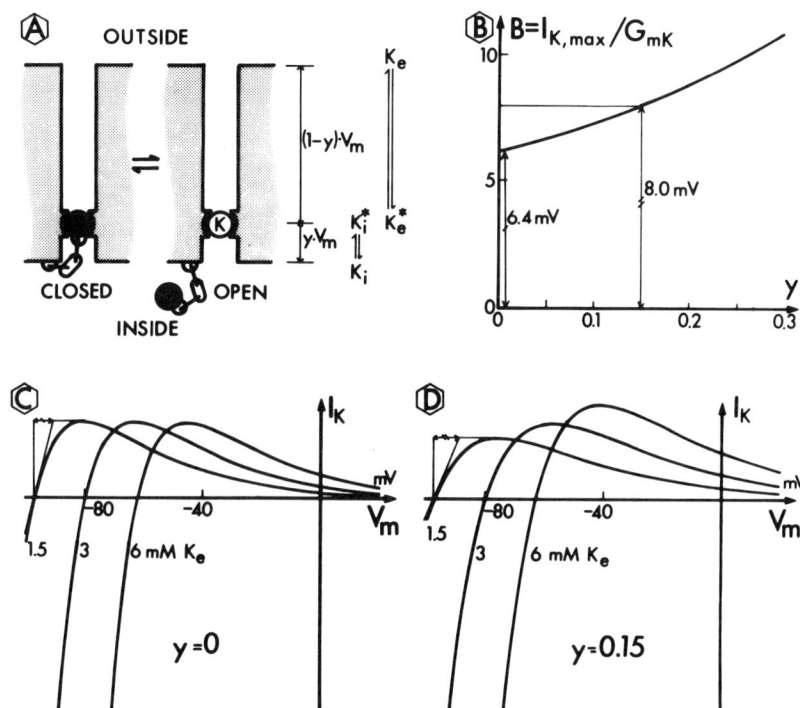

Fig. 7. Panel A shows a model for an inward rectifying
K channel. The model is described in the text. y is the
fraction of the membrane potential, V_m, subsiding between the
narrow section of the pore and the inside of the membrane.
The local K concentrations right inside, K_i^*, and outside, K_e^*,
the narrow section are in equilibrium with the intracellular
K concentration, K_i, and the extracellular K concentration,
K_e, respectively: $K_i^* = K_i \cdot exp(y \cdot V_m \cdot F/RT)$,
$K_e^* = K_e \cdot exp((y-1) \cdot V_m \cdot F/RT)$. The K current, I_K, is
calculated assuming that the fraction of open pores is
proportional to K_e^* and that the K flux through an open
channel is proportional to the local concentration gradient:
$I_K \sim K_e^* \cdot (K_i^* - K_e^*)$. Panel C and D show the shape of the
calculated current-voltage relations and the effect of
variations in K_e when y is 0 and 0.15 respectively. Panel
B shows how y may be estimated from the maximum value of the
current, $I_{K,max}$, divided by the conductance at the reversal
potential, G_{mK}.

narrow section is determined by the intracellular and
extracellular K concentration and by the electrical potentials
driving the K ions away from or towards the narrow section.
The model suggests that the rate-limiting processes occur at
a potential which is close to the intracellular potential.
A similar scheme has been suggested by Armstrong and
Bezanilla (Armstrong, 1971; Bezanilla & Armstrong, 1972) to
explain blockage of the K current in the squid axon by
internally perfused TEA and similar compounds.*

The shape of the calculated current-voltage relations in
the model described here depends on the exact "electrical"
location of the rate-limiting processes. The current-voltage
relations in panel C are calculated under the assumption
that they occur exactly at the inner surface of the membrane.
The current-voltage relations are inward-rectifying and have
a negative slope conductance at some membrane potentials.
Variations in the extracellular K concentration will just
shift the current-voltage relation along the potential axis.

Panel D shows the current-voltage relations calculated
from the assumption that the rate-limiting transitions are
removed from the inner surface of the membrane 15% in the
direction of the outer surface. The shape of the individual
current-voltage relations is altered slightly, and a
variation in the extracellular K concentration will now not
only shift the current-voltage relation along the voltage
axis, but also alter the magnitude of the maximal outward
current. The calculated increase in the magnitude of the
current is approximately 12% each time the reversal potential
depolarizes 10 mV.

Both cases give K-induced K current in a potential region
which approximately corresponds to the region of negative
slope conductances. The maximal outward current divided by
the conductance at the reversal potential was determined from
several preparations, and the values suggest that the shape
of the current-voltage relations is close to that shown in
panel D. The calculations require correction for
extracellular series resistance.

*Rojas & Rudy (1976) have observed that Na inactivation--
the h-process--is suppressed after internal perfusion with
pronase (alkaline proteinase b). They suggest that pronase may
cleave a short protruding polypeptide sequence with a function
similar to that ascribed to the ball on the chain. Hille and
Woodhull (e.g. Hille, 1975; Woodhull, 1973) have investigated
the selectivity and blockage of the Na channel in myelinated
axons and have proposed a blocking site with 27% of the
membrane potential residing between it and the extracellular
cell surface.*

The shape of the current-voltage relation for the initial membrane current is tested more rigorously in the experiment shown in Fig. 8. Here, TTX is applied to block the excitatory Na current, and the series resistance is removed functionally by using the chopped current clamp method first applied to cardiac muscle by Goldman & Morad (1977). The current is measured 10 and 50 msec after the beginning of the clamp pulse. Ten msec is more than sufficient time to achieve effective clamp control, and the extracellular K accumulation will still be insignificant after 50 msec. Below -30 mV, the two curves agree mutually, and they are well-approximated by the theoretical current-voltage relation suggested above. Above -30 mV, the membrane current apparently changes fairly rapidly, and neither of the measured current voltage relations are close to the theoretical current-voltage relation. A better fit in this potential range requires extrapolation from the slowly changing part of the current trace. This procedure is used in the experiments described below.

Fig. 9 shows how the initial membrane current changes in the presence of an afterpotential. The smooth curve through the black dots is measured at the beginning of clamp steps from rest. The curve approximating the open circles indicates the current-voltage relation measured after a 2.5-sec conditioning clamp pulse to -40 mV. The afterpotential develops rapidly around -40 mV, and the reversal potential is shifted 14 mV. A large part of the change in the current-voltage relation can be described by a shift along the voltage axis, but there is also a significant increase in the maximal outward membrane current. The increase is consistent with the suggested model, assuming again that approximately 15% of the membrane potential subsides between the K-gate and the intracellular compartment. The dashed curve is the membrane current measured at the end of 3.5-sec clamp pulses. It may be noticed that the conditioning clamp pulse to -40 mV results in a larger outward membrane current during the following test pulses to -30 to -10 mV than is obtained when the membrane is held continuously at these more depolarized potentials for an even longer time.

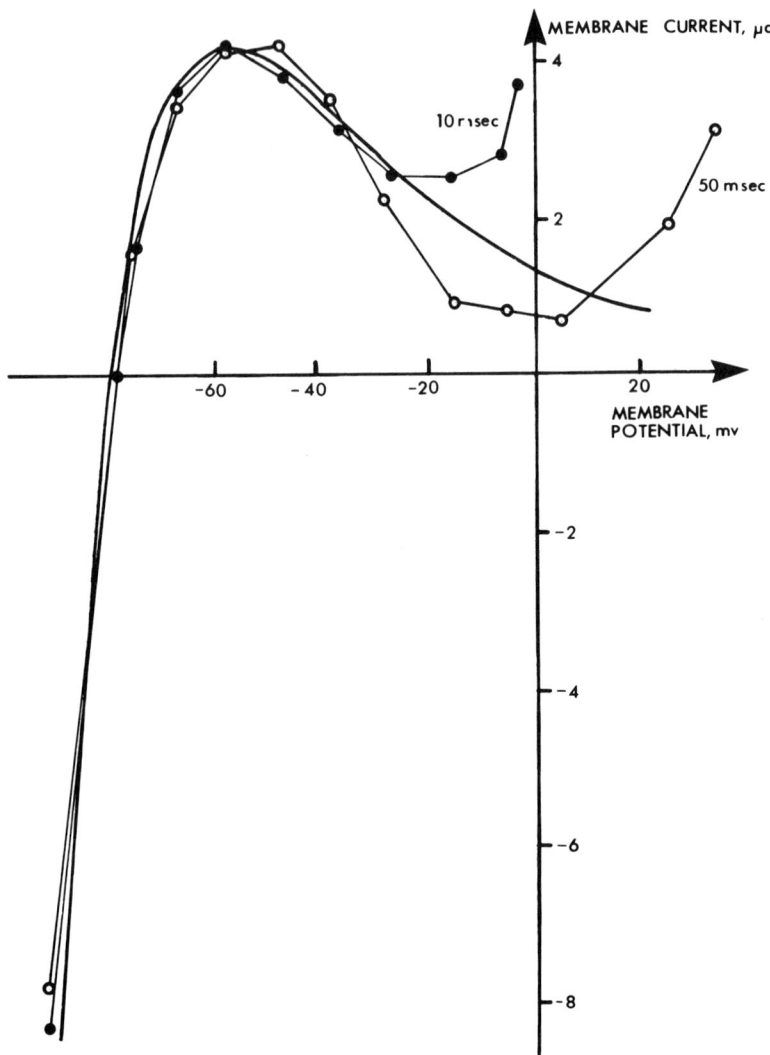

Fig. 8. Current-voltage relations measured 10 and 50 sec
after the beginning of the clamp pulses are compared to the
theoretical current voltage relation. The chopped current
pulse clamp method is used to achieve rapid clamp control
and to eliminate the effect of the extracellular series
resistance. $10^{-6}M$ TTX is added to the perfusate to block
the excitatory Na current.

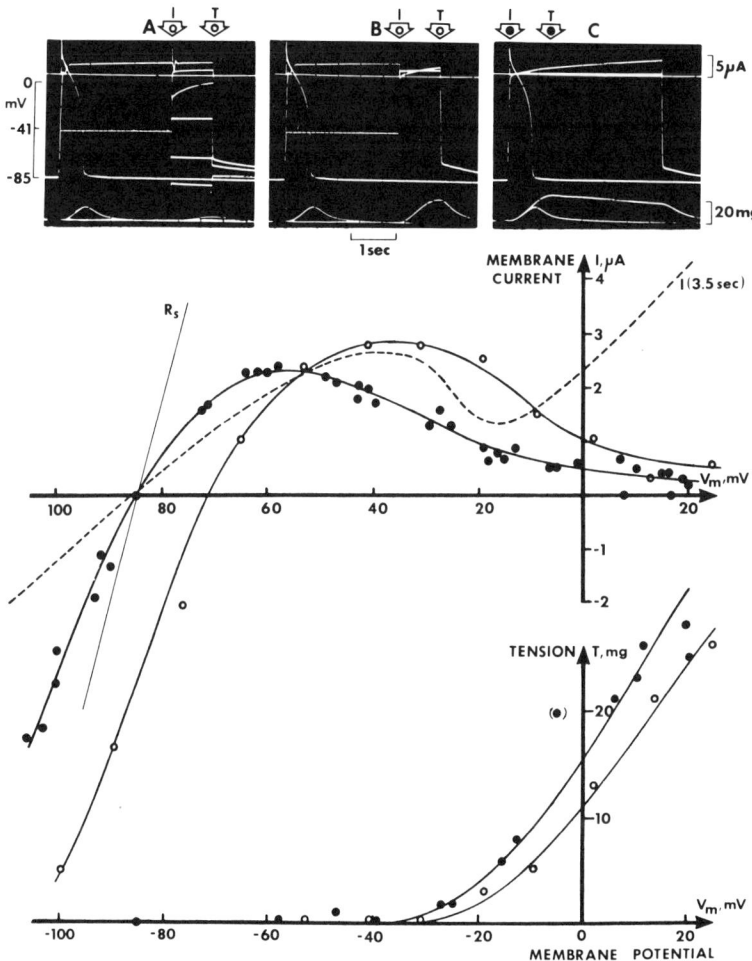

Fig. 9. *The effect of the extracellular K concentration on the current-voltage relation for the "initial" membrane current. The change in the extracellular K concentration is due to extracellular K accumulation during a 2.5 sec clamp pulse to -41 mV (Panels A and B). In the current-voltage diagram the series resistance, R_S, and the 3.5 sec current-voltage relation (dashed curve) are also indicated. The isometric tension plotted in the lower graph is measured 1 sec after the "initial" membrane current.*

EFFECT OF K^+ ACCUMULATION ON TIME DEPENDENT CURRENTS

Fig. 10 tests to what extent the slow current changes can be explained on the basis of extracellular K accumulation. The upper graph shows the membrane current at the beginning and at the end of 3.5-sec clamp pulses as a function of the clamped membrane potential. The lower curve is the afterpotential. The dashed curve is derived from the initial membrane current by shifting it along the voltage axis by the amount indicated by the afterpotential and at the same time scaling it slightly up or down by a factor which increases 12% for each 10 mV of depolarizing afterpotential. This transformation brings the initial membrane current very close to the final membrane current in the entire potential range below -20 mV. The current changes in the potential range below -20 mV can therefore be accurately related to the K-induced current resulting from extracellular K accumulation. The truly time- and voltage-dependent K permeability is limited to the potential region more positive than -20 mV.

EFFECT OF $[K]_o$ ON INWARD RECTIFICATION

In Fig. 11, the membrane current and the afterpotential were with this preparation measured first in a solution with 6 mM K and then in the normal Ringers solution with 3 mM K. The current-voltage relation for the initial membrane current is shifted along the voltage axis, and the maximum current is slightly increased as in the presence of a depolarizing afterpotential. In 6 mM K, there is significantly more outward current in the potential range from -40 to 0 mV. The final membrane current also indicates a region with increased outward membrane current, and the afterpotential indicates that the K efflux has increased in this range. The membrane current is, at positive potentials, unchanged both at the beginning and at the end of the clamp pulses. The afterpotential is somewhat smaller in 6 mM K, but this decrease is expected from the nonlinear relationship between the extracellular K concentration and the resting potential. When this is considered, the afterpotential indicates that the K efflux is unchanged at positive membrane potentials. The K-induced changes in the K current correspond well with the change in the shape of the action potential. There is little change in the initial part of the plateau which is, at positive potentials, where the time-dependent K current shows the same insensitivity to K. The changes observed at lower potentials towards the end of the action potential are consistent with current changes ascribed to the inward-rectifying K-sensitive K current.

K INDUCED K EFFLUX DURING A SINGLE ACTION POTENTIAL

Fig. 12 was obtained by Kline and Morad, who have measured the extracellular K accumulation using a double-barrel K-sensitive electrode. The accumulation,

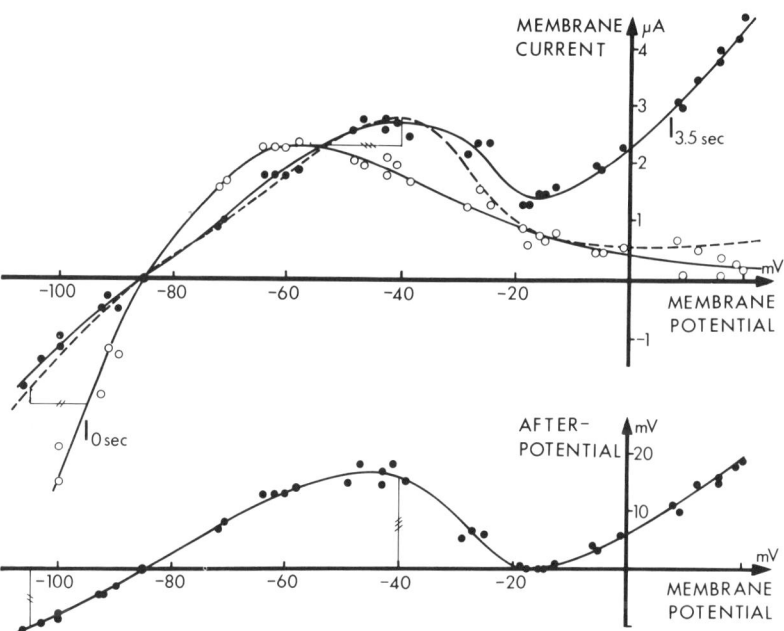

Fig. 10. The relationship between the initial membrane current (open circles), the final membrane current (filled in circles) and the afterpotential (lower graph). The dashed curve is calculated from the initial membrane current and the afterpotential using the model for inward-going K rectification. Below -20 mV the dashed curve closely approximates the final membrane current indicating that the current changes are due to extracellular K accumulation.

measured during a single action potential, is o.6 mM in
Ringers solution containing 3 mM K, and 0.4 mM in the
presence of 6 mM K. The accumulation of roughly proportional
to the duration of the action potential. The results
therefore indicate that the rate of K efflux during the

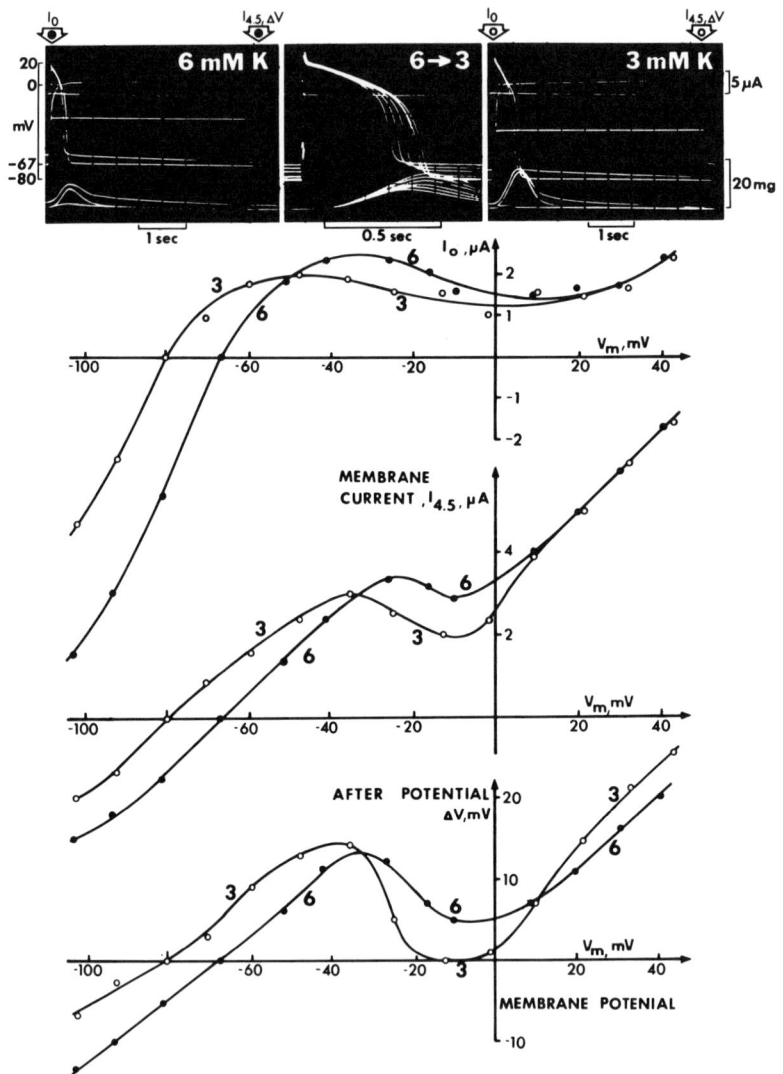

*Fig. 11.   The initial membrane current, the final
membrane current and the afterpotential with 6 mM and 3 mM
in the perfusing solution.*

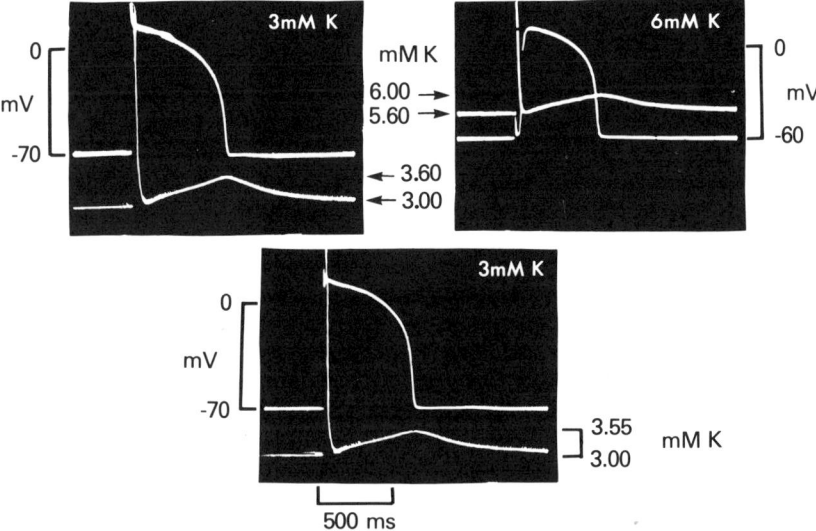

Fig. 12. The extracellular K accumulation measured during the action potential with a double barrel K selective microelectrode. The K concentration of the perfusing solution was changed from 3 mM to 6 mM and back to 3 mM. Experiment by R. Kline.

plateau of the cardiac action potential is unchanged. This observation is consistent with the preceding experiment, which suggests that the K-induced K current is most prominent in a potential range which includes only the last part of the plateau and the phase of rapid repolarization.

CONCLUSIONS

These experiments have supplied documentation for extracellular K accumulation in voltage-clamped frog ventricular muscle. K accumulation was indicated by the afterpotential, the K electrode, and the change in action potential duration. Further evidence for K accumulation was obtained from experiments which confirm the expected logarithmic relationship between the extracellular K concentration and the potential of the resting membrane.

The estimated size of the accumulation space and the time
constant for redistribution of K are fairly compatible with
the properties ascribed to the subendothelial fraction of
the extracellular space (Page & Niedergerke, 1972).

The extracellular K accumulation was used to investigate
both the voltage dependency and the time dependency of the K
current. It is concluded that the larger fraction of the
membrane current measured during long clamp pulses is carried
by K, and the accumulation studies confirm time-dependent
changes in the K current both below and above -20 mV. There
is, however, a marked qualitative difference between the
K current below and above this potential. There is
significant initial K current below -20 mV, while the K
current above -20 mV only develops slowly with time. A model
is suggested for an inward-rectifying K channel which gives
both negative slope conductance and K-induced current. The
shape of the calculated inward-rectifying current-voltage
relations have been tested, and the effect of extracellular
K has been confirmed both in experiments with clamp-induced
extracellular K accumulation and in experiments where the K
concentration of the perfusate was altered. It has been
verified that K accumulation can account quantitatively for
the slow time-dependent current changes in the entire
potential range below -20 mV.

The shortening of the action potential in solutions with
an elevated K concentration seems to be closely related to
the K-induced current if the inward-rectifying K current
plays a minor role during the first part of the plateau when
the membrane potential is positive. The dominant K current
is in this potential range the truly time-dependent K
current, and both voltage clamp experiments and the rate of
K efflux indicate that this current component is insensitive
to variations in the extracellular K concentration. The
inward-rectifying K current only plays a significant role
towards the end of the action potential.

The possibility exists that some of the time-dependent
current changes, which in previous studies have been
attributed to time-dependent permeability changes, in
reality are related to extracellular K concentration. The
current-voltage relations measured in frog atrial muscle by
Brown & Noble (1969 a,b), for instance, are quite similar to
the current-voltage relations measured in the present study,
but the current changes in the lower potential range are
associated with a time-dependent current component of the
type labelled I_{X1}. In light of the present investigation, it
may be suggested that the same current changes equally well

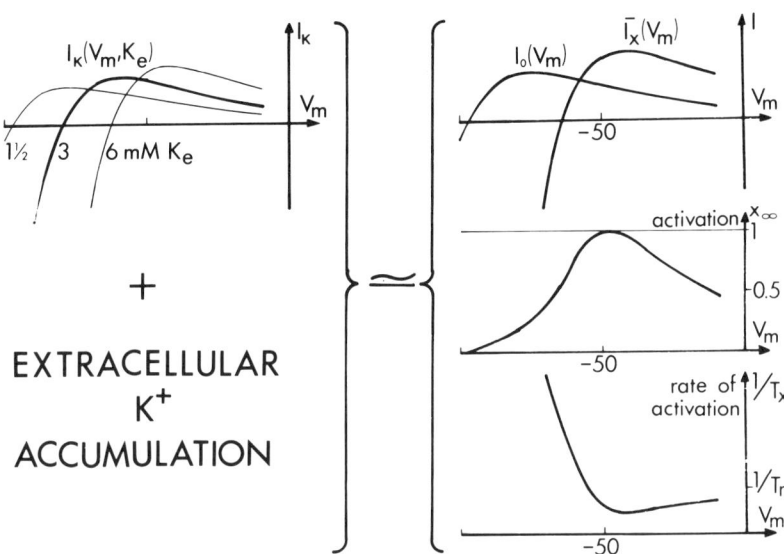

Fig. 13. Supplementary figure used during the discussion. Similar time dependent membrane currents may either result from extracellular K accumulation which influences a K sensitive current component, $I_K(V_m, K_e)$, (left side) or from time and voltage dependent membrane processes which determines the activation of a current component I_X, (right side). If the effect of the extracellular K concentration on the K sensitive current, I_K, can be described using a linear approximation and if the extracellular accumulation of K can be described by the volume of the accumulation space and the time constant for redistribution, T_r, then the equivalent time dependent current component, I_X, can be described using first order Hodgkin-Huxley kinetics and it is possible to calculate the current voltage-relation for the fully activated current, $\bar{I}_X$, the steady state activation parameter, $x\infty$, and the time constant for the activation, T_X. $I_o(V_m)$ is the current voltage relation observed when the time dependent current is fully deactivated corresponding to maximum extracellular K depletion.

could be caused by extracellular K accumulation. It has been suggested by Cohen, Daut & Noble (1976) that the K concentration in the narrow clefts of Purkinje fibers may deviate from the K concentration in the perfusate due to the activity of the Na-K pump. If the activity of the pump can cause K depletion, it is reasonable to expect that the passive K currents can cause accumulation. It is possible, therefore, that some of the measured current changes result from extracellular K accumulation.

DISCUSSION

   McNaughton: I'd just like to comment on somewhat similar results that we've obtained in Purkinje fiber. Using the Noble and Tsien terminology, I_{X2} has about the same time course as your accumulation current, but I_{X1} clearly does not. Would you agree that we're on fairly safe ground in saying that I_{X1} is not an accumulation current, although it is affected by accumulation, and that in fact accumulation occurs with a rather slower time course than the fast current changes in Purkinje fibers?
   Cleemann: It would be unwise for me to suggest without further evidence that any of the current components in Purkinje fibers are entirely related to potassium accumulation. But I think I could point out some of the properties you would expect from the potassium induced potassium current from accumulation. Fig. 13 shows that you can model all the time-dependent current changes, in the lower potential region, by saying that the potassium-induced potassium current plus extracellular potassium accumulation is exactly equivalent to Hodgkin-Huxley type current. You can balance the time-independent current, and the time-dependent current, with the activation curve and the rate constant shown in Fig. 13. The mathematical equations actually do not allow you to distinguish between the two possibilities. Only measurements with potassium electrode and changes of potassium concentrations make it possible to do that distinction. From this investigation you can determine the link between the shape of the activation curve, the rate of activation, and the shape of the current-voltage relations using the Hodgkin-Huxley formalism. It is conceivable that in Purkinje fibers a number of current components with various potential dependencies interplay in such a way as to hide at least to some degree potassium accumulation.

Tsien: I have a quick comment and a quick question. The comment is that according to the criteria you showed on the last slide (Fig. 13), namely, that there should be a relationship between the location of the current voltage curves and the activation curve, one can at least say that the pacemaker potassium current that Dr. Adrian talked about in his introductory speech is not largely due to accumulation.

Cleemann: That's true; I agree.

Tsien: I think with the plateau currents, one has to be much more cautious. The question I had concerns the very interesting results you showed right at the beginning, which I think were published in *Science*, where you show N-shaped curves for three different parameters: extracellular potassium measured with the electrode, afterpotential, and action potential duration. In those N-shaped curves, I've always wondered about the region underneath the voltage axis. And you said that you might explain that by an excess of potassium transport. Is that the explanation you believe, and if that's the case, does that mean that the sodium pump is voltage dependent?

Cleemann: The negative afterpotential, or the hyperpolarizing afterpotential, as I call it, is usually only a few millivolts. You could explain the negative or the hyperpolarizing afterpotential by having a sodium pump which is independent of voltage, when you consider that K^+ current turns off around -20 mV. Dr. Morad has recently explored the situation more carefully, and maybe he would like to comment on it.

Morad: We have done some experiments recently, and in fact, the results are very consistent with what Dr. Cleemann has said. It is really the fact that currents are very, very small around -20 mV that you get a chance to see the activity of the Na-pump. If you add adrenaline, which has been reported to activate the pump, we get to see an increase in this hyperpolarizing afterpotential; so I think we're fairly sure that it represents the pump. The problem is, we can't get it to increase in its activity with low concentration of oubain.

Adrian: I'm wondering to what extent in hearts which have an efficient coronary circulation, there will be potassium accumulation that you've been describing. I think in frog heart there is no coronary circulation.

Cleemann: I don't know much about the mammalian heart, but if you make calculations just based on the dimensions of for instance, Purkinje fibers, and consider the time it would take the ions to move in and out of that preparation through the narrow clefts, then you have a time constant in the order of seconds. I believe, therefore, it's a definite possibility that potassium might accumulate in mammalian heart.

Morad: We've been concerned about this problem for some time. The mammals are, in fact, more resistant to rate-induced accumulation, but remember that the pump is probably more effective. More experiments are required to determine the extent of K^+ accumulation in mammalian heart.

Page: This is again a reply that a mammalian trabeculae, like the frog, is surrounded by an endothelial layer. One does occasionally find capillaries within it.

REFERENCES

1. Armstrong, C.M. (1971). Interaction of tetraethylammonium ion derivates with the potassium channels of giant axons. *J. gen. Physiol. 58*, 413-437.

2. Bezanilla, F. & Armstrong, C.M. (1972). Negative conductance caused by entry of sodium and cesium ions into the potassium channels of squid axons. *J. gen. Physiol. 60*, 588-608.

3. Brown, H.F. & Noble, S. (1969a). Membrane currents underlying delayed rectification and pacemaker activity in frog atrial muscle. *J. Physiol. 204*, 717-736.

4. Brown, H.F. & Noble, S. (1969b). A quantitative analysis of the slow component of delayed rectification in frog atrium. *J. Physiol. 204*, 737-747.

5. Cohen, I., Daut, J. & Noble, D. (1976). An analysis of the action of low concentrations of ouabain on membrane currents in Purkinje fibres. *J. Physiol. 260*, 75-103.

6. Goldman, Y. & Morad, M. (1977). Measurement of the membrane potential and current in cardiac muscle: A new voltage clamp method. *J. Physiol. 268*, 613-654.

7. Haas, H.G. & Kern, R. (1966). Potassium fluxes in voltage clamped Purkinje fibres. *Pflügers Arch. ges. Physiol. 291*, 69-84.

8. Hille, B. (1975). Ionic selectivity, saturation and block in sodium channels. *J. gen. Physiol. 66*, 535-560.

9. McGuigan, J.A.S. (1974). Some limitations of the double sucrose gap, and its use in a study of the slow outward current in mammalian ventricular muscle. *J. Physiol. 240*, 775-806.

10. Morad, M. & Orkand, R.K. (1971). Excitation-contraction coupling in frog ventricle: Evidence from voltage clamp studies. *J. Physiol. 219*, 167-189.
11. Noble, D. (1975). *The Initiation of the Heartbeat,* Claredon Press, Oxford.
12. Noble, D. & Tsien, R.W. (1968). The kinetics and rectifier properties of the slow potassium current in cardiac Purkinje fibres. *J. Physiol. 195*, 185-214.
13. Noble, D. & Tsien, R.W. (1969a). Outward membrane currents activated in the plateau range of potentials in cardiac Purkinje fibres. *J. Physiol. 200*, 205-231.
14. Noble, D. & Tsien, R.W. (1969b). Reconstruction of the repolarization process in cardiac Purkinje fibres based on voltage clamp measurements of membrane current. *J. Physiol. 200*, 233-254.
15. Page, S.G. & Niedergerke, R. (1972). Structures of physiological interest in the frog heart ventricle. *J. cell. Sci. 11*, 179-203.
16. Rojas, E. & Rudy, B. (1976). Destruction of the sodium conductance inactivation by a specific pronase in perfused nerve fibres from *loligo*. *J. Physiol. 262*, 501-531.
17. Trautwein, W. (1973). Membrane currents in cardiac fibres. *Physiol. Rev. 53*, 793-835.
18. Walker, J.L. & Ladle, R.O. (1973). Frog heart intracellular potassium microelectrodes. *Am. J. Physiol. 225*, 263-267.
19. Weidmann, S. (1951). Effect of current flow on the membrane potential of cardiac muscle. *J. Physiol. 115*, 227-236.
20. Woodhull, A.M. (1973). Ionic blockage of sodium channels in nerve. *J. gen. Physiol. 61*, 687-708.

GENERAL DISCUSSION
ON "THE INWARD RECTIFIER"

Tsien: One thing which has intrigued me about inward rectification is that some people think about the rectification as being the shutting-off of a current with strong depolarizations, whereas the work of Almers, among others, suggests that the inward rectification turns on at more negative potentials. Is there any way of pulling all this together? Or de we have to think of it as being different ways of taking a linear thing and making it into a nonlinear function? I was hoping that Dr. Armstrong would comment.

Armstrong: I have a model which at least has the virtue of simplicity. It may not provide a rectifier characteristic that's sufficiently steep, but it's based on experiments in a squid axon with TEA. Those experiments can be explained simply by postulating that a potassium pore has a narrow outer portion, and a wide inner portion, which is not very selective, and that TEA can bind or stick in the inner portion of the channel, and that TEA is knocked out of the channel by a potassium ion that reaches this point. One can reverse the process by adding potassium to the outside of the fiber. So raising the external potassium shifts the equilibrium in such a direction as to open the channels. An open channel then has a linear instantaneous current voltage curve. It becomes nonlinear after a time-life which, in this case, is quite short. Inward rectification must explain two things: one is the fact that potassium permeability goes up as the external potassium is raised, and the other is that the potassium permeability increases as the voltage is made negative. And that's provided in this model simply by saying that the probability of finding a potassium in the channel depends on the membrane potential, and if this point is 90 or 100% of the way through the membrane field, then the probability will go up exponentially,

and will change e-fold in 25 mV. The net result of this
model, then, is that the fraction of the pores that are open
plotted as a function of membrane potential first diminishes
as a function of voltage, but raising the external potassium
concentration will tend to shift the curve in the depolarizing
direction.

Tsien: Is this the model where the TEA gets swept into
the channel when you depolarize?

Armstrong: No, it diffuses into the channel, and it is
swept out of the channel. Presumably, the rate of entry in
this model is totally independent of everything except the
concentration within the axoplasm. It's independent of the
membrane voltage, and it's independent of other ion
concentrations. The only thing that varies as one changes
the ionic concentration is the rate at which it's swept out
of the channels.

Tsien: Here's an experiment which may get at this point.
Assuming your model, suppose you were to clamp the membrane
at E_K, so you have zero current, and then step depolarize
by 100 mV, draw the time course of the current, step
hyperpolarize by 100 mV, and once again draw the time course
of the current. You're giving us the steady state, and I'd
like to know what is the dynamic behavior?

Armstrong: That depends on the rate constant for this
reaction. There is a time dependency in theory in the
closing and opening of the channels, and that can be very
slow if one uses some of the compounds that stick in the
channel very tightly. With TEA, the reaction is fast (a few
milliseconds) and still more rapid with cesium.

But you wanted the fiber clamped at E_K and then stepped
in both directions. Let's say that at E_K we happen to be at
a point on the curve which is equivalent to a state where
half of the channels are open, and we then step in the
positive direction. If the instantaneous current voltage
curve is linear, then it doesn't matter whether we go in the
positive or in the negative direction, the current will
jump, and the K will jump for a positive step, and it will
jump and then increase for a negative step. At low
concentration of TEA, this might take a half a msec, while
for other compounds, it can be much slower.

Adrian: You mentioned that you were uncertain about the
steepness of the voltage dependence of that model, which I
think you gave as e-fold change for 25 mV. I think when I
tried this on some data from skeletal muscle, expressing the
potassium current as a conductance function of voltage, the
change was something like e-fold for about twelve millivolts,
which would be considerably steeper than that model.

Armstrong: Yes. There is a way of getting this in skeletal muscle. Wolf Almers and I attempted to model some of his data, and we found that the instantaneous current voltage curve is not in fact linear, but already has a rectifier characteristic. One could steepen it to some degree by supposing that sort of a curve.

Cleemann: If you assume the process is taking place close to the inner surface of the membrane, you would automatically have introduced some inward rectification. In your I-V relations you have inward rectification where the current bends over, but the current just doesn't fall off. Then, adding TEA, which is removed by potassium, you get the fall off of the membrane current. That, at least, is the way it works in my model.

Hille: The question of steepness can be handled in still another manner. I have made models in close association with Dr. Armstrong and very much like Lars Cleemann's. Without a blocking particle, the current voltage relation of the channel looks more or less normal for a delayed rectifier. If you add one blocking particle, then the curve becomes steeper, more than e-fold for 25 mV. If you add external potassium, decreasing E_K, then you get recovery, which, in this particular model, happened to be tremendous. But it all depends upon how you write down the parameters. This model is one in which you have a channel, and the binding site is on the inside similar to that of Armstrong's C_9. The blocking particle itself has a univalent charge. If you wanted to increase the steepness, you could increase the charge of the blocking particle. Alternatively, if you knew that you had TEA or cesium as your blocking particle, then you could suppose that there are several sites for potassium ions coming from the outside. Now the process of blocking or of unblocking involves the movement of several particles going in a relay or as a single file mechanism. When one uses the single file mechanism, the voltage dependence of the flux through the channel, i.e. the voltage dependence of the probability of any one of these particles being in a certain position, and the voltage dependence of the time constants for transitions, all become steeper than e-fold for 25 mV. This is my view of the inward rectifier or, particularly, the blocking effects of cesium and sodium and TEA on the K^+ channel.

McNaughton: It occurred to me that quite a different type of model might fit in with the effects of various substances on the inward rectifier in heart muscle. First, if you remove the sodium from the outside, this inward rectifier is depressed. I'd be very interested if anyone

knows of similar effects in other tissues. The second point
is an old observation by Tsien and Noble, that when you
increase the potassium concentration, the reversal potential
shifts to a more positive value, as has been presented today;
but the turnover occurs at a much more positive position.
These two observations together make me wonder whether
somehow calcium inside the fiber is binding to the channel,
and blocking it at the inner site. This idea would fit in with
both these experiments, although there may be others which
it doesn't fit. When you reduce the sodium, it's well known
that the calcium rises inside the fiber, and this might
explain the dip across the whole range. Then, in the normal
condition, one would expect the dip to be due to the fact
that one is beginning to activate the calcium efflux, and one
is approaching the contractile threshold. Last of all, the
observation that Dr. Morad mentioned earlier on, is that
raising potassium has a negative inotropic effect. It may
have the effect of shifting the active threshold for the
calcium concentration, which is doing the blocking, to a more
positive potential.

IV

Structure and Function of
the Sacrotubular System

INTRODUCTORY REMARKS
STRUCTURAL BASIS OF E-C COUPLING

L.D. Peachey

Department of Biology
University of Pennsylvania
Philadelphia, Pennsylvania

Before I begin my paper, I would like to review some of the key problems in excitation-contraction coupling. Skeletal muscle cells, the topic of the first four papers, have an arrangement of membrane systems involved in excitation-contraction coupling, shown in Fig. 1 in simple diagrammatic form. The surface membrane of the muscle cell infolds at various places to form what we refer to as the transverse tubule system or T-system.* These tubules carry extracellular space and surface membrane from bands of openings spaced circumferentially around the fiber all the way to the center of the cell. These bands of invaginations are often spaced regularly along the muscle, one or two per sarcomere. A close association is found between this surface membrane system and the internal representative of the endoplasmic reticulum in the muscle cells, referred to as the sarcoplasmic reticulum or SR. This association is in the form of couplings whose exact structure varies from one kind of muscle to another. Commonly, in vertebrate muscles this coupling is a triad structure with two elements of sarcoplasmic reticulum flanking one T-tubule. Some specific morphological structures hold these together into a rigid

In subsequent publications, I will use the word helicoid for the geometric form of the T-system and striations as reported here. This is a more accurate term than spiral as employed in this paper, but the change will not be made here since spiral was the word used when the paper was presented.

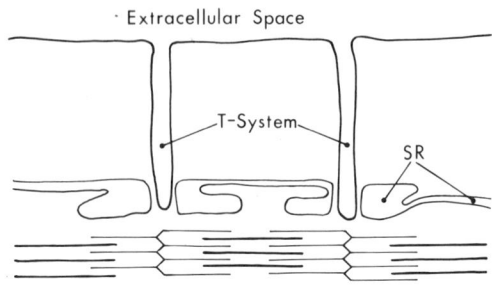

  *Fig. 1. Simplified diagram of the arrangement of membranes
in most vertebrate skeletal muscle cells. The surface
membrane, which separates extracellular space from
intracellular space, invaginates into the cell in networks of
tubules. These T-tubules associate with the sarcoplasmic
reticulum (SR) in three-part structures called triads. These
membranes lie adjacent to myofibrils, diagrammed at the bottom
of the figure. There is a relatively precise relationship
between the periodicity of the fibrils and the repeating
structural pattern of the membrane systems. This simple,
two-dimensional sketch fails to indicate that the membranes
are arranged in three-diamensional networks that envelope
fibrils throughout the volume of the muscle fiber.
Magnification of this representation is about 20,000x.*

unit. These tubules and the sarcoplasmic reticulum surround
all of the myofibrils in a complex three-dimensional network.
  Functionally, we know that the action potential, which
spreads along the surface of the muscle fiber, propagates in
along the transverse tubular system. It enters each
transverse tubular system in order, and reaches the center
of the fiber with a delay that, in a typical, say, frog
fiber at room temperature, could be of the order of 5 to 10
msec. This step is reasonably well established, but the next
is less certain. There is some form of coupling between the
action potential of the transverse tubules and the
sarcoplasmic reticulum. Perhaps this is related to the charge
movements discussed in earlier papers in this symposium.
The next step, the release of calcium from the sarcoplasmic
reticulum to activate the contraction, is also not well
understood. It seems that there must be a change in
permeability of the membrane of the sarcoplasmic reticulum,

but it is uncertain if there is an associated change in electrical potential across the membrane of the sarcoplasmic reticulum.

The first talk of this session is concerned with some three-dimensional aspects of the arrangement of these membrane systems. In the subsequent two talks, Drs. Baylor and Bezanilla will discuss attempts to learn about possible changes in potential across the sarcoplasmic reticulum using fluorescent dyes. There is evidence that when these dyes are placed in the membranes of the sarcoplasmic reticulum, an optical signal can be obtained, giving information about the potential across these membranes, which cannot be explored with microelectrodes because of the extremely small size of the SR. Dr. Rüdel's paper will deal with time course of release of Ca^{2+} with activation.

The coupling arrangement described above would apply with some minor variations to the skeletal muscle of mammals, amphibians, and most of the other vertebrates, and, with some further modifications, to many invertebrate muscles. The corresponding membrane arrangement in the heart has some similarities to and differences from skeletal muscle. The hearts of mammals do have the T-system and the sarcoplasmic reticulum, but the T-systems tends to be somewhat larger in caliber, the T-system tubules tend to be somewhat shorter, and the sarcoplasmic reticulum often is less developed. That is, there is less volume and area of sarcoplasmic reticulum in the cell on a per volume or on a per surface area basis. This difference probably does not limit excitation-contraction coupling, however, because heart muscle cells tend to be small, so there are shorter distances over which coupling has to take place. Furthermore, heart muscle cells generally are activated more slowly than most skeletal muscle cells.

One of the outstanding problems in the heart is to what extent the calcium that activates the heart muscle cell comes from the sarcoplasmic reticulum by internal release. This is the mechanism that is the dominant if not the only one in skeletal muscle. It is not completely clear to what extent calcium release from the SR is important in heart muscle in relation to calcium entry through the surface membrane of the cell. In amphibian cardiac muscle, where there is no T-system, it seems that calcium entry from the surface of the muscle cell can be the major source of calcium for activation of contraction. However, there is a sarcoplasmic reticulum even in these cardiac cells without a T-system. This SR lies under the surface membrane and forms

the same kind of specific couplings that occur deep in the
cell, but forms them with the fiber surface. Thus even in
these relatively simple cells there may be two sources of
calcium, one released from the sarcoplasmic reticulum
coupled with the surface membrane, and secondly, the calcium
entry from the external medium through the surface membrane.
We hope that the three papers after the intermission will
shed some light on this and other problems in cardiac muscle.

THREE-DIMENSIONAL STRUCTURE
OF MUSCLE FIBER

L.D. Peachey

Department of Biology
University of Pennsylvania
Philadelphia, Pennsylvania

Muscle tissue has always been attractive for structural studies. It has a highly regular arrangement of structural components and it is divided into a hierarchy of structures extending from the macroscopic and easily visible down to the extremely minute and only microscopically visible. This attraction was felt at least as early at the late 17th century by Leeuwenhoek in Holland, and it has survived to the present time because this fineness and regularity of structure persists down to structures resolved by current electron microscopes.

The two major components of muscle fiber, whose structure and function are directly related to the contraction of the fiber and its control, are the striated myofibrils and the membranes of the sarcoplasmic and transverse tubular system. The structures of these elements are quite familiar, both as seen in two-dimensional electron micrographs and in the more three-dimensional form shown in Fig. 1 (Page, 1965; Peachey, 1965). Most of the information shown in this kind of drawing has come from the examination of thin sections of embedded muscle tissue by electron microscopy. In addition to providing us with this kind of information, which is certainly useful, electron microscopy has had a second effect. While focusing our attention on very fine details of muscle cell structure, it has led us to ignore some three-dimensional aspects of muscle cell structure at a coarser level. I intend to draw attention to some of these today, and to look

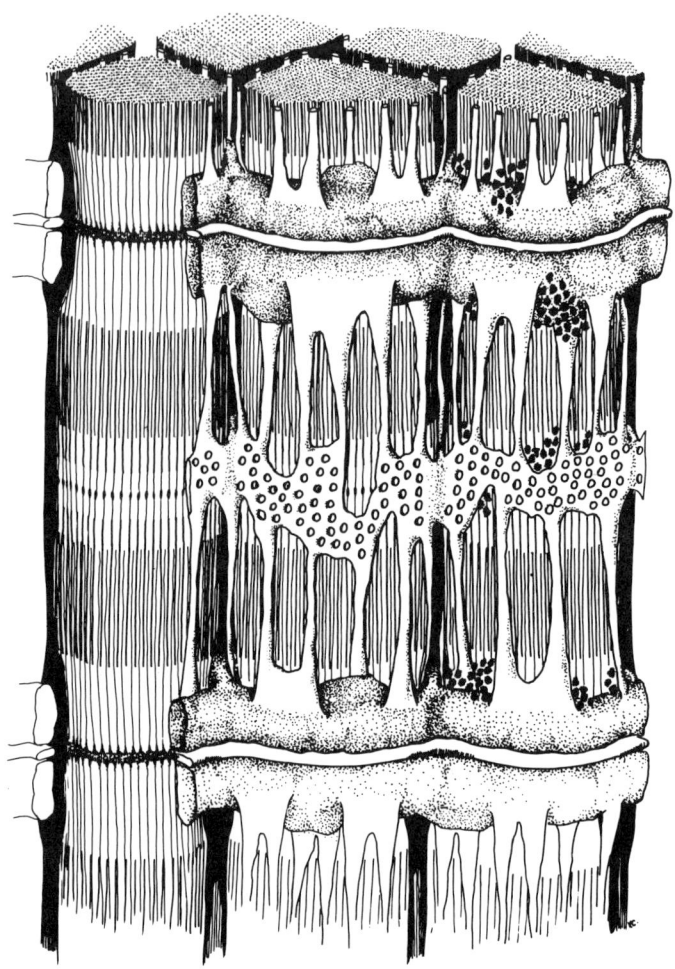

Fig. 1. Artist's reconstruction of the transverse
tubules and sarcoplasmic reticulum surrounding a few
myofibrils of a frog twitch skeletal muscle fiber. One
sarcomere occupies the center of the figure in the vertical
direction. T-tubules are oriented horizontally and form the
center elements at triads opposite the myofibrillar Z-lines
at the top and bottom of this sarcomere. The lateral elements
of the triad, above and below the T-tubules, are dilated
terminal cisternae of the sarcoplasmic reticulum, which are
connected across the middle of the sarcomere by a network of
tubules and flattened cisternae. (Reproduced, with permission,
from the Journal of Cell Biology: Peachey, 1965).

back in history to see where some of these ideas originated,
since some very accurate ideas of muscle cell structure in
three dimensions were known long ago, but have been lost by
more modern generations.

One can begin this discussion with a quotation from
William Bowman's comprehensive and critical paper *On the
Minute Structure and Movements of Voluntary Muscle*, published
in the *Philosophical Transactions* of the Royal Society of
London for the year 1840. On the cross striations of muscle,
Bowman writes:

> Their existence was doubtless known to Hooke, and
> Leeuwenhoek has given more than one very accurate
> description of them, as well as made constant
> reference to them in his letters. He believed, during
> the early years of his inquiry, that they were
> circular bands or girths, surrounding a bundle of
> fibrillae; but at a later period he regarded them
> as of a spiral shape, and endeavored to show, by a
> fancied analogy with an elastic coil of wire, that
> they were in some manner the originators of motion.

Bowman did not doubt the existence of striations as
Leeuwenhoek described them on muscle fibers in 1712. In
fact, Bowman goes on to point out the reality of
cross-striations, and the fact that they can be observed on
the individual fibrils (fibrillae) when a muscle fiber is
mascerated and the fibrils separate. Fig. 2 shows, in
Bowman's Fig. 17, heart muscle of the ox. The individual
fibrils, separated out of the muscle fiber, are drawn with a
certain periodicity, indicating that the striations are
present not only on the surface of the muscle fiber, as
suggested by Leeuwenhoek, but also are present in the fibrils,
an interpretation with which we agree today.

Where Bowman's description did differ from that of
Leeuwenhoek is in the form of the striations when examined at
the level of the whole fiber. Bowman described the
separation of adjacent striations into disks, saying that
this is one form of natural cleavage of the muscle fiber; the
other form of cleavage is into fibrils, as mentioned earlier.
Fig. 2 also shows Bowman's Figs. 21 and 22 from his paper of
1840, of the separation of pig and human muscle fibers into
a series of planar disks. In fact, such an arrangement was
implied considerably earlier by John Hunter in his 1781
Croonian lecture on muscular motion, in which he discussed
the rearrangement of the component parts of a muscle fiber
that he felt must occur during contraction. Without having

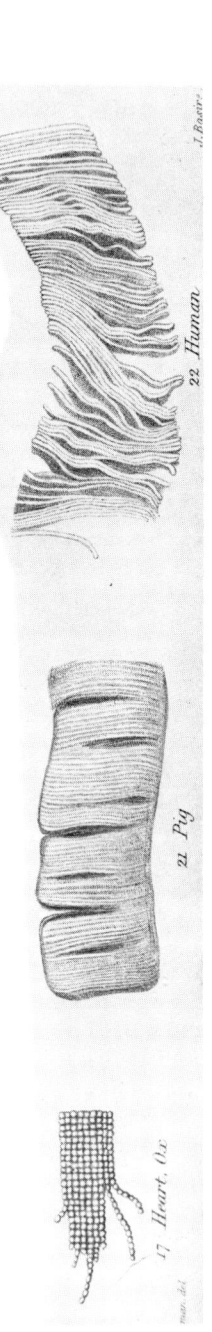

Fig. 2. Three figures from Bowman's paper of 1840. His Fig. 17 is a fragment of a
macerated ox heart and shows periodically beaded myofibrils. His Figs. 21 and 22 show skeletal
muscle fibers from two species separating into transverse disks along the striations.

seen these units, Hunter suggested that the fiber must be made up of repeating parts along its length, and that these parts must become both broader and shorter when the muscle fiber contracts.

While he does not explicitly deny the accuracy of Leeuwenhoek's interpretation of spiral bands, Bowman's emphasis on disks or plates had the effect of distracting attention from the spirals. From that time onward, with a few notable exceptions, the English language literature emphasizes the transverse striations as disklike rather than in the form of spirals. One notable exception is in the papers of the late Australian entomologist O.W. Tiegs, who described observation of spiral bands as widespread throughout vertebrate and invertebrate skeletal and cardiac muscles, first in a series of papers in the 1920's, and later in 1955 in an extensive review on insect muscle fibers (Tiegs, 1922, 1924, 1955). But this spiral arrangement of bands was largely ignored by morphologists and physiologists interested in muscle tissue. For example, in his 1902 description of the reticular networks in skeletal muscle cells, Veratti made no reference to spirals. He drew his figures as shown in Fig. 3, as if the transverse reticulae were in the form of planar bands extending all the way across the fiber.

The importance of these facts for our discussion today lies in the close association of the transverse tubular and sarcoplasmic reticulum networks with the striations of the myofibrils. Certain features of the three-dimensional arrangement of the striations must somehow be reflected in the arrangement of the membrane networks. In particular, if the striations are spiral, and if the transverse tubular networks maintain close alignment with the myofibrillar striations, as we believe they do, then the transverse tubular networks themselves must lie in planes that are spiral and not lamellar or diskoid. Some recent high-voltage electron microscopic studies of muscles that Drs. Brenda Eisenberg, Clara Franzini-Armstrong, and I have stained selectively to show the transverse tubules in unusually thick electron microscopic preparations have given us a chance in the last few years to check on some of the points I have been discussing. In particular, we have been able to examine the spiral arrangement of the T-system and of the bands of the muscle fibers, and have been able to confirm it (Peachey & Eisenberg, 1975, 1978).

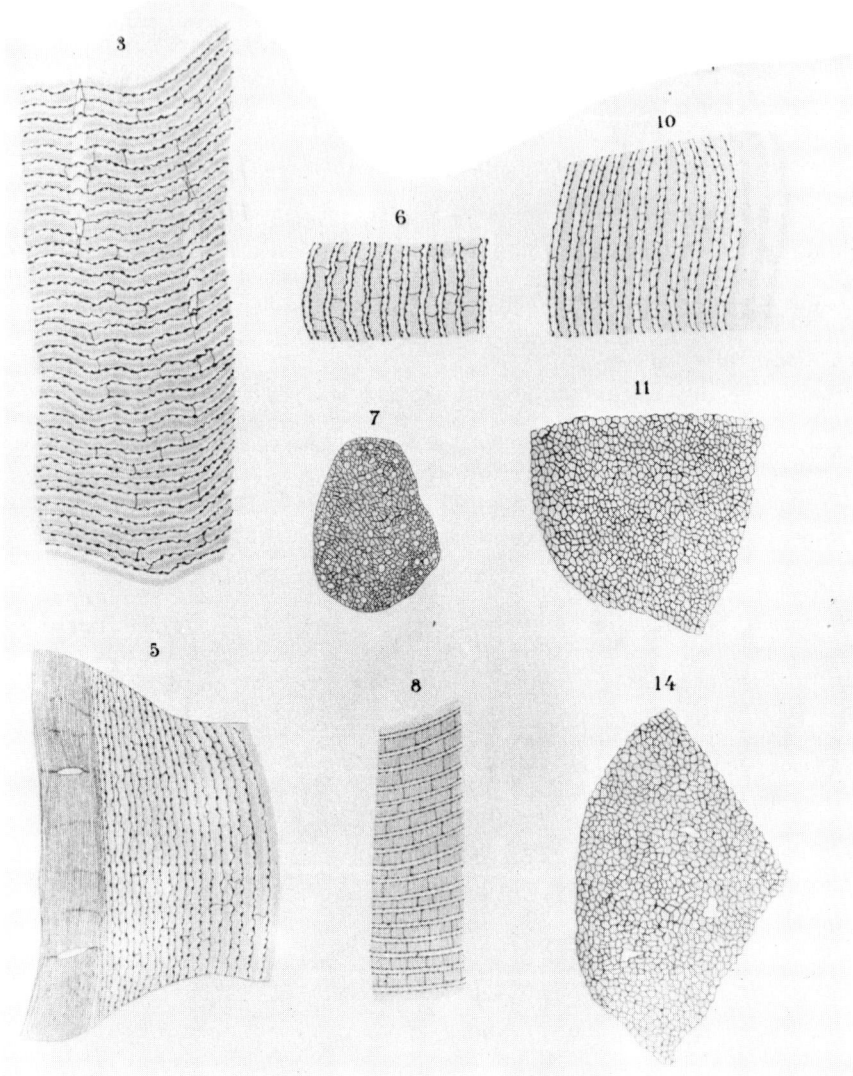

Fig. 3. Some of the drawings of muscle fiber reticula published by Veratti in 1902. His Figs. 3, 5, and 6 show muscle fibers of a mouse, Figs. 7 and 8 are from a bat, Figs. 10 and 11 are from a pigeon, and Fig. 14 is from a lizard.

Fig. 4 shows the ability of the high-voltage electron microscope, when the transverse tubules are selectively stained, to produce good images through relatively thick tissue slices and to show certain features of the transverse tubular system which were extremely difficult to find when

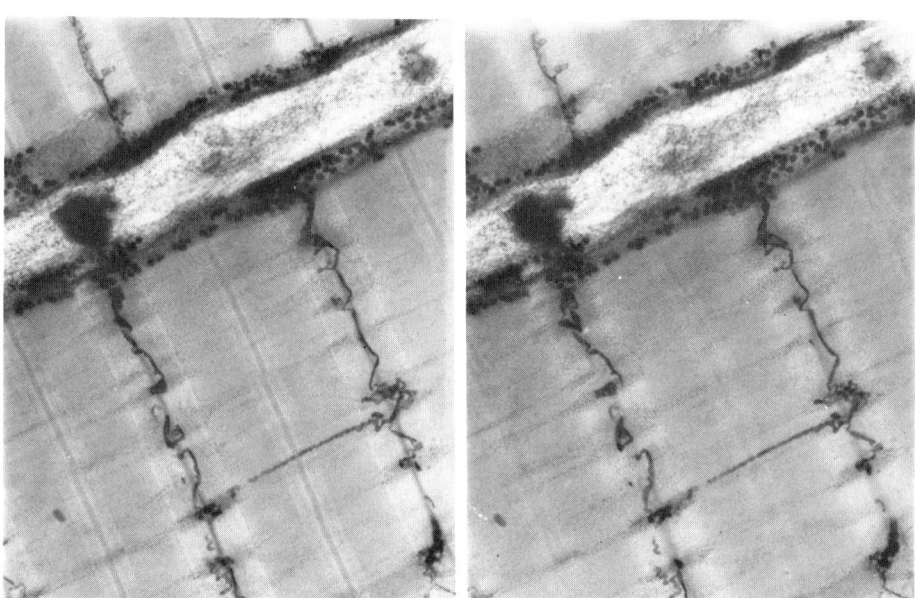

Fig. 4. Longitudinal slice about one µm thick of frog skeletal muscle. This and all subsequent figures are from the author's work in collaboration with Dr. Clara Franzini-Armstrong and are from sartorius muscles of Rana pipiens. *Stereoscopic pairs of micrographs can be viewed with stereo viewers, such as those supplied by Hubbard Scientific Co., Northbrook, Ill., or Abrams Instrument Corp., Lansing, Mich. They also can be viewed without a viewer if the reader can diverge or cross his eyes to superimpose the two images and keep them in focus at the same time. The depth dimension will be reversed from that intended in the case where the eyes are crossed.*

This preparation was stained using the lanthanum procedure of Revel and Karnovsky (1967), which fills the T-tubules and the caveolae connected to the surface of the fibers. Two adjacent fibers are shown, with extracellular space between. Where the surface membrane passes slightly obliquely through the slice, groups of caveolae can be seen. In one place, in the lower fiber at the left, a narrow connection of a T-tubule into a caveola is seen. Micrograph taken at 1000kV, x15,000.

much thinner sections were used. This slice is one μm thick,
about ten times the thickness ordinarily used in electron
microscopy. The tissue is stained with lanthanum, which
stains the transverse tubules especially selectively. They
can be seen approaching the surface of the fiber. In many
cases, they connect into the caveolae or small inpocketings
in the surface membrane present at the surface of the cell.
The myofibrils are present in these preparations, but they
can be seen only faintly because of the high selectivity of
this particular stain. The connection of the transverse
tubule on the surface often shows a very narrow constriction
of the tubule as it enters into the caveola, and from there
into the surface, as seen at the left in Fig. 4; this is a
very common finding in these micrographs.

Fig. 5 is a somewhat thicker slice, showing the T-system
at lower magnification. Several transverse tubular
networks can be seen standing up rather like a row of fences.
There is some distortion because the third dimension has
been exaggerated by a somewhat excessive tilt when these
micrographs were made. In the region where the surface

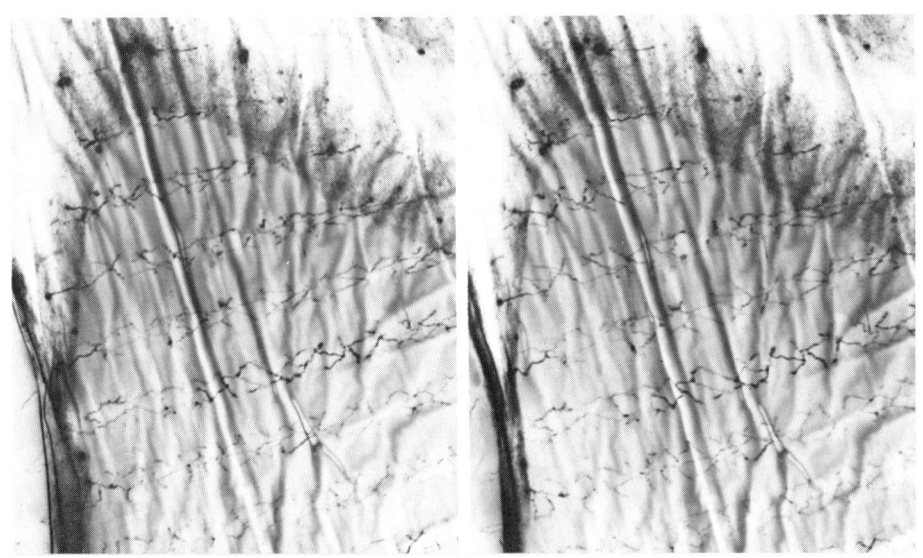

*Fig. 5. Approximately longitudinal, 2 μm thick slice of a
lanthanum stained muscle showing portion of T-system networks
at several sarcomeres. Some T-tubules approach the surface
membrane, but it cannot be seen if a patent connection is made.
The irregular surface seen at the rear of the stereo image is
a rough surface of the slice produced when it was cut; 1000kV,
x5000.*

membrane goes obliquely out of the section, there are some possible points of connection of the transverse tubules with the surface membrane. The thickness of this slice, about 2 μm, is sufficient to show the transverse tubules surrounding several different myofibrils, but no spiral is visible in this region.

Fig. 6 shows a similar slice at somewhat higher magnification. We are outside the fiber, looking in. Note the transverse tubular networks standing up and the surface membrane going obliquely through the slice. At this magnification and in this slightly oblique slice, we can begin to identify the elements of the network of the transverse tubular system where particular myofibrils pass. For example, there is a five-sided tubular polygon that can be identified in part in three successive networks (arrow). There are other examples of places where we can find the corresponding shape opening in successive networks, indicating the portion of the network through which an individual myofibril passes.

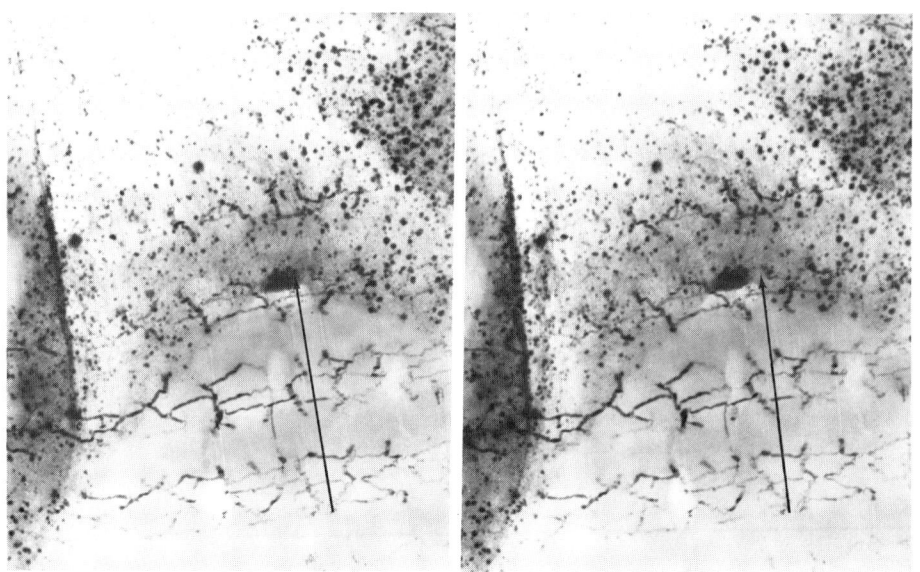

Fig. 6. Higher magnification of the same kind of field as Fig. 5. The arrow indicates openings in the T-system networks where a particular fibril passes; 1000kV, x8000.

In Fig. 7 the view is from inside the muscle fiber,
looking out at a rather large expanse of undulating surface
membrane. Three successive Z-lines are shown. The caveolae,
inpocketings of the surface membrane, are arranged in patches,
and at various places, the stubs of T-tubules come toward
the observer. The circumferential distribution of T-tubules
as they approach the surface along the muscle fiber can be
determined from such pictures, and the lateral spacing
between these is about 1-2 μm. This is somewhat closer than
what was suggested in the local stimulation experiments
described by Professor Huxley (Huxley & Taylor, 1958).

The next figures are high-voltage pictures made using a
different stain, the Golgi black reaction used by Veratti in
1902, but slightly modified to better preserve the tissue.
This stain is very good for extremely thick slices; it leaves
a dense deposit in the form of small spheres which give very

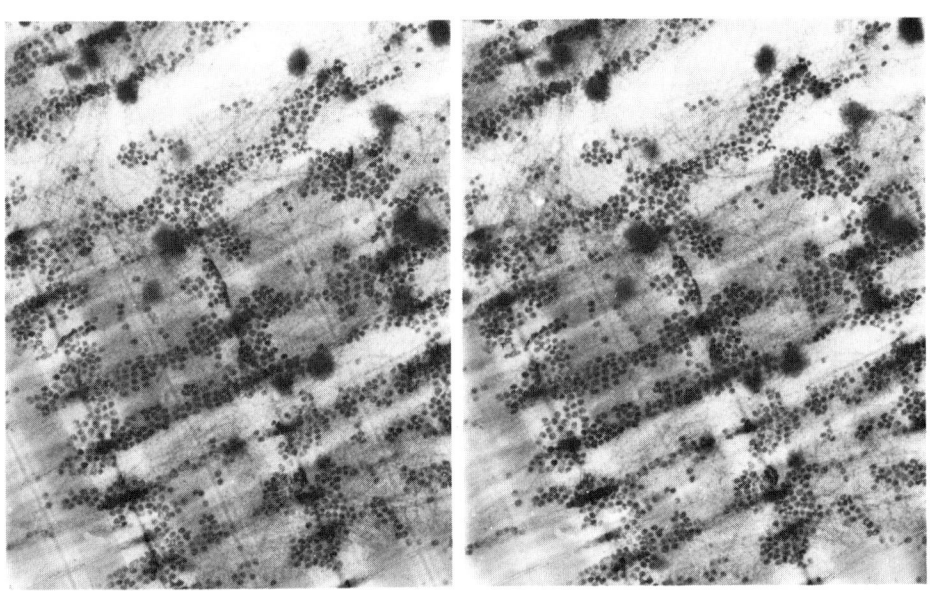

*Fig. 7. Another field with the same stain. The
stereoscopic view is from inside the fiber looking out at
the sarcolemma. Groups of sarcolemmal caveolae and stubs of
T-tubules are filled with the dense stain; 1000kV, x12,000.*

high scattering in the electron microscope. Fig. 8 is a micrograph of such a preparation 5 micrometers thick, a not quite transverse slice from an area of the muscle fiber that is perhaps 30 microns square. There are four successive T-systems extending somewhat obliquely. Fig. 9 shows, using this method, a direct verification of Leeuwenhoek's original spiral description. In addition to the spiral, there are dislocations in the band pattern; for example, there is a sharp dislocation where there is a step in the T-system. Because the T-system remains so closely associated with the Z-lines of the myofibrils in frog muscle, the bands of the muscle fiber must make the same step and must have the same spiral.

This new approach, using thicker than ordinary preparations and the high-voltage electron microscope, gives us an opportunity to extend electron microscopic analysis. In the past we could examine only very restricted regions of

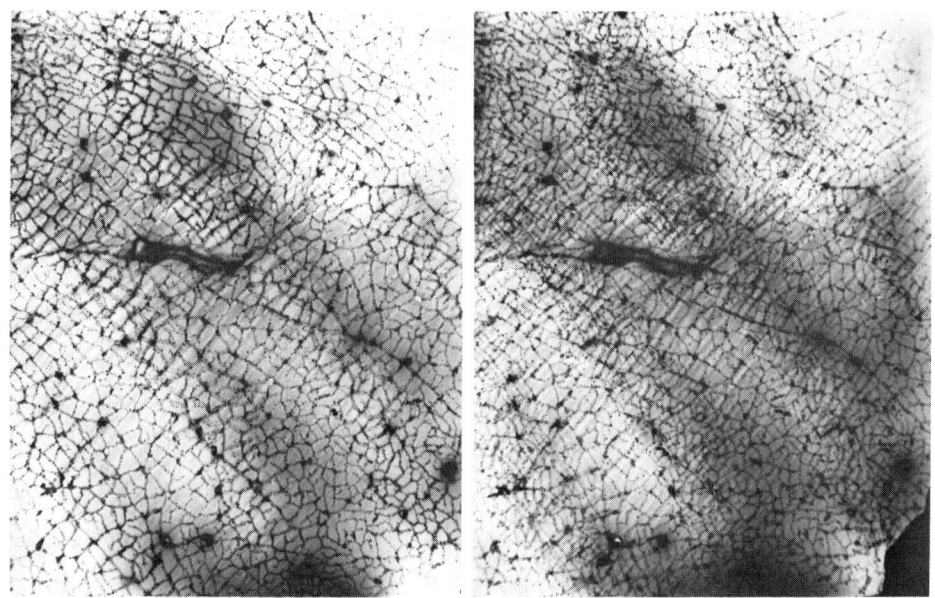

Fig. 8. Low magnification of a transverse slice of frog sartorius muscle stained using the Golgi black reaction. Slice thickness is about 5 μm. Five layers of T-system network are seen passing obliquely through the slice thickness; 1000kV, x3000.

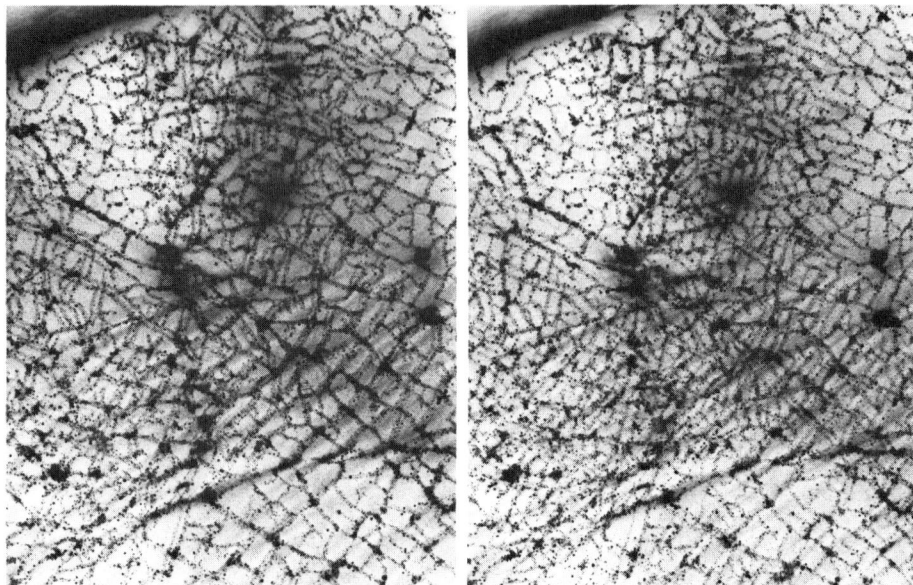

*Fig. 9. A similar view, but in a region where a spiral
is present in the T-system. The axis of the spiral is near
the center of the figure. The T-system enters the slice at
the left, and passes below center, up to the right, across
the top, and behind itself on the left. It continues in this
way, completing almost two full turns of the spiral before
leaving the slice at the rear, as seen stereoscopically;
1000kV, x5000.*

the cell, although at very good resolution. Now we are
beginning to return to the view of the light microscopists,
and to examine fairly large coarse regions of the cell,
but of course we retain the better resolution of the electron
microscope. As we do this, we begin to discover that some
of our notions about the three-dimensional arrangements of
muscle fibers may have been excessively simplified.

DISCUSSION

Eppling: Are you suggesting that the spiral transverse
tubules extend throughout the length of the muscle cell in
one continuous spiral?

Peachey: Well, we don't know that. Tiegs says that this is true, but I'm not sure. In principle we could serial slice a whole fiber from end to end. More practically, we could use slices about 10 microns thick and have the possibility of serially slicing, in some reasonable number of sections, through several hundred microns fiber length. My suspicion is that individual spirals are not going to extend from end to end. This suspicion is based on some reconstructions from serial longitudinal sections, but the results were not so good that I would want to be very definite at this time.

Page: I wonder, were these from sartorius or semitendonosus? Might there be a different spacing of the openings in semitendonosus, which was probably the muscle of the Huxley and Taylor experiments?

Peachey: Yes, that's an excellent point. We have looked a little at semitendonosus, but the particular pictures that I showed were of the sartorius.

Page: Are the openings different in semitendonosus? My impression from thin sections is that the T-tubules are less complete in some of the semitendonosus.

Peachey: I can't answer that yet because we've just recently got the stains the way we want them. Now we can begin to do the sort of analysis that you suggest.

Huxley: These high-voltage micrographs are extremely beautiful, but as regards seeing the arrangement in the whole fiber, I wonder about using stereo viewing with a light microscope. Golgi preparation ought to make the tubules visible as they were in Veratti's preparation, and there's no difficulty in using an oil immersion objective aperture with a free working distance of the order of a millimeter, and having excellent stereo arrangements with Polaroids and right angles under the two halves of the condenser and in the eyepiece. You should be able to see the arrangement in the entire fiber, and move it around, and get a direct answer to these things without going through the laborious business of serial stereo pictures in high voltage.

Peachey: Yes, that's an excellent idea and I thank you for the suggestion. We have started to look at some of the Golgi preparations using the light microscope. We first used Nomarski optics. Our particular Nomarski microscope has only a high dry objective, so the depth of the field was too great to be able to see the spirals in transverse slices. Stereoscopy might help.

Winegrad: I wonder if you could tell us what you think the functional implications of the spiraling of the transverse tubular system and the banding pattern might be.

Peachey: Well, I'd like to pass that question right back
to the audience, because I've not been able to think of any
significant functional implications. For example, if you
try to think of a functional implication of the spiral
arrangement with respect to the spread of electrical activity
into the muscle fiber, you might think that the electrical
impluses would go longitudinally along the spiral. Well,
that is unlikely, because the velocity of conduction of the
surface action potential is such that the two longitudinally
adjacent T-tubule networks are synchronous with each other
within a matter of a few tens of microseconds: the signal
is propagating in at both levels of the spiral almost
simultaneously. So it's not going to go in one and spiral
to the next sarcomere. I don't think there's any major
electrical implication to the spirals. I'm not sure if
there's any mechanical implication. The fibrils still run
straight and parallel to each other, and the spiral
arrangement really only means that the bands are not exactly
placed adjacent to each other. I don't see how this would
dramatically change the mechanical activity.
   Maisami: I wanted to ask you about one of the slides
you showed, about the T-tubules. I noticed that there were
many interruptions of this stain; it didn't look continuous.
I wonder whether this is an artifact of the preparation,
or was it functionally significant?
   Peachey: In many cases the lanthanum stain is not very
reliable. It does not stain as large regions as does the
Golgi stain, which in this way is more reliable. In many
cases where a tubule seemed to be missing, on close
inspection you really see it, but it's quite lightly
stained. So that part is an artifact. The true part of it
is that even when it is very well stained, with all of the
stains, including peroxidase and the Golgi stain, we find
a good many tubules that have dead ends. There are more of
these than we thought from thin sections, because we could
never be quite sure when we saw a tubule ending in the thin
section whether it was really ending or going out of the
plane of the section.
   Maisami: Regarding Dr. Winegrad's question, could it
be that this spiral is just a matter of embryology of the
muscle and the T-tubules, instead of having an obvious
functional significance?
   Peachey: Yes, this is exactly the thought we've had on
this. As the muscle fiber forms in myogenesis, we know
that many fibrils are formed with many repeating bands, and
they're somewhat irregularly arranged, but then the muscle

cell seems to get things sorted out. It's just possible that it does not sort it out always precisely in planes, but that some regions of the cell end up in spirals. Then, having formed in spirals, it's stuck with these spirals for at least a considerable length of time.

REFERENCES

1. Bowman, W. (1840). On the minute structure and movements of voluntary muscle. *Phil. Trans. Roy. Soc. 130*, 457-501.
2. Huxley, A.F. & Taylor, R.E. (1958). Local activation of striated muscle fibres. *J. Physiol. 144*, 426-441.
3. Leeuwenhoek, A. van (1712). A letter from Mr. Anthony van Leeuwenhoek, F.R.S. Containing his observations upon the seminal vesicles, muscular fibres, and blood of whales. *Phil. Trans. Roy. Soc. 27*, 438-446.
4. Page, S. (1965). A comparison of the fine structure of frog slow and twitch muscle fibres. *J. cell Biol. 26*, 477-497.
5. Peachey, L.D. (1965). The sarcoplasmic reticulum and transverse tubules of the frog's sartorius. *J. cell Biol. 25*, No. 3, Part 2, 209-231.
6. Peachey, L.D. & Eisenberg, B.R. (1975). The T-systems and striations of frog skeletal muscle are spiral. *Biophys. J. 15*, 253a.
7. Peachey, L.D. and Eisenberg, B.R. (1978). Helicoids in the T-system and striations of frog skeletal muscle fibers seen by high voltage electron microscopy. *Biophys. J. 22*, in press.
8. Revel, J.P. & Karnovsky, M.J. (1967). Hexagonal array of subunits in intercellular junctions of the mouse heart and liver. *J. cell Biol. 33*, C7-12.
9. Tiegs, O.W. (1922). On the arrangement of the striations of voluntary muscle fibres in double spirals. *Trans. Roy. Soc. S. Austr. 46*, 222-224.
10. Tiegs, O.W. (1924). On the mechanism of muscular action. *Austr. J. exp. Biol. med. Sci. 1*, 11-29.
11. Tiegs, O.W. (1955). The flight muscles of insects--their anatomy and histology; with some observations on the structure of striated muscle in general. *Phil. Trans. Roy. Soc. B238*, 221-348.

OPTICAL INDICATIONS
OF EXCITATION-CONTRACTION COUPLING
IN STRIATED MUSCLE

S.M. Baylor
W.K. Chandler

Department of Physiology
Yale University School of Medicine
New Haven, Connecticut

INTRODUCTION

As first shown on axons (Cohen, Hille & Keynes, 1968; Tasaki, Watanabe, Sandlin & Carnay, 1968) small optical signals can be detected which accompany the action potential. Many of these signals appear to monitor changes in membrane potential rather than membrane current or permeability (see review by Cohen & Salzberg, 1978). More recently a number of investigators have begun to look for similar kinds of signals in striated muscle. Such signals might be associated with potential changes across interior structures such as the transverse tubular (T-system) or sarcoplasmic reticulum (SR) membranes. Thus optical methods may be useful for studying the mechanisms involved in excitation-contraction (E-C) coupling.

This report briefly reviews some previously reported optical experiments on intact, single skeletal muscle fibers (Baylor & Oetliker, 1975; Oetliker, Baylor & Chandler, 1975; Baylor & Oetliker, 1977a,b,c) and presents findings from recent experiments in which activity was initiated by either the action potential mechanism or by voltage clamp techniques. Many of these results are generally similar to work reported by other laboratories (Bezanilla & Horowicz, 1975; Nakajima, Gilai & Dingeman, 1976; Kovacs & Schneider, 1977; Bezanilla & Vegara, this book).

EXPERIMENTAL PROCEDURES

A single, intact twitch fiber from frog (*Rana temporaria*) is isolated from either the semitendinosus or ilio-fibularis muscle and mounted horizontally in a narrow chamber having glass sides to transmit the light. The fiber is held isometrically at the tendon ends by two vertical rods, one of which is the extended arm of a sensitive tension transducer. Using an optical bench arrangement, light from a 100-watt tungsten-halogen bulb is focused onto and collected from a small length of fiber (300 μm or less) by two symmetric long-working-distance objectives.

Light measurements can be made in one of three modes: birefringence, transmission, or fluorescence. The birefringence measurement is an "intrinsic" one (in the sense of not requiring the addition of an external chemical), whereas the "extrinsic" transmission and fluorescence signals discussed below depend on dyes being added to the Ringer solution in order to stain the fiber. Various optical components are inserted in the light path to make the three types of measurements:

(1) birefringence: a linear polarizer located between the light source and the preparation, oriented at +45° with respect to the fiber's long axis; and a second linear polarizer (analyzer) positioned between preparation and photodetector, oriented at -45° to the fiber axis.

(2) transmission: a bandpass interference filter located between light source and preparation (primary filter)

(3) fluroescence: a primary filter, as for the transmission measurement, plus a longer wavelength cut-on filter (secondary filter) located between preparation and photodetector, permitting passage of primarily fluorescent light to the photodetector.

In all three modes, a relatively large light intensity (I) is measured in the resting state, and small changes in light intensity (ΔI) occur during activity. The active signals are plotted as fractional changes, $\Delta I/I$.

   The composition of the normal Ringer's solution was
120 mM NaCl, 2.5 mM KCl, 1.8 mM $CaCl_2$, 2.15 mM Na_2HPO_4 and
0.85 mM NaH_2PO_4, pH 7.1. Movement artifacts were minimized
or eliminated by working either in mechanically paralyzing
solutions (e.g., strongly hypertonic Ringer) or by
stretching a fiber in isotonic solution. Propagated action
potentials were initiated by a brief (0.2 to 0.5 msec) shock
from two platinum electrodes positioned locally near the
fiber.

   For the voltage-clamp experiments Na^+ was replaced by
TEA^+ (tetraethylammonium) and TTX (tetrodotoxin), 1 µg/ml,
was used to block the sodium channels. The fiber rested
on a supporting pedestal and was impaled from above by
voltage-sensing and current-passing electrodes, located 50
to 100 µm away from the site of optical recording. Voltage
decrement in the region of optical recording was minimized
in one of two ways, either by working at a fiber end, where
the longitudinal voltage gradient is small, or by working
in the middle of a fiber and restricting the optical field
to a narrow (50-60 µm) transverse slit. Further details
of some of the experimental methods are given in Baylor &
Oetliker (1977*a*).

RESULTS

   The optical signals described below fall into two main
classes according to time-course:

   Class 1: Time-course close to that of surface and/or
T-system action potential.
   Class 2: Time-course distinctly slower than class 1 but
faster than the development of positive tension.

The two-component birefringence record shown in Fig. 1
(trace 0) demonstrates both classes of signals on a fiber
in three-times hypertonic Ringer. Following the stimulating
shock (downward arrow), there is a 2-3 msec delay before
onset of the "first component" of the birefringence signal
(trace 0, peak labelled 1). A similar delay is seen
preceding onset of the action potential (trace *V*, recorded
by an internal micro-electrode). This delay represents the
time required for the action potential to propagate to the
region of optical and electrical recording, 7 mm from the
stimulating cathode. During most of the action potential the

optical and electrical traces superimpose. Moreover, the
polarity of the optical signal indicates a decrease in
optical retardation, which is the same change as that found
on the squid giant axon accompanying the action potential
(Cohen, Hille & Keynes, 1970). On this basis the first
component has been attributed to the surface action potential
(Baylor & Oetliker, 1975).

The "second component" of the birefringence record
(Fig. 1, trace 0, plateau labelled 2) is a larger and more
slowly developing signal. Both its magnitude and kinetics
are influenced by the tonicity of the bathing solution. In

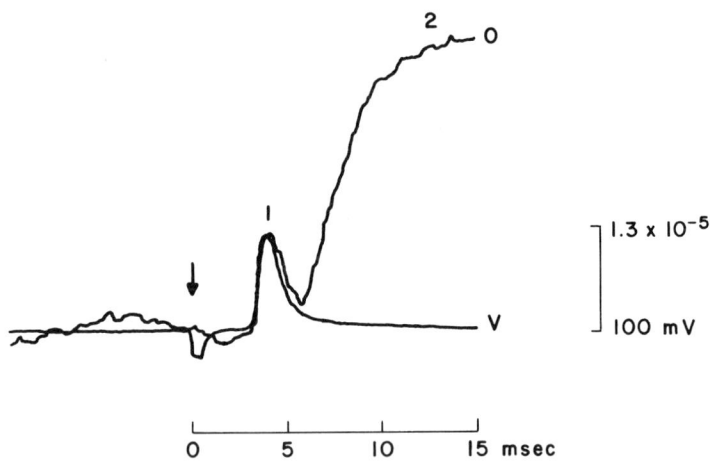

Fig. 1. *Action potential experiment showing optical
(trace 0) and intracellular voltage (trace V) recordings
from a fiber in Ringer solution of 3 times normal tonicity
(370 mM instead of 120 mM NaCl). The optical trace has
been inverted (i.e., an upward signal is a decrease in
intensity) and scaled to facilitate comparison with the
action potential. Optical peak labelled "1" refers to the
"first component" whereas the plateau labelled "2" refers
to the "second component", both signals being a decrease in
light intensity. Both traces are the average of 300 sweeps,
taken simultaneously. Resting potential of the fiber
decreased from -70 mV at the beginning of the average to
-65 mV at the end. Fiber diameter 65 μm; field of optical
recording, 500 μm; optical and electrical recording 7 mm
from the stimulating cathode; stretch unrecorded. Downward
arrow marks the moment of the shock. 20° C.*

isotonic Ringer the magnitude is usually 50 to 100 times
larger than the signal shown in Fig. 1 (e.g., see Fig. 2, trace
B_1). This increase plus a faster rate of rise obscure the
first component in isotonic Ringer (Baylor & Oetliker, 1975).
The "second component" and related class 2 signals are
further discussed in connection with Figs. 4-11.

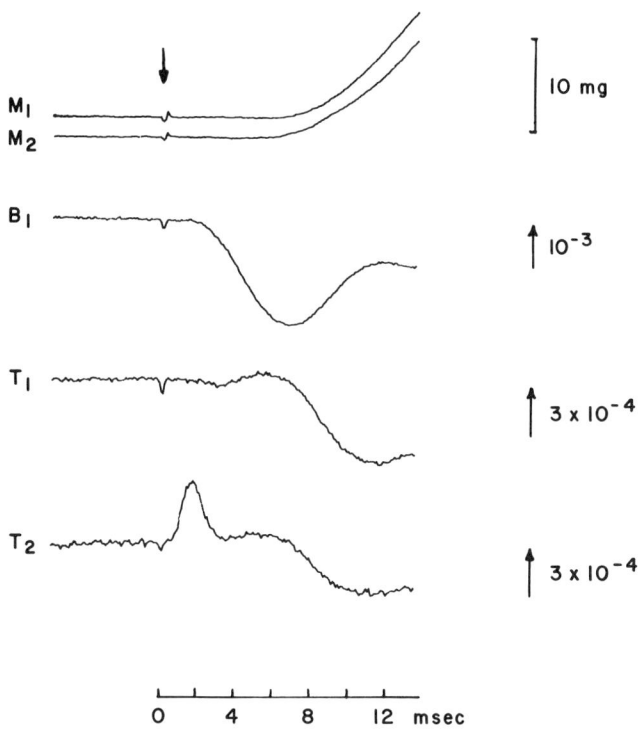

Fig. 2. *Action potential experiment in normal Ringer
comparing tension (M), and second component of the
birefringence signal (B), and the transmission change (T).
Subscripts 1 and 2 refer to traces taken before and after
the addition of a merocyanine-rhodanine dye (10 μg/ml). All
optical traces were taken using a primary filter passing
light of 570 ± 30 nm. Fiber diameter, 104 μm; field of
optical recording, 300 μm; fiber stretched 1.54 times slack
length; optical recording 1.5 mm from cathode; 4 sweeps per
trace. In this and subsequent figures the direction of the
light calibration arrow indicates an increase in light
intensity. 20° C.*

The Transmission Signal of Merocyanine-Rhodanine

Ross, Salzberg, Cohen, Grinvald, Davila, Waggoner &
Wang (1977) have shown that in squid giant axons stained with
a merocyanine-rhodanine dye (XVII of Ross *et al.*), changes
in dye absorption provide a rapid, linear monitor of changes
in membrane potential. At a dye concentration causing no
measurable toxicity, these absorption changes can be detected
with a large signal-to-noise ratio (Ross *et al.*, 1977).
Since this dye does not appear to penetrate nerve membranes
(due to the presence of a "localized" charge, Ross *et al.*,
1977), in muscle it might be expected to stain only surface
and T-system membranes if simply added to the bath.

Fig. 2 demonstrates a class 1 signal in muscle due to
the presence of merocyanine-rhodanine. Subscripts 1 and
2 refer to traces recorded before and after the addition of
dye. The mechanical records (M_1 and M_2) show that the
residual twitch response was not significantly affected
by the presence of the dye. The birefringence trace (B_1)
shows the second component in isotonic solution to be a
relatively large signal, reaching an apparent peak when
significant tension appears on the mechanical record. A
comparison of the transmission records for light of
wavelength 570 nm (traces T_1 and T_2) shows that there is an
early, transient signal (increase in transparency) that
depends on the addition of dye. The onset of the signal is
similar to that expected for the surface action potential;
however, its duration is probably somewhat longer. The
half-width of the dye signal is 2.5 msec, whereas the
half-width of the action potential at the same temperature
is 1-1.5 msec (Hodgkin & Nakajima, 1972). A discrepancy in
this direction would be expected if the transmission signal
monitors potential changes of both surface and T-system
membranes, as reported by Nakajima, Gilai & Dingeman (1976)
who used a related dye, merocyanine 540.

Fig. 3 shows that the polarity of the transmission
signal depends strongly on wavelength. An increase in
transparency is seen using light of wavelength 570 nm (trace
O_1), a decrease at 690 nm, and an increase again at 750 nm.
A similar triphasic wavelength dependence has been observed
in the squid giant axon (Ross *et al.*, 1977). In squid,
however, the triphasic signal was seen only with linearly
polarized light oriented perpendicular to the axon axis;
signals recorded with unpolarized light (as in Fig. 3) or
light polarized parallel to the axon axis showed a monophasic

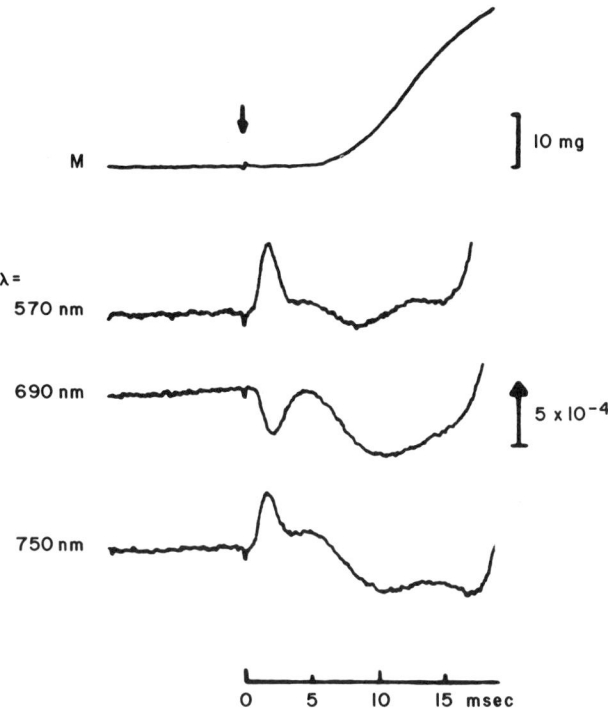

Fig. 3. Action potential experiment showing the
wavelength dependence of the merocyanine-rhodanine transmission
change. Wavelengths used for the transmission records were
570 ± 30 nm (upper optical), 690 ± 30 nm (middle optical), and
750 ± 30 nm (lower optical). Same fiber as in Fig. 2. 20° C.

dependence on wavelength. Although the basis for this
difference between squid axon and frog muscle is not
understood, it is possible that the different orientations
of surface versus T-system membranes play a role.

Class 2 Signals

Other types of optical signals can be obtained in muscle
using dyes that permeate membranes. Dyes of this kind that
give large membrane-potential related signals in axons
might provide a means for determining whether changes in SR
voltage are associated with Ca^{+2} release. Dye signals

possibly related to SR events were reported by Bezanilla &
Horowicz (1975), who used Nile Blue A, and subsequently by
Oetliker, Baylor & Chandler (1975), who used an
indodicarbocyanine dye (XIX of Ross et al., 1977), hereafter
simply called the "cyanine" dye.

Oetliker et al. (1975) showed that the early time course
of the second component of the birefringence signal in
hypertonic Ringer was nearly identical to that of the
fluorescence signal observed using the cyanine dye. Fig. 4
shows a similar experiment carried out on a fiber in isotonic
Ringer. Movement was minimized by stretching the fiber to
1.9 times slack length, giving an average measured sarcomere

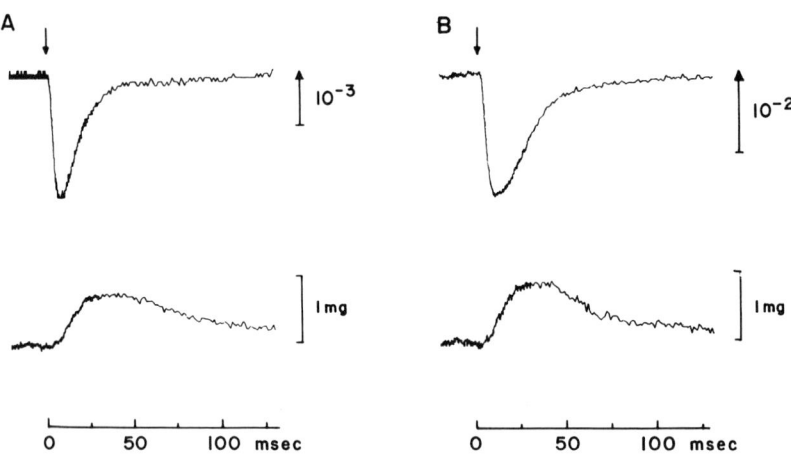

*Fig. 4. Action potential experiment in normal Ringer
comparing the second component of the birefringence signal
(part A) and the fluorescence change of the
indodicarbocyanine dye (part B) on the same fiber. Upper
traces are optical, lower traces are mechanical. Birefringence
trace was taken using 570 ± 30 nm light before the addition
of dye (average of 2 sweeps); fluorescence trace was taken
10 min after the addition of dye (1 µg/ml; single sweep),
using a primary filter passing 570 ± 30 nm light and a
secondary filter passing light longer than 695 nm. Fiber
diameter, 85 µm; sarcomere spacing was 3.9 µm in the field
of optical recording (middle of fiber) which was 300 µm,
at the cathode. 17° C.*

spacing of 3.9 μm in the field of optical recording. The two signals (A, birefringence; B, fluorescence) almost superimpose to time of peak but differ slightly at later times. However, the general similarity of the two signals supports the conclusion that they may reflect closely related events.

The nature of these events has not been established. However, the signals are most likely not "movement artificats", e.g., fiber material moving out of, into or twisting within the field of view (Baylor & Oetliker, 1977a; also see below). Rather, the time course of the optical signals, following the action potential and preceeding the main tension response, is suggestive of an underlying event related to excitation-contraction coupling. Table 1 lists three possibilities for this event and suggests possible mechanisms for associated birefringence and fluorescence signals. The first possibility, and the one we consider most likely, is based on an analogy with nerve, namely that a membrane potential change can give rise to both intrinsic birefringence and cyanine fluorescence signals. Under this hypothesis the properties of the signals (see below) strongly suggest that the membranes of the sarcoplasmic reticulum are the responsible structure.

A second possibility is that the signals are somehow related to small volume changes (Baskin & Paolini, 1966) that might take place as calcium is released from the SR. If, for example, calcium movements were accompanied in an obligatory

TABLE I

| Basis | Possible Mechanism for | |
	Birefringence	Fluorescence
Potential change across SR	similar to nerve	similar to nerve
SR or myoplasmic volume change	change in form birefringence	change in dye concentration
Contractile filament activity	change in birefringence of actin or myosin	change in dye bound to filaments

way by movement of an anion in the same direction, or by two
monovalent cations in the opposite direction, changes in
osmotic pressure would result, requiring volume readjustments.
In the birefringence case this might give rise to a change
in form birefringence of structures such as the SR or
contractile proteins; in the case of fluorescence, changes
in dye concentration might occur, which might cause changes
in fluorescence.

A third possibility is that the signals may reflect
activation of the contractile proteins. Since a muscle fiber
has a large resting birefringence, due primarily to the
contractile filaments, contractile protein activity might
be considered a likely source for a birefringence signal.
For the fluorescence case, one might suppose that the
fluorescence properties of dye bound to contractile
filaments are altered during activation. Hopefully, future
experiments will help decide whether one of these or some
other mechanism underlies these optical signals.

Birefringence Signals Under Voltage Clamp

In order to learn more about the physiological properties
of the optical signals, experiments were carried out using
a two-microelectrode voltage clamp. Results from a
birefringence experiment on a highly stretched fiber in
isotonic Ringer are shown in Fig. 5 (upper traces, voltage;
middle traces, birefringence; lower traces, tension). For a
depolarizing pulse to -60 mV little or no optical signal
occurred (upper birefringence record), whereas following a
larger depolarization a decrease in light intensity was seen.
The signal developed after a marked delay, achieved a
maximum rate of decrease, then reached a plateau or peak;
following repolarization, the intensity level returned to
the baseline with a final time constant of 25-35 msec (10° C).
The maximum rate of change and peak magnitude of the signal
were markedly dependent on voltage.

An important question which arises in this kind of
experiment concerns the location of the voltage threshold
for a just detectable contraction. The threshold is not
apparent from Fig. 5, since the experiment was carried out
in the middle of a highly stretched fiber. However, other
experiments in less stretched fibers showed similar decreases
in light intensity first occurring 5-10 mV below the
mechanical threshold, suggesting that these small signals
may be related to small, subthreshold amounts of activation.

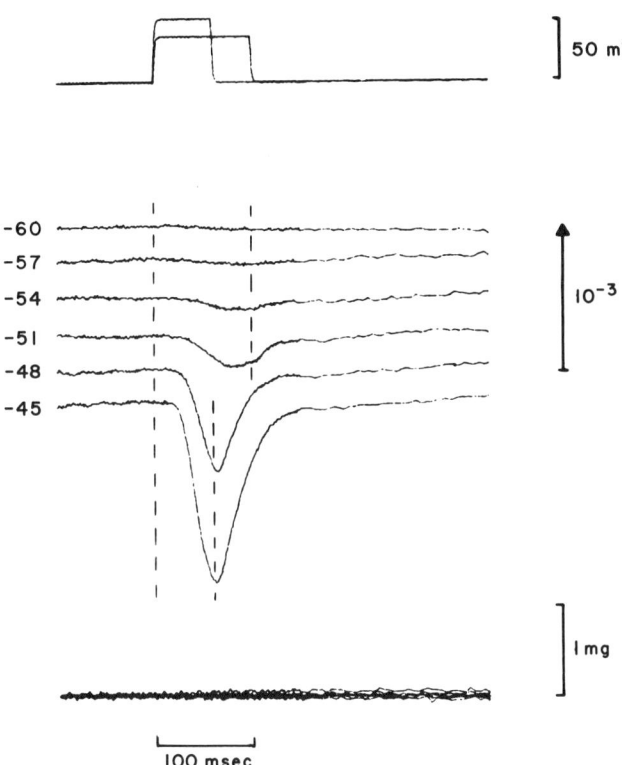

Fig. 5. *Birefringence signals near the contractile threshold in a voltage-clamp experiment in a highly stretched fiber. Upper traces show two voltage pulses, 100 msec and 60 msec in duration; middle traces show birefringence records; lower traces show superposed tension records. All traces are single sweeps. Dashed vertical "timing lines" mark beginning and end of the pulse. Sarcomere spacing was 3.9 μm at the site of optical recording in the middle of the fiber; illumination was with a narrow transverse slit (about 60 μm wide) of white light positioned about 50 μm from the voltage-sensing microelectrode. Resting potential of the fiber was -90 mV; holding potential was -100 mV. Numbers to the left of the optical traces indicate membrane potential (in mV) during the pulse. 10° C.*

A somewhat general hypothesis concerning the origin of the birefringence signal is that it is somehow related to the amount of calcium that is released by the SR. For instance, if the signal reflects a potential change across the SR membranes, the early part of the signal might be proportional to the integral of calcium current across the SR, since the initial current would be expected to charge SR membrane capacity according to the equation $Q = CV$. On this basis the early amplitude of the optical signal would correspond to the amount of calcium released and the rate of change of the optical signal to the rate of calcium release. A similar relationship between the optical signal and Ca release might also hold under alternative explanations of the signal (for example, the volume hypothesis discussed in connection with Table 1).

Fig. 6 examines the kinetic features of the birefringence signal in more detail, showing the voltage pulse, the optical signal, and the rate of change of the optical signal. Not only the original optical signal, but also its rate of change, follow a sigmoid time course. In particular, the initial rate of change of the optical signal does not even approximately follow first order kinetics, a finding similar to that reported by Kovacs & Schneider (1977) for the kinetics of the intrinsic transmission signal observed in "cut" fibers under voltage clamp. This kinetic feature is of interest since it is not predicted by the simplest version of a recent model proposing how a membrane charge movement might regulate calcium release (Schneider & Chandler, 1973; Chandler, Rakowski & Schneider, 1976). In this model each Ca^{+2}-release site in the SR is assumed to be gated by a single charged group in the T-system membrane. The individual charged groups are distributed between resting and activating positions according to tubular membrane potential. Changes in distribution, and therefore changes in the rate of SR Ca^{+2}-release, are assumed to follow first-order kinetics.

Another property of the birefringence signal is also inconsistent with the simplest version of the charge movement model, namely the steepness of the relationship between the birefringence signal and voltage. Fig. 7 shows the (absolute value of the) maximum rate of change of the optical signal plotted on a logarithmic scale versus voltage. The points are well fitted by a straight line, with a slope corresponding to an e-fold change for each 2.5 mV. A similar potential dependence is obtained if the peak amplitude rather than the maximum rate of change is used. This

remarkably steep dependence on voltage has been consistently
seen in other experiments, with e-fold values generally
falling between 3 and 4 mV. The comparable relationship
between charge movement and voltage is less steep, with
e-fold values of 8 to 12 mV (Schneider & Chandler, 1973;
Adrian & Almers, 1976; Chandler *et al.*, 1976; Schneider &
Chandler, 1976). If the charge movement process is involved
in "gating" the optical signal, it is clear that the simplest
model fails and additional features will be required.

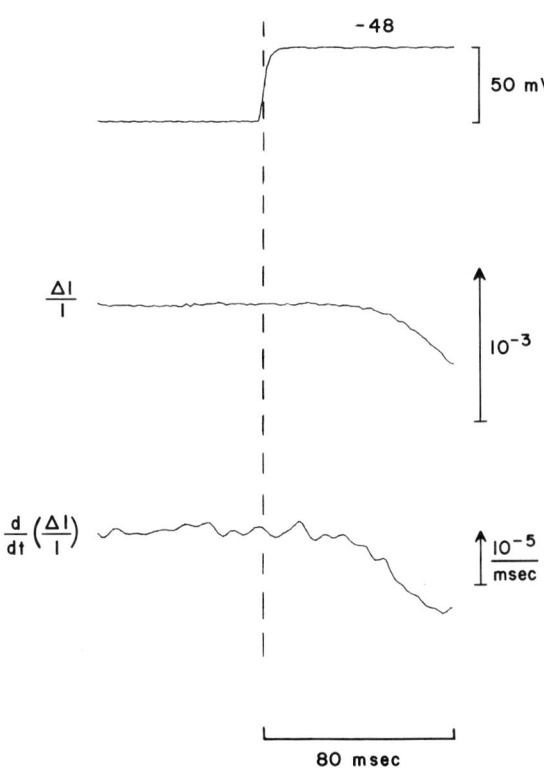

*Fig. 6. Delay in onset of the rate of change of the
birefringence signal. Same experiment as Fig. 5, showing
birefringence signal (middle trace) during the pulse to
-48 mV (upper trace), and the rate of change of the
birefringence record (lower trace) obtained by taking first
differences of the digitized birefringence data and
performing a smoothing algorithm.*

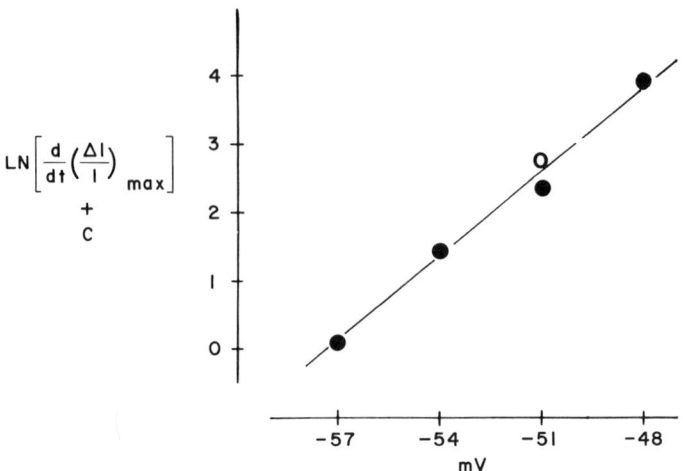

Fig. 7. *Logarithmic plot of the maximum rate of change of the birefringence signal versus voltage. Solid circles, from the records in Fig. 5; open circles, from a bracketing record during the run. The straight line was drawn by eye; its slope corresponds to an e-fold change per 2.5 mV. The scale for the ordinate has been arbitrarily shifted.*

Fluorescence Signals Under Voltage Clamp

Fig. 8A shows records from a voltage-clamp fluorescence experiment. The fiber was soaked for 10 min in Ringer plus cyanine dye (1 µg/ml), then washed in dye-free Ringer prior to impalements with the microelectrodes. In this experiment a hyperpolarizing pulse to -160 mV showed little or no fluorescence change (upper optical record, Fig. 8A), although the dye would be expected to stain surface and T-system membranes. Evidently any potential-dependent signal from these membranes is lost in the noise of a single sweep. Other experiments, however, in which larger concentrations of dye were used for longer soaking periods, have shown fluorescence changes on hyperpolarization that presumably reflect potential changes across surface and T-system membranes.

The optical signals (Fig. 8A) associated with depolarizations near the contractile threshold closely resemble the birefringence signals already described. The signal develops after a marked delay, with an amplitude and

maximum slope that is strongly voltage dependent. In
addition, the signals are first seen below the threshold
for detectable movement. For the three largest
depolarizations in Fig. 8A, the maximum rate of change
of the optical signal increased e-fold for about 3.5 mV.
Similar e-fold values, falling between 3.5 and 4.5 mV, have
been consistently observed in other fluorescence experiments.
  Some of the fluorescence signals in squid (Cohen,
Salzberg, Davila, Ross, Landowne, Waggoner & Wang, 1974)
and in red blood cells (Sims, Waggoner, Wang & Hoffman, 1974)

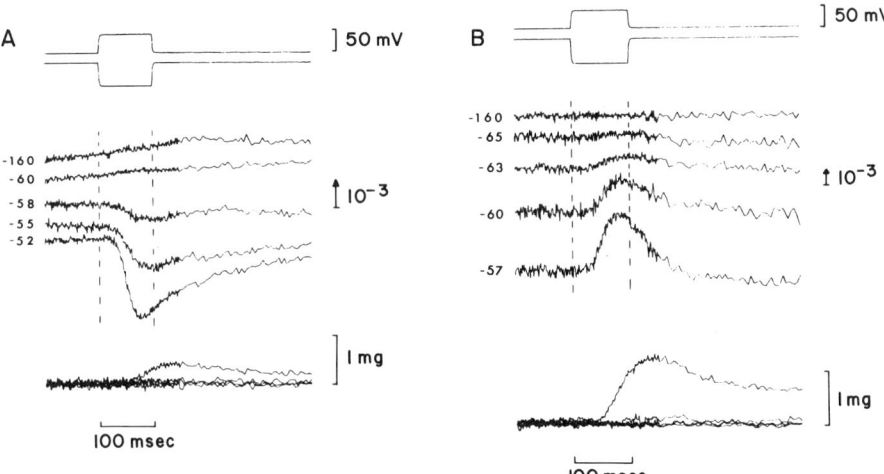

Fig. 8. *Fluorescence signals near the contractile
threshold. Intensity changes were from a 300 μm length at
the end of a highly stretched fiber. Records in A were taken
after a 10-min soak in a 1 μg/ml concentration of cyanine dye;
those in part B were taken after a 10-min soak in 0.2 μg/ml.
Upper traces show a single voltage pulse, middle traces show
fluorescence records (570 ± 30 nm primary filter, 665 nm
secondary filter), and lower traces show tension. The visible
tension change corresponds to the largest depolarizing step.
Numbers to the left of the fluorescence records indicate
membrane potential during the pulse (in mV). Holding
potential in each experiment was -100 mV and resting
potentials were -95 mV (part A) and -96 (part B). Average
sarcomere spacing was 3.0 μm (A) and 3.4 μ (B). Temperature
was 8.5° C in both experiments. The fluorescence signal
changes e-fold for each 3.5 mV in A and 4.0 mV in B.*

have been found to reverse polarity at low dye concentrations.
We were therefore interested to see how the cyanine signal in
muscle depended on dye concentration. In one experiment a
fiber was exposed to a low dye concentration, 0.2 μg/ml for
a 10-min soaking period. Fig. 8B shows voltage clamp records
taken after removing the dye. No signal was seen during
hyperpolarizations. However, depolarizations resulted in
fluorescence signals similar to those in Fig. 8A, but of
opposite polarity. The mechanism(s) responsible for the
reversal of polarity in this and other preparations is not
understood.

Birefringence Signals for Large Depolarizations

     Fig. 9 shows birefringence signals from the same
experiment as Fig. 5 but using larger depolarizations. As
the voltage was made progressively more positive, the

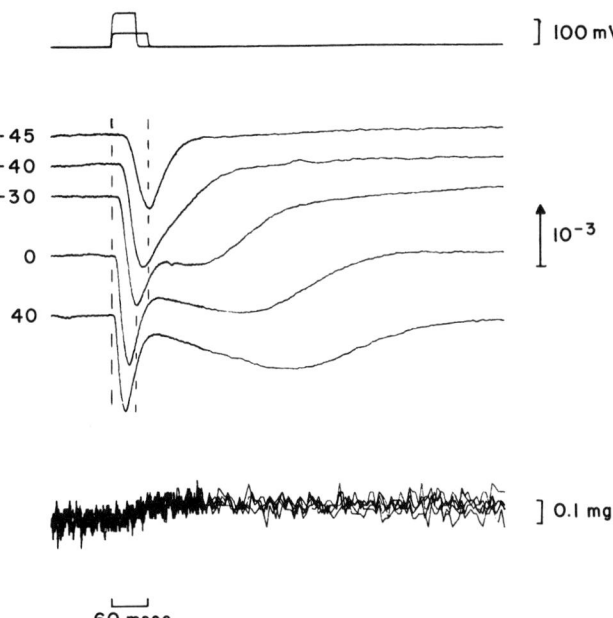

*Fig. 9. Birefringence signals associated with large
depolarizations. Same experiment as Fig. 5.*

magnitude of the signal increased and the delay in onset decreased. A very small amount of tension (Fig. 9, lower records) was also observed in each record, although it is not known which part of the fiber was responsible for this change. At strong depolarizations a late component was seen in the birefringence traces. This was not analyzed in detail.

Dependence of the Birefringence Signal on Voltage

In Fig. 10 the maximum rate of change of the early optical signals in Figs. 5 and 9 is shown on a linear scale as a function of voltage. The data fall on a smooth sigmoid-shaped curve that might, as discussed above, approximate the relationship between the maximum rate of calcium release and the potential across surface and T-system membranes. The activation curve appears to be asymmetric, showing a steep "foot" at negative potentials but only a gradual approach to saturation at positive potentials. These features may provide clues as to how membrane potential regulates calcium release from the sarcoplasmic reticulum.

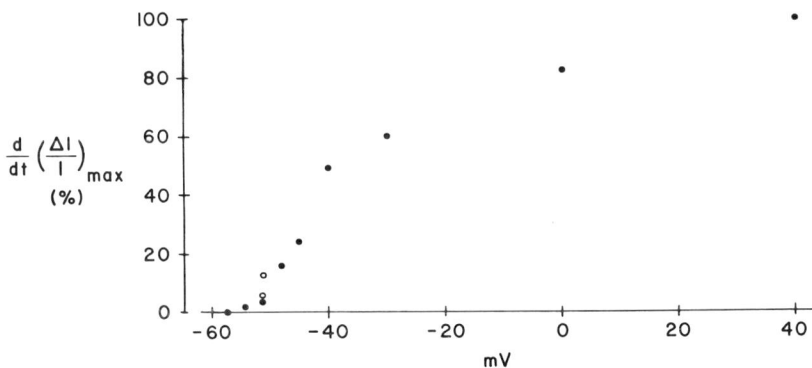

Fig. 10. Maximum rate of change of the early birefringence signal vs. voltage. Solid circles, from Figs. 5 and 9; open circles, bracketing measurements taken during the experiment. (No brackets were obtained following the 3 largest potential steps, as the fiber gave out following the step to +40 mV.)

CONCLUSIONS

Among class II signals, changes in birefringence and dye fluorescence can be easily measured from small regions of single muscle fibers in single sweeps. These signals are remarkably similar, whether they are elicited by an action potential or under voltage clamp. The similarities suggest that the signals arise from a common underlying process or closely related processes. The time course and voltage dependence are consistent with the idea that the underlying event(s) is a step(s) involved in E-C coupling. Hopefully, future experiments will identify the step and thereby increase the usefulness of optical methods for studying problems in muscle activation.

DISCUSSION

Hille: Did you use a compensator with different retardations? If so, the amount of light coming through depends upon the net retardation of fiber plus compensator, and you have a muscle whose retardation varies because it has A- and I-bands.

Baylor: In the ideal, yes, but for typical-sized fibers this longitudinal variation should be small, because the myofibrils cannot be expected to be in register throughout an average path length. That is, I think in our preparation we see most of the fiber's retardation averaged out longitudinally over A- and I-bands.

Hille: My suggestion was that perhaps if you took a ribbon-shaped muscle and stretched it out considerably, then you might be able to distinguish the A- and I-band contributions. For example, you could be operating in some parts of the fiber on one slope of the intensity-retardation relationship and in other parts on another slope. If there were birefringence changes in a fully compensated region of the fiber you shouldn't see intensity changes, whereas you would in others. Then, by adding a polarizer, you could move the relationship over so that the situation was reversed. Using that kind of technique, it seems to me one might be able to further get a geometrical resolution of where these signals are coming from.

Baylor: I think that's a good observation, and we are aware of the possibility of doing that kind of experiment. However, I'm afraid it may be technically difficult in a typical-sized frog twitch fiber. Yet we may be able to get fibers that are small enough and stretched out far enough so that we can do that kind of experiment. I've also been inquiring of the anatomists to suggest a preparation better suited for that experiment. I've been told that the A- and I-bands, even though very long, really don't line up that well either--that in fact there's more stagger in the A- and I-bands there than in a frog twitch fiber. But it may be possible to use them. If successful, that kind of experiment might provide some very useful information, because if the signal does come from either the A-band or I-band alone, such a finding would make an SR hypothesis less attractive and a contractile filament hypothesis more likely, whereas if the signal comes from the entire sarcomere length more or less uniformly, it would make the SR hypothesis more attractive.

Hille: Another related point is that in the beginning you said it's possible that there is an artifact in the birefringence trace not returning to zero. To test for this optically, one could change the compensation and see if this signal exactly inverts.

Baylor: I think that experiment does help somewhat. We know that throughout their time course, including the return to the baseline, the birefringence signals I've described do behave as a change in retardation. Namely, with a predictable amount of compensation they exactly invert or almost exactly invert. Yet that doesn't really tell us whether any given signal is explainable as all one process or not. For example, any movement complication in the signal that depended on a change in retardation should also invert in such a compensation experiment.

Morad: Steve, do you measure at all any surface membrane potential change with the indodicarbocyanine dye?

Baylor: Yes. We have in voltage clamp experiments where we've used a larger concentration of the dye for a longer soaking period, something like 2 micrograms per ml for 20 minutes.

Morad: Why didn't we see any in the records you showed; do you think it's just too small to be observed?

Baylor: Yes. I think that if we signal-averaged, we probably would see a signal. But in a way, for some records it's nice not to have the complication of a surface and

T-system signal coming in. In fact, when we have seen the
surface and T-system signal coming in, it makes it more
difficult to tell where that process leaves off and the
later process begins.

Morad: Another point I wasn't quite clear about. If
you did a fluorescence versus voltage plot did the
fluorescence signal also give you the same thing as a
birefringence?

Baylor: The same general shape yes. It has a steep
foot that's quite asymmetric compared with the gradual
approach to the saturation level for large depolarizations.

Morad: And you think that the signal is actually
calcium moving out from the SR under some kind of an
electrochemical driving force?

Baylor: That would be one hypothesis, yes.

Morad: And if so, you can calculate the reversal
potential for the process, etc.?

Baylor: No. There's no way to get that from our data.
For one thing we don't have any reliable calibration for
either the fluorescence or birefringence signals to know,
if it is a potential change, how much the change might be.

Tsien: Steve, you drew attention to the long delay
after the beginning of the voltage pulse. If you apply two
voltage pulses in fairly quick succession, the second pulse
coming after the signal has declined back to the baseline,
do you see any difference in the time course of the optical
signals with the second pulse relative to the first?

Baylor: We haven't done that experiment yet.

REFERENCES

1. Adrian, R.H. & Almers, W. (1976). Charge movement in
    the membrane of striated muscle. *J. Physiol. 254*,
    339-360.
2. Baylor, S.M. & Oetliker, H. (1975). Birefringence
    experiments on isolated skeletal muscle fibers suggest
    a possible signal from the sarcoplasmic reticulum.
    Nature, Lond. 253, 97-101.
3. Baylor, S.M. & Oetliker, H. (1977a). A large
    birefringence signal preceeding contraction in single
    twitch fibers of the frog. *J. Physiol. 264*, 141-162.

4. Baylor, S.M. & Oetliker, H. (1977*b*). The optical properties of birefringence signals from single muscle fibers. *J. Physiol. 264*, 163-198.

5. Baylor, S.M. & Oetliker, H. (1977*c*). Birefringence signals from surface and T-system membranes of frog single muscle fibers. *J. Physiol. 264*, 199-213.

6. Baskin, R.J. & Paolini, P.J. (1966). Muscle volume changes. *J. gen. Physiol. 49*, 387-404.

7. Bezanilla, F. & Horowicz, P. (1975). Fluorescence intensity changes associated with contractile activation in frog muscle stained with Nile Blue A. *J. Physiol. 246*, 709-735.

8. Chandler, W.K., Rakowski, R.F. & Schneider, M.F. (1976). A non-linear voltage dependent charge movement in frog skeletal muscle. *J. Physiol. 254*, 245-284.

9. Cohen, L.B., Keynes, R.D. & Hille, B. (1968). Light scattering and birefringence changes during nerve activity. *Nature, Long. 218*, 438-441.

10. Cohen, L.B. & Slazberg, B.M. (1978). Optical measurements of membrane potential. *Rev. Physiol. Biochem. Pharmac.*

11. Cohen, L.B., Salzberg, B.M., Davila, H.V., Ross, W.N., Landowne, D., Waggoner, A.S. & Wang, C.H. (1974). Changes in axon fluorescence during activity: molecular probes of membrane potential. *J. Membr. Biol. 19*, 1-36.

12. Hodgkin, A.L. & Nakajima, S. (1972). Analysis of the membrane capacity in frog muscle. *J. Physiol. 221*, 121-136.

13. Kovacs, L. & Schneider, M.F. (1977). Increased optical transparency associated with excitation-contraction coupling in voltage-clamped cut skeletal muscle fibers. *Nature, Lond. 265*, 555-560.

14. Nakajima, S., Gilai, A. & Dingeman, D. (1976). Dye absorption changes in single muscle fibers: an application of an automatic balancing circuit. *Pflug. Arch. 362*, 285-287.

15. Oetliker, H., Baylor, S.M. & Chandler, W.K. (1975). Simultaneous changes in fluorescence and optical retardation in single muscle fibers during activity. *Nature, Lond. 257*, 693-696.

16. Peachey, L.D. (1965). The sarcoplasmic reticulum and transverse tubules of frog's sartorius. *J. cell Biol. 25*, 209-231.

17. Ross, W.N., Salzberg, B.M., Cohen, L.B., Grinvald, A.,
     Davila, H.V., Waggoner, A.S. & Wang, C.H. (1977).
     Changes in absorption, fluorescence, dichroism and
     birefringence in stained giant axons: optical
     measurement of membrane potential. *J. Membr. Biol.*
     33, 141-183.
18. Schenider, M.F. & Chandler, W.K. (1976). Effects of
     membrane potential on the capacitance of skeletal
     muscle fibers. *J. gen. Physiol. 67*, 125-163.
19. Sims, P.J., Waggoner, A.S., Wang, C.-H. & Hoffman, J.F.
     (1974). Studies on the mechanism by which cyanine dyes
     measure membrane potential in red blood cells and
     phosphatidylcholine vesicles. *Biochemistry, 13*,
     3315-3330.
20. Tasaki, I., Watanabe, A., Sandlin, R. & Carnay, L. (1968).
     Changes in fluorescence, turbidity, and birefringence
     associated with nerve excitation. *Proc. natn. Acad.
     Sci. U.S.A., 61*, 883-888.

FLUORESCENCE SIGNALS
FROM SKELETAL MUSCLE FIBERS

F. Bezanilla
J. Vergara

Department of Physiology
and the Brain Research Institute
University of California at Los Angeles
Los Angeles, California

We know from the work of Cohen, Salzberg, Davila, Ross, Landowne, Waggoner & Wang (1974), that nerve membranes stained with quite a variety of fluorescent dyes exhibit changes in extrinsic fluorescence in response to membrane potential changes. Furthermore, most of these dyes produce fluorescence signals whose magnitudes are linearly related to potential, at least for voltage pulses no longer than 100 msec. Some of these dyes seem to penetrate the surface membrane and thus seem able to distribute into internal membranes as well; therefore, if we apply these dyes in order to indicate the membrane potentials in muscle, we should obtain different kinds of signals depending on the type of dye (Landowne, 1974; Bezanilla & Horowicz, 1975; Oetliker, Baylor & Chandler, 1975; Vergara & Bezanilla, 1976; Nakajima, Gilai & Dingeman, 1976; Vergara & Bezanilla, 1977).

In our experiments we tried two types of dyes: the nonpenetrating Merocyanine-540 (Vergara & Bezanilla, 1976) and the penetrating Nile Blue A (Bezanilla & Horowicz, 1975; Vergara & Bezanilla, 1977). The Merocyanine experiments will be discussed first because they give information about the early step of the excitation-contraction process; that is, the spread of the action potential along the surface and T-system membranes. Subsequently, we shall discuss experiments with Nile Blue A that are thought to give information about depolarization of the sarcoplasmic reticulum or some closely related process.

I.   TECHNIQUES

     Fig. 1 shows the setup used for the experiments in bundles
of muscle fibers obtained from semitendinosus or sartorius
muscles. A light source (B) through a slit (D) is focused
onto a segment of the bundle. We measure tension, and we
detect the light at 90°. The two important features of the
optical setup are the interference filter (F) (a bandpass
filter) and the cutoff filter (N). In the case of
Merocyanine, we excited the dye at 550 nm with the cutoff at
610 nm. For Nile Blue we excited a 625 nm with the cutoff
at 645 nm. The particular experimental methods used on
bundles of muscle fibers have been published elsewhere
(Vergara & Bezanilla, 1976) and will not be repeated here.

II.   MEROCYANINE SIGNALS IN BUNDLES OF MUSCLE FIBERS

     Fig. 2 shows the type of result obtained from a bundle
of fibers, stained with Merocyanine-540, stimulated to
propagate an action potential. Traces a and b show
fluorescence and tension signals, respectively, recorded
after a single stimulus pulse. The light trace has a
biphasic shape. The end of the light trace seems to be
partly altered by tension or movement. To eliminate this
movement, we used hypertonic solutions and stretched the
fibers. The results are shown in Fig. 3, where the upper
signal corresponds to the fluorescence change and the lower
signal is an action potential recorded with a microelectrode
positioned in the illuminated segment of one of the fibers.
There is a certain relation between the two signals in terms
of time. The fluorescence signal (Fig. 3a) seems to be
slower than the action potential (Fig. 3b). If we
consider the random variation of propagation velocities and
record action potentials from several different fibers, we
find that the delays measured could account for the slower
rising phase of the fluorescence signal; however, the falling
phase is probably too slow. Also, during the rising phase of
the fluorescence signal, shown in Fig. 3a, a small "hump" is
seen which is almost coincident with the action potential
shown in Fig. 3b. This hump can be interpreted as the
contribution of the surface action potential to the total

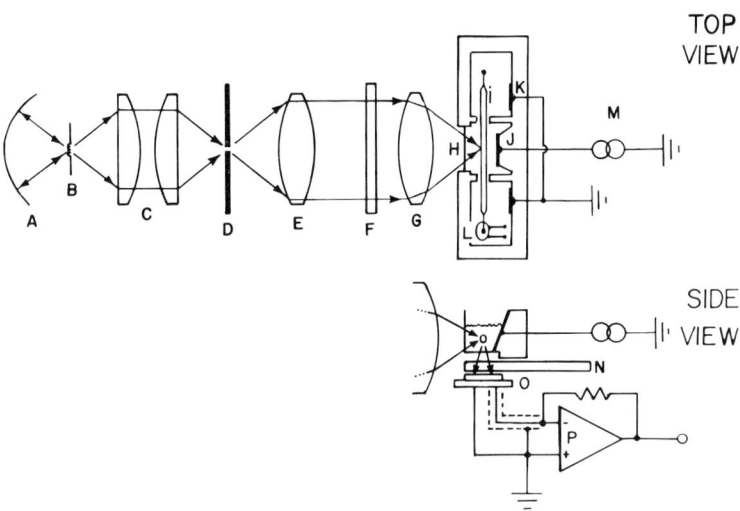

Fig. 1. Schematic drawings of the experimental setup
used to record optical signals from bundles of muscle fibers.
Top View: A. Spherical mirror.
            B. Bulb filament from a 100W tungsten-halogen
                lamp.
            C. Condenser lens system.
            D. Slit.
            E. Colimating lens.
            F. Interference filter.
            G. Condenser. Focuses light into a 1 mm
                spot on the bundle.
            H. Glass window made of coverslip glass.
            I. Muscle bundle. Experiments ranged from
                a few fibers (30-60) to about 1/4 of a
                whole muscle.
            J. Central current platinum plate platinized
                with the Pt black. Current was passed
                between this plate and the lateral plates
                (K) to depolarize or hyperpolarize TTX
                treated bundles.
            K. Lateral current plates.
            L. Tension transducer.
            M. Pulse generator.
Side View: N. Colored glass cut-off filter.
            O. Photodiode.
            P. Current to voltage converter.

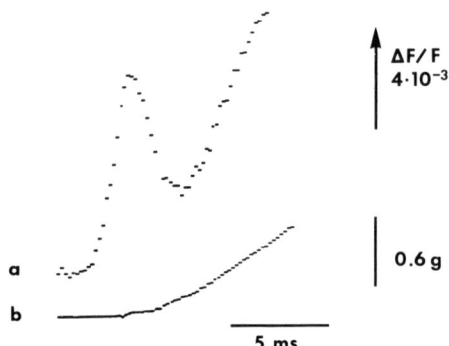

Fig. 2. Merocyanine fluorescence change and tension from
a bundle of about 60 fibers dissected from the semitendinosus
muscle of the Chilean frog, Calyptocefaleya gayi.

a.   Fluorescence signal
b.   Tension

The bundle was stained with 10^{-4} M Merocyanine 540 for
6 minutes. The arrow indicates increment in fluorescence.
Perfused with Ringer's solution at 21° C.

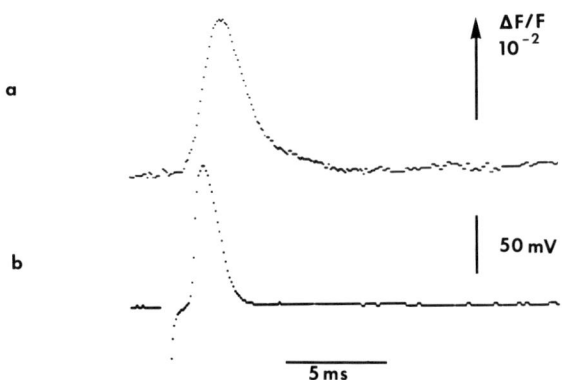

Fig. 3. Fluorescence signal and action potential.

a.   Change of fluorescence from a bundle from the
semitendinosus muscle.
b.   Action potential recorded by a microelectrode impaled
in a fiber with a resting potential of -80 mV in the center
of the illuminated region of the central compartment.

Stimulus duration was 0.2 ms. The bundle was stretched
to 1.4 times its slack length and was immersed in 345 mM
NaCl hypertonic Ringer's solution.

signal. If the depolarization of the surface membrane is
assumed to be identical to that of the T-system, the ratio
(area of T-system membrane)/(area of surface membrane) is
calculated, from measurements in Fig. 3, to be about 7/1.
This ratio is in agreement with morphological measurements
(Peachey, 1965) and also with the detubulation experiments
described later.

In the experiment shown in Fig. 4, tetrodotoxin was
applied to block the active sodium conductance. The fibers
of a bundle were depolarized by passing current between the

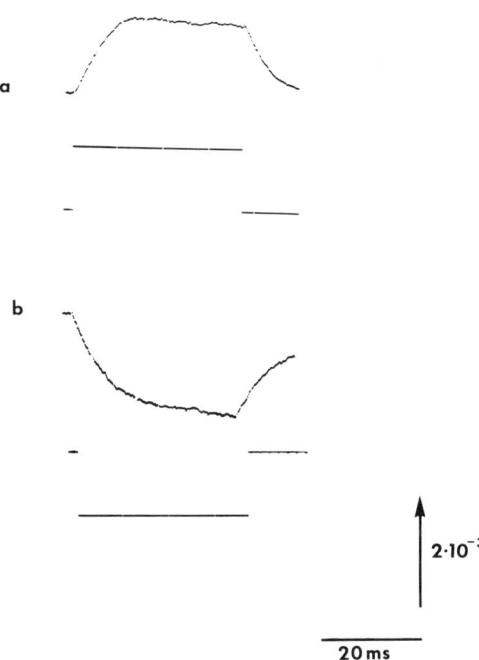

$2 \cdot 10^{-3}$

20 ms

*Fig. 4. Fluorescence changes during prolonged stimulation
of a TTX treated bundle from sartorius muscle. The bundle
was stretched to 1.3 times its slack length and it was
immersed in 345 NaCl hypertonic Ringer's solution with 1 μM
TTX.*

*a. Fluorescence intensity changes during a depolarizing
current pulse.*
*b. Fluorescence changes during a hyperpolarizing pulse.
Both records are the average of 10 sweeps.*

The stimulus pulses are shown below each trace.

central compartment and the two lateral compartments of the
experimental chamber, shown in Fig. 1. A positive change in
fluorescence is observed for a depolarizing pulse, and a
negative change in fluorescence for a hyperpolarizing pulse.
Both fluorescence signals show a rising phase with a time
constant of about 18 msec, which is very similar to that
observed with microelectrodes. Note that with the
hyperpolarizing pulse, one even sees the "creep" that is also
observed with microelectrodes and that has been explained in
terms of anomalous rectification. As suggested above, the
origin of the fluorescence signal seems to be related to the
surface and T-system membranes. Since the T-system has much
more membrane area than the surface, it was expected that
most of this fluorescence signal arises from the T-system.

Fig. 5 shows the results of a detubulation experiment
(Eisenberg, Howell & Vaughan, 1971). The upper trace
corresponds to the fluorescence signal recorded before the
detubulation procedure. We elicited propagated action
potentials by stimulating a small bundle of about 60 fibers
with a short current pulse (arrows in Fig. 5) applied 9 mm
away from the light recording spot. The time elapsed between
the stimulation and the rising phase of the fluorescence
signal is due mainly to propagation delay. Fig. 5*b* shows the
fluorescence signal recorded after detubulation. Conduction
velocity was increased about 40% by the glycerol treatment.
This is in good agreement with results obtained with
microelectrodes and it should be expected from the decrement
in capacitance due to the detubulation. Fig. 5 also shows
that the falling phase of the fluorescence signal is
significantly faster in detubulated fibers than in intact
fibers. In addition, Fig. 5 shows that the magnitude of the
fluorescence signal is diminished considerably by the
glycerol treatment. In the experiment shown in Fig. 5,
the signal is reduced to less than 5% of the control. In
other experiments the percentage fluctuates between 5% and
10% (Vergara & Bezanilla, 1976).

The results obtained in detubulation experiments are
consistent with the interpretation that most of the
Merocyanine fluorescence signal arises at the tubular
membranes. The surface membrane depolarization is expected
to contribute to the overall signal, but only in proportion
to the amount of surface membrane area relative to the
T-system membrane area. Even though the ratio obtained
with the detubulation procedure is in agreement with
morphological studies, one must be cautious, because glycerol
treatment is known to depolarize the muscle fibers. With

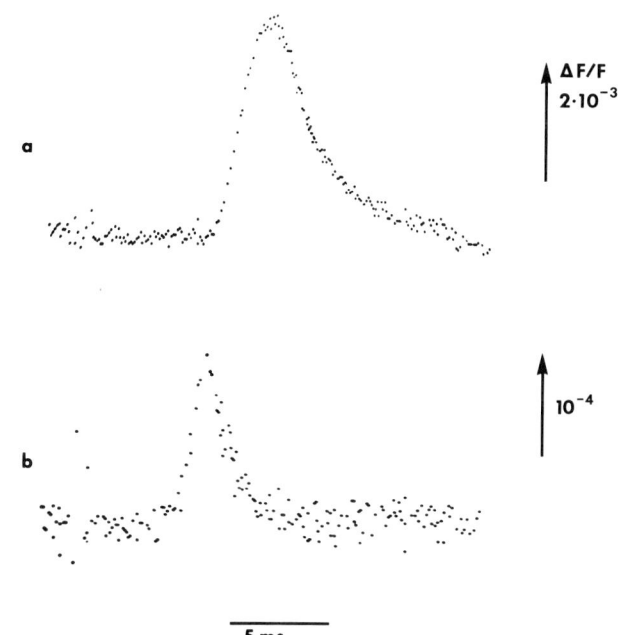

Fig. 5. *Propagated fluorescence signals before and after detubulation. a and b were obtained from the same bundle of about 40 fibers from the semitendinosus muscle immersed in 345 NaCl hypertonic Ringers solution.*

a. *A single sweep before detubulation.*
b. *Average of 40 sweeps after detubulation.*

The arrow indicates the time at which the stimulus pulse was applied. Temperature 20° C.

the procedure of Eisenberg *et al.* (1971) used here, however, a better survival of the fibers is expected, and the loss of signal can be ascribed mainly to the dissappearance of the T-system contribution.

III. NILE BLUE SIGNALS IN BUNDLES OF MUSCLE FIBERS

Fig. 6 shows the light signal and tension recorded from a bundle of fibers stained with Nile Blue A when the bundle is allowed to contract slightly. Tension was suppressed by stretching the bundle and bathing the fibers in hypertonic Ringer's solution. The twitch tension is a few milligrams

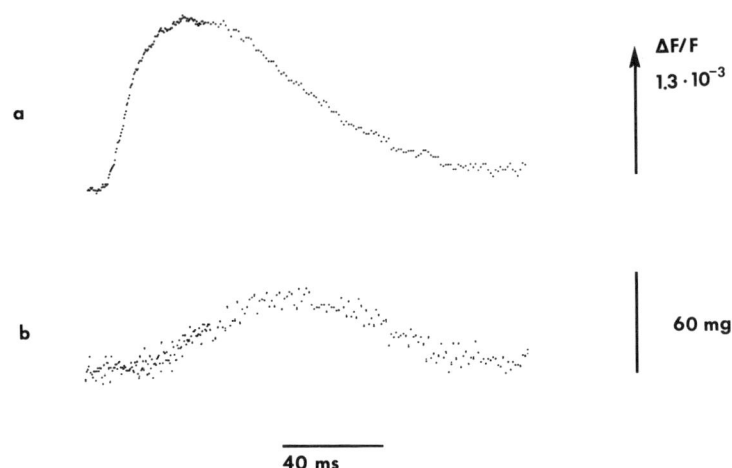

Fig. 6.   Time relation of Nile Blue fluorescence signal
to mechanical response.

  a.   Fluorescence signal in response to single stimulus.
  b.   Tension response.

and is shown in order to compare its time course with the
Nile Blue signal. The time course of the fluorescence signal
(upper trace) is obviously faster than the tension trace.
One difference that can be readily noticed between the Nile
Blue signal and the Merocyanine signal is that the former
is about 3 times slower than the latter.
      In order to stress this difference, an experiment was
performed in which we stained a bundle of fibers with both
Merocyanine 540 and Nile Blue A. The result of that
experiment is shown in Fig. 7. The excitation light used
was 550 nm, which is specific for Merocyanine but which was
able to excite Nile Blue somewhat. There are an early and
a late peak in the fluorescence trace shown in Fig. 7; they
can be ascribed to Merocyanine and Nile Blue, respectively.
The late peak is slightly larger than the early peak, even
though the illumination is more efficient for Merocyanine.
The first peak disappeared when the exciting light was
changed to that specific for Nile Blue (625 nm). Although
we cannot make quantitative comparison between the two signals
in this experiment, we can conclude that the Nile Blue signal
is larger than the Merocyanine signal. This is a
peculiarity of muscle preparations, since Merocyanine is
known to give about 7 times larger signals than Nile Blue in

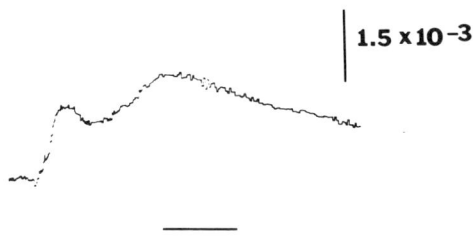

1.5 x 10^{-3}

10 msec

Fig. 7. Light signal from a bundle of about 1/5 of the semitendinosus muscle. The bundle was stained first with Nile Blue A and later with Merocyanine 540.
   The trace corresponds to an average of 20 sweeps. Temperature 17° C. See text.

the squid axon preparation (Cohen *et al.* 1974). We believe that Figs. 6 and 7 suggest that the Nile Blue signal is monitoring a physiological event of the excitation-contraction coupling which: (1) occurs later than the T-system depolarization, (2) has a slower time course, and (3) if it corresponds to a membrane depolarization of a reasonable magnitude, arises from a compartment with a larger area than the T-system.

IV. NILE BLUE SIGNALS IN SINGLE MUSCLE FIBERS

   In order to obtain optical signals under voltage clamp conditions, we essentially used a modification of Hille and Campbell's vaseline gap voltage clamp technique (Hille & Campbell, 1976). This procedure enabled us to do optical experiments in a single fiber preparation that is not only electrophysiologically but mechanically active. In these experiments, we were able to prevent muscle movement by intracellular perfusion with a solution containing 2 mM EGTA, which apparently did not interfere with the calcium release from the sarcoplasmic reticulum. The fibers were stained by external perfusion with a solution containing Nile Blue A for 10-15 minutes. Fluorescence records were obtained by collecting light across the single fiber at 180° with respect to the illumination.

Fig. 8 shows records of an action potential and a Nile
Blue fluorescence signal displayed with two different time
bases. The upper traces correspond to a membrane action
potential recorded from a 180 μm muscle segment (Pool A:
Hille & Campbell, 1976). The fluorescence signal (lower
traces) is very much like the one observed in bundles. The
signal shown was obtained when the fiber was stimulated only
once; no signal averaging was necessary. Comparing Fig. 8
(upper and lower traces), it can be noted that the optical
signal peaks during the decaying phase of the action potential
when the negative after potential is observed in the
electrical record. Fig. 8*b* emphasizes the delay of about
3-5 msec between the foot of the light signal and the foot
of the action potential. The records shown in Fig. 8 give
evidence, in addition to previous records in bundles of
muscle fibers, that the Nile Blue signal is related to

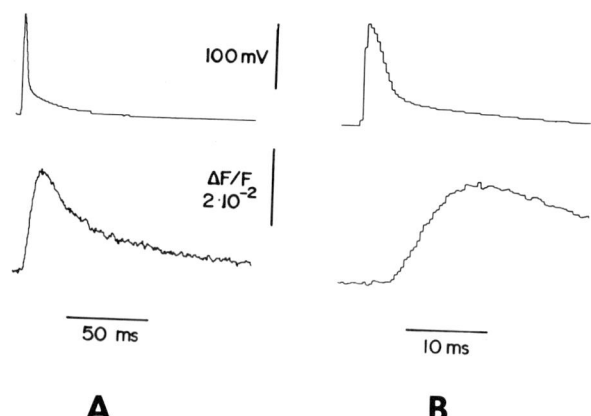

A   **B**

*Fig. 8. Action potential and Nile Blue fluorescence
signal from a current clamped single muscle fiber. Fibers
were dissected from the semitendinosus muscle of the frog
Rana catesbiana and were stained for 10 minutes with a
solution containing 0.001 mg/ml of Nile Blue A. A and B are
the same records displayed at different speeds. Top traces
show the action potential and bottom traces show the
fluorescence signal. Fluorescence was recorded at 180° with
an exciting light of 625 nm and a cut-off filter of 665 nm.
The fluorescence trace corresponds to an average of 5 sweeps.
Temperature 13° C.*

physiological events much slower than the action potential.
As stated before, movement was abolished in this preparation
by the use of EGTA. Visual inspection with a 40x objective
was used to check that there was no movement before the
experiments were started. In experiments in which
movement was not completely abolished, we could see changes
in the shape of the signal, always occurring late in the
decaying phase.

    An interesting feature of the Nile Blue signal is shown
in Fig. 9. When the fiber is stimulated repetitively at
50 per sec and a temperature of 15°, the signals did not add
up, but each one started from where the other ended,
reaching the same height. This result is different from
what is expected for tension or movement. Another feature
is that the relaxation of the last signal has very much the
same time course of the single signal shown in Fig. 8.
As a final piece of evidence against movement artifacts,

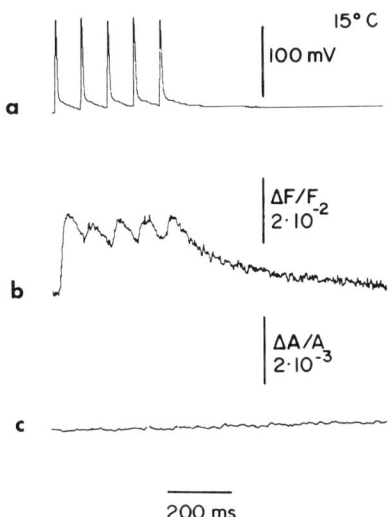

Fig. 9. *Action potentials and fluorescence and
absorption traces during a repetitive stimulation of a single
muscle fiber.*

    a. *Action potentials.*
    b. *Nile Blue fluorescence trace.*
    c. *Absorption, control trace.*

*Fluorescence trace corresponds to a single sweep. Absorption
record obtained with monochromatic light of 625 nm. Frequency
of stimulation 20 Hz. Pulse duration 0.5 ms.*

trace 9*c* shows what occurs when the cutoff filter is removed.
No absorption change is observed even at an order of magnitude
higher sensitivity. This result demonstrates that the signal
shown in Fig. 9*b* is a real fluorescence signal. The signal
is obviously produced by Nile Blue: unstained fibers do not
give a fluorescence signal.

     Fig. 10 shows results from a two pulse experiment. The
interval between stimuli is shown below every record. Upper
traces show the membrane action potentials and lower traces
correspond to the optical records. When the two stimuli are
very close (20 msec), the second signal is slightly smaller
than the first one, but as they are separated, the second
signal seems unaffected by the first. When the interval is

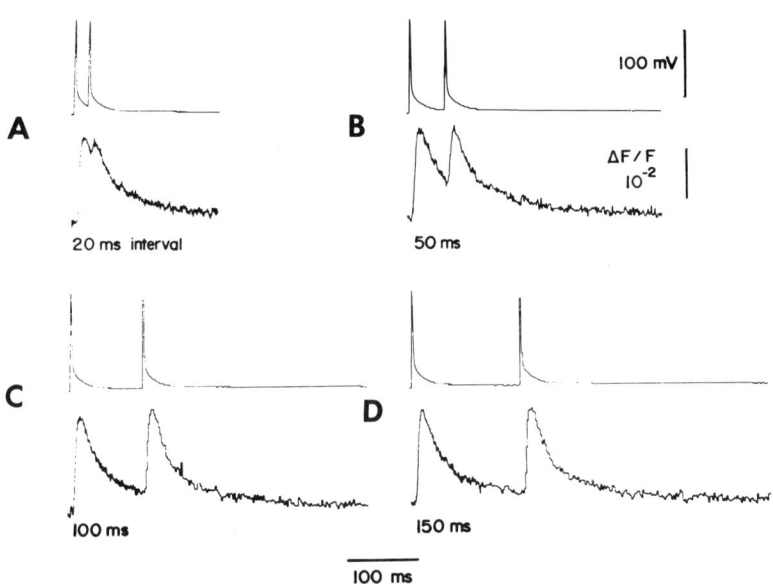

*Fig. 10. Action potentials and fluorescence records from
a single fiber stimulated with two consecutive pulses. A, B,
C, and D are obtained with pulse separations of 20, 50, 100,
and 150 ms, respectively. Top traces show action potentials
and bottom traces show the fluorescence changes.
Fluorescence traces correspond to a single sweep. Temperature
17° C.*

very long (100 msec), the second signal is possibly slightly larger than the first, which could suggest a certain degree of potentiation. When we modify the action potential by the action of tetraethylammonium (TEA) outside, as shown in Fig. 11, the optical signal is slightly prolonged (Fig. 11b), but it is not larger than the control (Fig. 11a). An interesting feature observed with two pulses in a TEA-containing solution (Fig. 11c) is that starting from a potential that is not fully repolarized to -90 or -100, there is no second peak in the fluorescence signal (Fig. 11c), but the signal widens.

Fig. 12 shows a voltage clamp experiment using tetrodotoxin (TTX) to eliminate the active sodium current. With a 100 mV depolarizing pulse, we observe a sustained

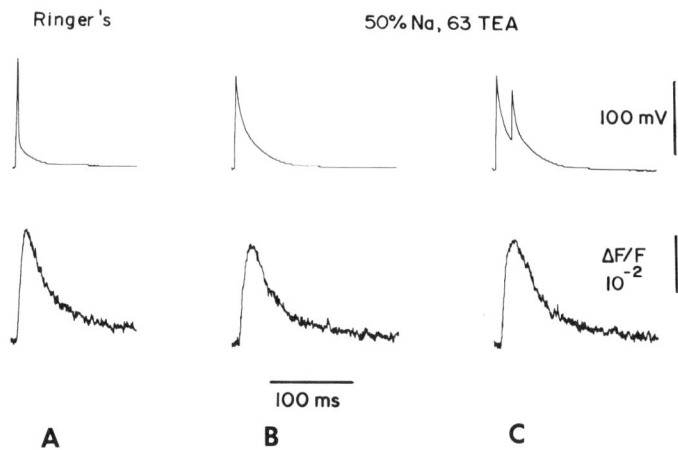

Fig. 11. *Effect of TEA on the action potential and the Nile Blue fluorescence signal.*

A. *Action potential and fluorescence signal in Ringers solution.*
B and C. *Action potentials and fluorescence signals in a solution in which 50% of the NaCl was replaced by 63 mM of TEA-Cl.*

Top traces show the action potentials and bottom traces show the fluorescence signals. Record C corresponds to a two pulse experiment. The two pulses were applied with a 20 ms interval. Note that even though the second action potential is smaller than the first one, it is still larger than 100 mV. Temperature 17° C.

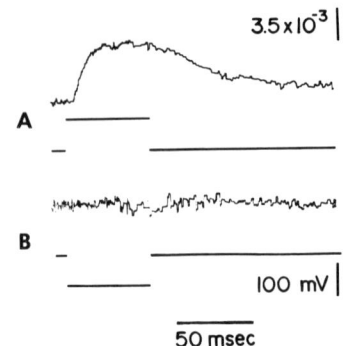

Fig. 12. Changes in fluorescence in a voltage clamped single muscle fiber.

*   A. Top trace, fluorescence. Bottom trace, 100 mV depolarizing pulse.*
*   B. Top trace, fluorescence. Bottom trace, 100 mV hyperpolarizing pulse.*

Estimated holding potential -90 mV. Fluorescence traces correspond to an average of 10 sweeps. Temperature 14° C.

signal while the pulse is applied to the fiber. After the end of the pulse, the signal decays with a time constant slower than the rising phase. For the hyperpolarizing pulse, however, there is no detectable signal. This rectification property of the Nile Blue signal is probably the strongest evidence suggesting the involvement of the sarcoplasmic reticulum in the genesis of the signal. It was first observed by Bezanilla & Horowicz (1975) in whole muscles depolarized or hyperpolarized by currents applied with triangular fluid electrodes and was later verified by us in bundles of fibers using the chamber described above (Fig. 1) and in voltage clamped single muscle fibers as shown in Fig. 12.

Fig. 13 shows the results of an experiment in which the K currents were blocked by TEA. The effects of TTX on the fluorescence signals are shown for pulses of 70, 80, and 130 mV. The current traces are shown above each light trace. Without TTX (left column), there is a net inward current and a noticeable difference in the time course and size of the signal as compared to records with TTX (right column). For a small depolarization (70 mV) in the presence of TTX, the signal is very small and has a long lag, as previously shown by Baylor and Chandler (this volume). But when we do not

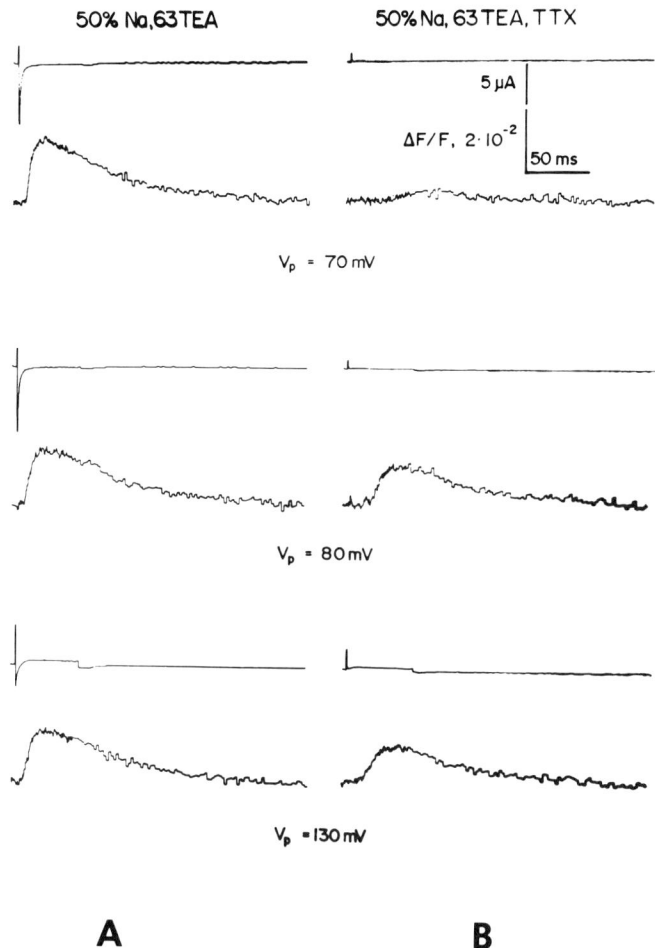

A   **B**

Fig. 13. Effect of TTX on the fluorescence signal in TEA treated voltage clamped single muscle fiber.

A. Fluorescence and current records without TTX.
B. Fluorescence and current records with TTX.

Above every fluorescence trace, the corresponding current trace is shown. Magnitude of stimuli (V_p) shown in the medial line of the figure. Fluorescence traces correspond to single sweeps. Holding potential estimated at -90 mV. TTX concentration 100 nM. Temperature 15° C.

block the inward current, there is a big optical signal at
the same depolarization in the same fiber (Fig. 13a). To
explain these results, one must consider that the membrane
potential can be controlled at the surface, but there might
be an uncontrollable potential in the tubules when an
inward current is present. In other words, we cannot control
the potential of a network that has series resistance and an
access resistance if there is a nonlinear conductance
(Adrian & Peachey, 1974). We would expect that the tubules
undergo a tubular action potential that charges the membranes
of the tubules much faster than in the presence of TTX,
presumably producing a better transfer of information to the
triad.

Fig. 14a shows graphically the voltage dependence of the
Nile Blue signal without TTX. The plot illustrated the peak
change of fluorescence as a function of membrane potential,
which is indicated here by the pulse amplitude. The value
of the holding potential in this preparation is not certain
but it is probably very close to -90 mV because we obtained
a steady state inactivation curve of the sodium current before
setting the holding potential. There is a marked steepness,
practically a threshold, in the appearance of the signal.
Actually, for the curve that was fitted with an e-fold change
for 2 mV, the Boltzman distribution is not steep enough.
Again, the tubular action potential appears to be producing
the steepness by activating the tubules practically in an all
or none process. Fig. 14b shows the same experiment with
TTX. The voltage dependence of the signal appears much more
smooth. An e-fold change for 7.5 mV is obtained. This
value is not too far from that observed in the charge
movement associated with contraction (Schneider & Chandler,
1973).

Another interesting feature of the Nile Blue signal is
shown in Fig. 15. The time to peak and the time to half-peak,
which are measurements of the time the signal takes to
appear, seem to be a function of membrane potential, being
different for experiments with and without TTX. The inset
shows how these measurements were done. The solid circles
show results with tetrodotoxin; the open symbols show the
case without tetrodotoxin. Both parameters (time to peak
and time to half-peak) indicate that the time course of the
signal is faster when sodium current is present. When
tetrodotoxin is added, the signal slows down, probably
because of the time it takes to charge the tubules by the

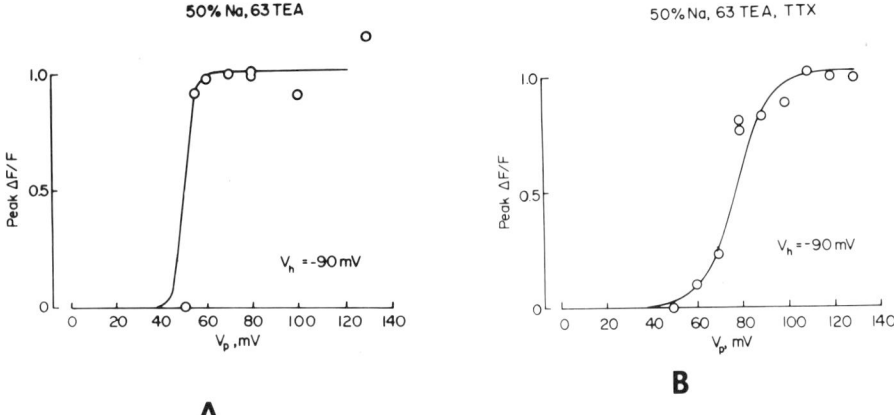

A

B

Fig. 14. *Voltage dependence of the peak Nile Blue fluorescence signal in a voltage clamped single muscle fiber.*

A. *Without TTX.*
B. *With TTX.*

These curves were obtained from the same experiment shown in Fig. 13. Both graphs correspond to normalized fluorescence changes vs. pulse magnitude. The maximum value of $\Delta F/F$ in Fig. 14a is 2×10^{-2} and in Fig. 14b is 1.2×10^{-2}. The continuous curve fitting the experimental points (open circles) in Fig. 14b was obtained from the equation:

$$Normalized \ (\Delta F/F) = \frac{1}{1 + exp \ [-(V-78)/7.5]}$$

In Fig. 14a the continuous curve corresponding to the equation

$$\frac{1}{1 + exp \ [-(V-50)/2]}$$

is given as a reference.
     V_h is an estimated value of the holding potential.

clamp system. It is also clear, in this case, that there
is a strong dependence on the voltage. The more positive
the voltage becomes, the faster the signal gets, although
there is a tendency to saturate.

Even though there is no conclusive evidence that these
changes in fluorescence are related to the sarcoplasmic

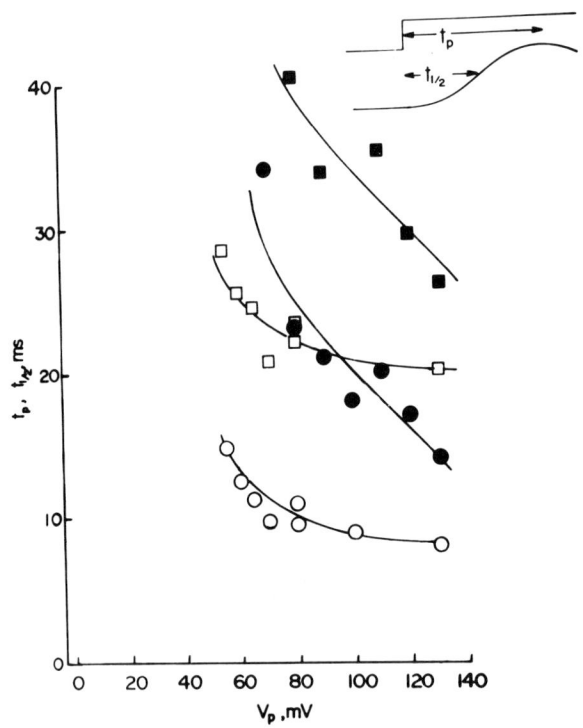

Fig. 15. *Voltage dependence of the time to peak and time
to half-peak of the fluorescence signal in a voltage clamped
fiber. The results were obtained from the same experiment of
Figs. 13 and 14. The insert in the upper corner of the figure
defines the terms: time to peak (tp) and time to half-peak
(t½). The data points obey the follow nomenclature
according to the experimental conditions:*

    Open squares: tp, 1/2 Na - TEA
    Filled squares: tp, 1/2 Na - TEA + TTX
    Open circles: t½, 1/2 NA - TEA
    Filled circles: t½, 1/2 Na - TEA + TTX

*The continuous curves were drawn by eye through the data
points.*

reticulum membrane, we have used this assumption as a working
hypothesis. In any case, this signal is not concomitant with
the action potential, and it does not behave as if it arises
from the tubular membrane. It is not linear, whereas one
would expect to see a linear signal associated with membrane
potential. Its time course is too slow. Among the different
possible events occurring after the tubular depolarization,
we are tempted to think that the Nile Blue signal arises
from depolarization of the sarcoplasmic reticulum primarily
because Nile Blue is a voltage sensing dye.

V. EFFECTS OF TETRACAINE ON THE NILE BLUE SIGNAL

Tetracaine is known to eliminate contraction (Caputo,
1976; Almers & Best, 1976) without affecting the displacement
currents in muscle fibers (Almers & Best, 1976). The latter
authors suggest that tetracaine inhibits the calcium release
from the SR. Fig. 16 shows the result of an experiment in
a single muscle fiber that is voltage clamped and treated
with TTX. It can be observed that only 0.3 mM tetracaine is

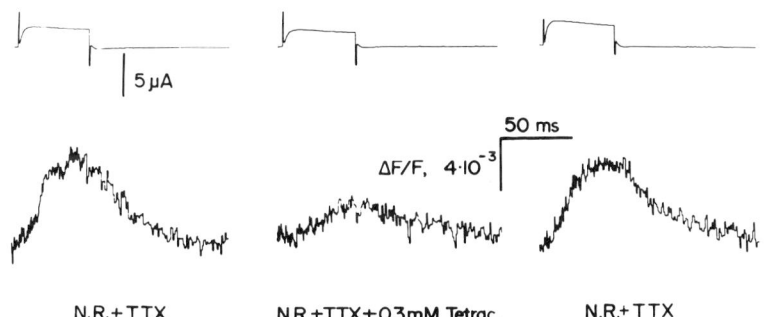

N.R.+TTX   N.R.+TTX+0.3mM Tetrac.   N.R.+TTX

*Fig. 16. Effect of tetracaine on the Nile Blue
fluorescence signal of a voltage clamped single muscle fiber.*

Upper traces, membrane current.
Lower traces, fluorescence.

*The fluorescence and current records were obtained by
averaging 10 sweeps. Holding potential -90 mV. Stimulus
150 mV, depolarizing pulse of 50 ms duration. Temperature
22° C.*

necessary to block at least 65% of the signal with an almost
complete recovery after 20 min of washing. With 1 mM
tetracaine the signal is completely eliminated, but the
recovery is less complete. Comparison of this result with
those of Almers & Best (1976) suggests that the Nile Blue
signal depends somehow on the Ca^{++} release processes. The
reversible blockage of the Nile Blue signal by tetracaine
can be paralleled by 10 mM procaine.

Fig. 17 shows results of another experiment in which a
bundle of muscle fibers was perfused with Ringer's + 500 nM

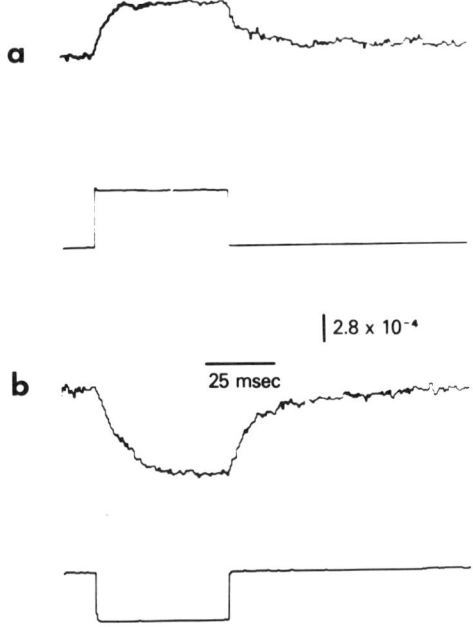

Fig. 17.   Linearization of the Nile Blue signal by
tetracaine in a bundle of muscle fibers stimulated with long
current pulses. The bundle was stretched to 1.4 times its
slack length and was immersed in 345 mM NaCl hypertonic
Ringers solution with TTX and tetracaine added. The tension
record was flat. The experimental conditions and organization
of the figure is similar to Fig. 4.

a.   Fluorescence intensity change during a depolarizing
current pulse.
b.   Fluorescence intensity change during a hyperpolarizing
pulse.

The current pulses are shown below each trace.   Temperature
18° C.

TTX and 0.5 mM tetracaine. It demonstrates that when the SR signal is blocked and the gain of the light recording system is increased, there is a linear signal obtained with Nile Blue that is identical to the Merocyanine signal (Fig. 4). It can be observed that there is a signal for hyperpolarization and depolarization. They are more or less symmetrical except for the "creep" that is observed in the hyperpolarizing direction. Note that the magnitude of the linear signal is very small with a peak $\Delta F/F$ of the order of 5×10^{-4}. This value is significantly smaller than the peak value of the SR signals shown in Figs. 6 and 7. It is also smaller than the linear Merocyanine signal shown in Fig. 4, which is not surprising since it is known that Nile Blue is about 7 times less efficient than Merocyanine 540. In fact, if that value is used to scale the records shown in Fig. 17 to those in Fig. 4, it is found that they match almost perfectly not only in their time course, but also in their magnitude.

VI.   CONCLUSION

It is not unreasonable, then, to think that Nile Blue A is acting as a voltage sensing dye, mainly giving a signal that is related to the membrane potential of the SR. Using this interpretation as a working hypothesis, however, we can gain some knowledge of how the sarcoplasmic reticulum works during calcium release. It is possible to suggest, for example, that the release of calcium produces a membrane potential change across the sarcoplasmic reticulum and that this is the signal that we are observing. Alternatively, it is possible that the external depolarization is inducing a depolarization of the SR membranes after a gating process is activated. In this case, the Ca^{++} release would be a consequence of the SR depolarization.

To decide between these possibilities, we will continue making use of our preparation with cut fibers. In principle, it should be possible to study ionic dependence of the Nile Blue or the indodicarbocyanine signals (or any other dye) in this preparation. Presumably we would end up studying the properties of the sarcoplasmic reticulum membrane, which is inaccessible by other means. Both the indodicarbocyanine dyes, as demonstrated by Baylor and Chandler (this volume), and the Nile Blue dye produce very similar signals. The most important common property that the dyes share is that they are penetrating membrane potential indicators. Clearly,

however, we must do more experiments with other dyes. If we
use several membrane potential indicating dyes, either
penetrating or nonpenetrating, in the end pools so that they
can diffuse into the single fibers' sarcoplasm, and if all
of them produce a very similar signal, we will have a much
firmer basis for our belief that the signal arises at the
sarcoplasmic reticulum's membranes.

DISCUSSION

    Rougier: We have performed a voltage clamp experiment on
semitendinosus muscle of frog in the double sucrose gap
system, measuring the contraction and the current together
under different conditions, and I want to make some
comparisons with your work. The main result that we obtain
is the following. If we plot the contraction vs.
depolarization we find that depolarizing to 120 or 140, which
is above the sodium equilibrium potential, we obtain
contractions which are all or none and are independent of
the membrane potential. We interpret this response in the
same way as the optical signal you obtained in sodium Ringer.
In other fibers in sodium Ringer, we obtain smooth
tension-voltage relations which show strong dependence on
membrane potential. If we add tetrodotoxin, we observe a
shift in the mechanical threshold and a smooth increase of
contraction independent of the resting potential. Under
these conditions the membrane potential of the tubular
system does not escape as it does under other conditions,
and perhaps it is better controlled.
    Peachey: Do you have any idea what is the basis of
this difference between these two kinds of fibers? Have they
normal membrane potentials and so on?
    Rougier: The only possibility is that the geometry of
the fibers is not the same, and perhaps the tubular system
is deeper in some fibers than in others. Perhaps also the
access resistance of this kind of fiber is very small.
    Baylor: When we plotted the maximum rate of change of
the optical signal vs. voltage, we saw e-fold increase for
about 3 to 4 mV. You plotted the peak, and got an e-fold
increase for 7.5. Have you plotted the maximum rate of
change for the optics?
    Bezanilla: No, not yet.

Armstrong: In your conditions, how could there still be calcium in the sarcoplasmic reticulum to produce potential change--to accord with one of the hypotheses--that it simply reflects the amount of charging of the sarcoplasmic reticulum membrane? You have EGTA and I would think there would be no calcium left in the sarcoplasmic reticulum after a very short time.

Bezanilla: Well, I think that Endo could make a comment about that, because he has measured how long the load of the SR under EGTA conditions lasts.

Endo: We have done many experiments with skinned fibers, and we even loaded SR with calcium and then washed with a relaxing solution containing say, 10 mM EGTA. After 30 min or an hour a significant amount of calcium still remains in the SR, so I don't think that the calcium in the SR is gone in your experiments.

Armstrong: How many pulses do you apply, and how much calcium do you suppose that you release in each one? I presume this is in the absence of any sort of stimulation of the SR.

Bezanilla: We don't apply too much stimulation if possible. All the traces I showed were single traces. As a matter of fact, the signal goes down after a while, although the electrical properties are still good. Also, in the experiments with cut end fibers, we use ATP and magnesium in the perfusing solutions. So perhaps we still can transport back some of the calcium.

Morad: Since this is a cardiac muscle symposium, it's only fair to say that we haven't been able to measure any SR type of signal in cardiac muscle, although we have been able to measure perhaps even better or bigger signals from the surface membrane. And we've tried rat ventricular tissue which is fairly similar to skeletal muscle: with a short action potential, a lot of SR, and a lot of T-system. Reducing the extracellular calcium concentration often abolished this signal. We've also tried hypertonic solutions, which are very effective in the rat, but not very effective in the frog heart. I draw two conclusions from the hypertonic solution studies; one, that when the tension really went down, there was no signal, although action potential was there, and two, that in the frog, since hypertonic solution didn't work, one must be somehow affecting the T-tubular SR junction. I wonder if all these pharmacological manipulations or the use of hypertonic solution don't somehow alter the E-C coupling processes in such a way that the signal you measure is not really very meaningful.

Bezanilla: Well, all these signals that I presented at the end were not in hypertonic solutions.

Reuter: What is known about the ionic composition in the SR? I heard that Somlyo had done some electron probe studies and found similar ionic composition in the SR as in the myoplasm. Wouldn't this be a rather strong blow against the hypothesis that the SR membrane can be depolarized at all?

Bezanilla: I would prefer that Saul Winegrad comment on the concentration of different ions made by Somlyo, because he's more familiar with their work.

Winegrad: The information that the Somlyo's have found using electron probe microanalysis of frog skeletal muscles is in a way preliminary,* but essentially this is what it consists of. In the isotonic perfusion medium, there is no detectable gradient within the limits of their system for sodium or for chloride, across the SR membrane. However, when the cells are bathed in hypertonic solution, 2.2 or 2.5 times normal tonicity, then one does see a gradient of these two ions. The system is not well suited for measuring sodium gradients, however, because of a very unfavorable signal to noise ratio. So, one can't say that the composition is the same; one can only say that as far as

*Editors note, submitted by A.V. Somlyo. The results of electron probe x-ray microanalysis of cryosections of frog muscle, showing that Cl is not compartmentalized in the SR and that its concentration does not differ from that of the cytoplasm (Somlyo, Shuman & Somlyo, J. cell Biol. 74:828-857, 1977), have since been confirmed in the swimbladder of the toadfish (Somlyo, Shuman & Somlyo, Nature 268:556-558, 1977). In the later experiments the Na concentrations were also measured and showed that the Na content of the SR and adjacent cytoplasm (34 ± 4.2 vs. 30 ± 4.3, mM/Kg dry wt ± S.E.M.) are also not significantly different. These findings indicate that the SR is not an extracellular compartment and do not support the existence of a Cl or Na potential across the SR membrane in resting muscle. The NaCl containing vacuoles along the Z line of muscles treated with hypertonic solutions (Somlyo, Shuman & Somlyo, J. cell Biol. 74:828-857, 1977) have recently been shown by freeze substitution to represent swollen regions of the T-tubule system (Franzini-Armstrong, Heuser, Reese, Somlyo & Somlyo, submitted for publication).

those two ions are concerned, in isotonic solution, within
the limits of error of the system, there is no difference,
but there is a difference in the hypertonic solution.

Bezanilla: If there is no resting potential difference, if
calcium goes out, for example, it will charge the membrane
capacitance and consequently it will produce a potential change
assuming that it goes out unaccompanied by another cation
going in or unaccompanied with anion.

Endo: I'd like to add to Dr. Winegrad's comment. He
told you that in hypertonic solution, the Somlyo's found
much chloride in some structures, but they recently wrote me
that they now believe that those structures are dilated
T-system and not the SR.

REFERENCES

1. Adrian, R.H. & Peachey, L.D. (1973). Reconstruction of
   the action potential of frog sartorius muscle.
   J. Physiol. 235, 103-131.
2. Almers, W. & Best, P.M. (1976). Effects of tetracaine
   on displacement currents and contraction of frog skeletal
   muscle. *J. Physiol. 262*, 583-611.
3. Bezanilla, F. & Horowicz, P. (1975). Fluorescence
   intensity changes associated with contractile activation
   in frog muscle stained with Nile Blue A. *J. Physiol.
   246*, 709-735.
4. Caputo, C. (1976). The effect of caffeine and
   tetracaine on the time course of potassium contractures
   of single muscle fibres. *J. Physiol. 255*, 191-207.
5. Cohen, L.B., Salzberg, B.M. Davila, H.V., Ross, W.N.,
   Landowne, D., Waggoner, A.S. & Wang, C.H. (1974).
   Changes in axon fluorescence during activity: Molecular
   probes of membrane potential. *J. membrane Biol.
   19*, 1-36.
6. Eisenberg, R.S., Howell, J.N. & Vaughan, P.C. (1971).
   The maintenance of resting potentials in glycerol-treated
   muscle fibres. *J. Physiol. 215*, 95-102.
7. Hille, B. & Campbell, D.T. (1976). An improved vaseline
   gap voltage clamp for skeletal muscle fibers. *J. gen.
   Physiol. 67*, 265-293.
8. Landowne, D. (1974). Changes in fluorescence of skeletal
   muscle stained with merocyanine associated with
   excitation-contraction coupling. *J. gen. Physiol.
   64*, 5a.

9. Nakajima, S., Gilai, A. & Dingeman, D. (1976). Dye absorption changes in single muscle fibers: An application of an automatic balancing circuit. *Pflügers Arch. 362*, 285-287.

10. Oetliker, H., Baylor, S.M. & Chandler, W.K. (1975). Simultaneous changes in fluorescence and optical retardation in single muscle fibres during activity. *Nature, Lond. 257*, 693-696.

11. Peachey, L.D. (1965). The sarcoplasmic reticulum and transverse tubules of the frog's sartorius. *J. cell Biol. 25*, 209-231.

12. Schneider, M.F. & Chandler, W.K. (1973). Voltage dependent charge movement in skeletal muscle: A possible step in excitation-contraction coupling. *Nature, Lond. 242*, 244-246.

13. Vergara, J. & Bezanilla, F. (1976). Fluorescence changes during electrical activity in frog muscle stained with merocyanine. *Nature, Lond. 259*, 684-686.

14. Vergara, J. & Bezanilla, F. (1977). Nile blue fluorescence signals in frog single muscle fibers under voltage or current clamp conditions. *Biophys. J. 17*, 5a.

ACKNOWLEDGMENTS

We thank Drs. B.M. Salzberg and R. Orkand for their hospitality, loan of equipment, and helpful advice throughout the course of part of this work. Tetracaine experiments were done in collaboration with Dr. Carlo Caputo. We also thank Dr. Sally Krasne for reading this manuscript. Supported by grants from the Muscular Dystrophy Association of America and from the Oficina Tecnica de Desarrollo Cientifico of the University of Chile.

THE AEQUORIN SIGNAL DURING ACTIVATION
OF MUSCLE

Reinhardt Rüdel

Physiologisches Institut der Technischen Universität
Biedersteiner Strasse 29, D-8000 München 40
Federal Republic of Germany*

One of the most successful methods for determining rapid
changes of cytoplasmic calcium concentration has been the
intracellular application of the calcium-sensitive
bioluminescent protein aequorin (Blinks, Prendergast & Allen,
1976). This substance has the advantage that it reacts
rather specifically with calcium and that no other co-factors
are required for the light-producing reaction. One limitation
of the indicator is that once each aequorin molecule has
reacted with calcium it is inert or "discharged" (Shimomura,
Johnson & Saiga, 1962).

Although the multiple injection of small amounts of
aequorin into groups of cardiac cells in conjunction with the
use of signal averaging techniques has recently led to
recording aequorin luminescence from heart muscle (Allen &
Blinks, personal communication), most of the information on
intracellular calcium transients during muscular activity has
so far come from experiments with skeletal muscle cells from
barnacle (Ashley & Ridgway, 1970) and various amphibia
(Taylor, Rüdel & Blinks, 1975; Rüdel, Taylor & Blinks, 1978).
The larger volume of skeletal muscle cells makes it
relatively easy to inject 1 to 3 ml of an isotonic KCl
solution containing 1 to 3 mg/ml aequorin, e.g., into a
single frog semitendinosus fiber (volume about 100 ml). This

*The experiments described in this article were performed
in the Department of Pharmacology of the Mayo Medical School,
Rochester, Minn. 55901, USA, in collaboration with Drs. John
R. Blinks and Stuart R. Taylor.*

amount is sufficient to give, on subsequent stimulation, calcium-mediated luminescent responses with a signal to noise ratio of better than 30.

This report deals with the physiological correlates of such luminescent responses.

I.   METHODS TO RECORD THE AEQUORIN SIGNAL

Fig. 1 shows a diagram of the setup used for aequorin injection and subsequent experimentation with the injected cell. An isolated frog muscle fiber is mounted in a temperature-controlled (5-25° C) chamber on two stainless steel hooks. A glass micropipette containing aequorin with a tip diameter < 1 μm is inserted into the fiber. Aequorin is then injected by applying air pressure above the solution in the pipette for some 10 sec while the site of impalement is inspected through a microscope. A slight transient swelling of the fiber indicates that aequorin solution has entered the cell. After termination of injection, the pipette is withdrawn, leaving no visible damage to the fiber. The intracellular presence of aequorin does not seem to influence the contractile performance of the muscle fiber, as aequorin-injected fibers function normally in subsequent experimentation for over ten hours.

Two platinum plate electrodes extending more than the length of the fiber are provided in the chamber for massive stimulation with square-wave pulses of 0.5 msec duration.

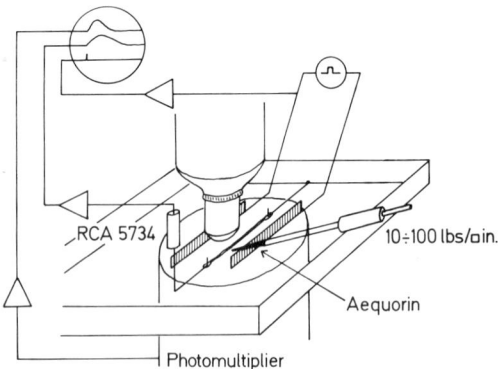

Fig. 1. Schematic drawing of the setup for the injection of aequorin into a single muscle fiber and for subsequent experimentation.

One of the two steel hooks is connected to a mechano-electric transducer for the recording of isometric force. Aequorin luminescence is monitored by means of a photomultiplier tube having a 45 mm wide photocathode with shutter, which is mounted directly under the chamber. The outputs of force transducer and photomultiplier are displayed on an oscilloscope screen and photographed. To reduce noise on the records, a computer of average transients (C.A.T.) is used when necessary.

The sensitivity of the light recording apparatus is such that a calcium concentration of 3×10^{-7} mol/l within the muscle cell within reasonable time produces a response just significantly above the dark count level. We were not able to detect luminescence from a resting aequorin-injected muscle fiber, at least in the investigated temperature range of 5 to 15° C, as photon counting for up to 30 min gave equal results with the shutter closed or open. Thus, in a resting muscle fiber the calcium level must be below 3×10^{-7} mol/l.

II. THE AEQUORIN SIGNAL DURING A SINGLE ACTIVATION

When a fiber is stimulated it emits a light flash, the aequorin response. The intensity but not the time course of this flash depends on the amount of aequorin injected. The time course of the aequorin response to a single stimulus is illustrated in Fig. 2, together with the simultaneously

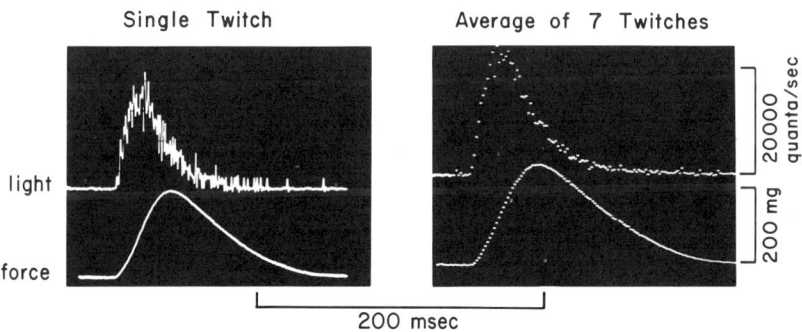

Fig. 2. *Rana temporaria, tibialis anterior, 15° C, 2.1 μm. The aequorin signal in response to a single stimulus. The left-hand panel shows a record of light and force during a single rested-state contraction. The right-hand panel shows a C.A.T. average of seven such twitches. Calibrations apply to right-hand panel only.*

recorded isometric force. In the left-hand panel a single
event is shown. To reduce the photomultiplier shot noise and
improve the quality of the light trace, in the right-hand
panel seven such responses have been averaged on a C.A.T.
computer. The light signal starts to rise nearly
simultaneously with force, but reaches its peak much earlier,
at about the time when the rate of force development is
maximal. It then declines exponentially with a half time of
about 10 msec at 15° C, so that it has fallen to less than
half of its peak value by the time force attains its maximum.

This signal reflects the transient change of intracellular
calcium concentration occurring on activation of the muscle
fiber. To test the contribution of an influx of
extracellular calcium to the intracellular calcium transient,
we removed all the extracellular Ca^{2+} from the bathing fluid.
For up to an hour fibers were kept in a Ca-free solution
containing 3 mmol/l Mg^{2+} and 2 mmol/l EGTA. No significant
changes were detectable in luminescent and mechanical
responses during this period. It thus appears that virtually
the entire luminescent reaction arises from intracellularly
released calcium.

Thousands of aequorin responses can be elicited from a
fiber during the 8 to 12 hours following a successful
injection. The time course and the amplitude of the optical
response will be virtually unchanged as long as the fiber has
been rested for at least 2 min before stimulation.* This
seems paradoxical because with each single event the
light-emitting aequorin molecules are irreversibly
inactivated. An explanation for the resiliancy of the
response is that only a fraction of the calcium-activated
molecules are used up in each activation. A rough estimate
of the aequorin consumption can be obtained by integrating
the number of photons registered per twitch, multiplying by
the efficiency of the counting prodecure, and dividing the
result into the total of the some 10^{10} molecules injected.
This number indicates that only about 10 ppm of the injected
molecules are inactivated at one time.

This low fraction is also one of the reasons why the
presence of aequorin does not interfere with contraction by
altering free calcium concentration. Assuming that three

*As will be shown in Fig. 6, this interval is just enough
to abolish all detectable influence of previous activity.
This is by definition the condition for a rested-state
response (Blinks & Koch-Weser, 1961).

calcium binding sites for each aequorin molecule have to be
filled for luminescence to occur (Allen, Blinks &
Prendergast, 1977), one can show that 3×10^{-8} mol/l calcium
will be bound by aequorin in a twitch. In a tetanus this
figure will go up by a factor of 10, but this is still three
orders of magnitude lower than the concentration of calcium
bound to fully activated myofibrils (2×10^{-4} mol/l, Weber
& Herz, 1963).

For these calculations we had to know the local
concentration of aequorin attained in the fiber. Thus it was
necessary to determine in what volume fraction of the muscle
fiber the injectate was distributed. Such estimates were
made by scanning the light intensity of the aequorin response
along the length of frog tibialis anterior fibers (5 to 8 mm).
A 0.5 mm wide fiber-optic probe coupled to a second
photomultiplier was mounted on a micromanipulator and moved
along the fiber in 0.2 mm steps. At each step the fiber was
stimulated. The peaks of the aequorin responses obtained
in two scans carried out 2.3 and 5.3 hours, respectively,
after the injections are plotted in Fig. 3. They show that
although aequorin spreads from the site of injection, its
distribution is far from uniform over the whole length even
several hours after injection. A coefficient for
longitudinal diffusion of aequorin in muscle fibers of

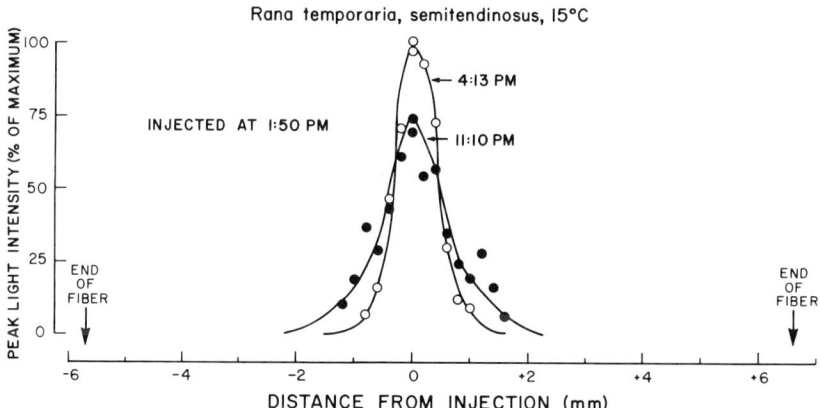

Fig. 3. *Rana temporaria, semitendinosus*, 15° C.
*Diffusion of aequorin from the site of injection. Muscle
fiber was injected with aequorin at a single point near its
middle. Two scans along fiber of light intensity during
activation were carried out, as described in the text, 2.3
(open circles) and 5.3 (filled circles) hrs after the
injection.*

$D = 0.5 \times 10^{-7}$ cm^2 sec^{-1} could be calculated from these
results, in good agreement with the value for the diffusion
of aequorin in barnacle myofibrils (Ashley, Moisescu & Rose,
1974).

Fortunately, nonuniformity in the intracellular aequorin
distribution has little influence on the aequorin response
of an activated muscle cell, for the light intensity from a
solution of given calcium concentration is directly
proportional to the concentration of active aequorin. Thus,
as long as the total amount of aequorin remains unchanged,
its spreading should not influence the aequorin response, as
in fact is observed.

Nonuniformity of calcium concentration, on the other
hand, has a distinct influence on the amplitude of the
aequorin response, because the light intensity from a given
amount of aequorin varies exponentially with [Ca^{2+}] over a
wide range (Allen *et al.* 1977). During a single twitch,
spatial gradients of calcium concentration will certainly
exist within the sarcomere. The release of calcium takes
place very rapidly, much faster than the rise of the
aequorin response indicates, the latter being limited by the
kinetics of the calcium-aequorin reaction (limiting rate
constant 75/sec at 15° C, Hastings, Mitchell, Mattingly,
Blinks & van Leeuwen, 1969). Although distances are small for
diffusion of calcium from the site of release to calcium
binding sites, the duration of a single activation is too
short for equilibrational distribution to be attained.
The highest concentration occurs at the sites of release,
probably the terminal cisternae of the sarcoplasmic reticulum
(Winegrad, 1968). This highest concentration should dominate
the aequorin response; therefore, the peak amplitude of the
aequorin response seems primarily determined by the amount
of calcium released. The latter quantity is only one of
several factors determining the calcium concentration at
the site of troponin. Thus, one should not expect a strict
relationship between peak aequorin response and force
developed (see below).

The decline of the aequorin response is slower (rate
constant 50/sec) than would be required by the kinetics of
the calcium-aequorin reaction itself (75/sec). Analysis of
the decline can therefore give information about the fall
of sarcoplasmic calcium concentration. Quantitative aspects
of calcium sequestration, however, are not easily deduced,
because of several other factors like diffusion, binding,
and the nonlinear relationship between [Ca^{2+}] and light

intensity. If sarcomere length is held constant, diffusion and binding of calcium should be stable, and a change in the time course of the declining phase of the aequorin response might then be taken as an indicator of altered calcium uptake rates by the sarcoplasmic reticulum.

III. THE AEQUORIN SIGNAL DURING MULTIPLE ACTIVATION

The aequorin signal during a tetanus is determined by several factors which can be best understood from records at low, and then progressively higher stimulation frequency. When an aequorin-injected fiber is stimulated with a train of low-frequency stimuli, each stimulus elicits a luminescent response. During the train, these responses undergo progressive changes, the most prominent being a descending staircase which leads gradually to a lower steady level of peak light and a progressive slowing of the declining phase. This is illustrated in Fig. 4 with a train of 25 pulses of 3 Hz frequency. The length of the staircase increases with falling temperature; at 15° C it is complete within about 12 contractions. This figure is roughly independent of stimulation frequency; i.e. with decreased stimulus interval the steady state of peak value occurs after the same number of contractions, but in a shorter time. The level of the steady state peak light amplitude falls with increasing stimulation frequency until a rate is reached where summation of successive responses begins to occur (about 10 Hz at 5° C, Fig. 5). At higher frequencies summation may cause the luminescent response to reach much higher amplitudes than the peak of the rested-state single response (Fig. 5, 40 Hz, Fig. 6).

The summation process is complicated by the second progressive change in the aequorin responses during repetitive stimulation, the slowing of the decline. The latter is already detectable at low stimulation frequency, as illustrated in Fig. 4 on an expanded time base. At an intermediate frequency, when successive aequorin responses begin to fuse, the slowing of the decline becomes evident as a progressive rise in the minimum light level between stimuli (Fig. 5, 20 Hz). At high frequencies, when the aequorin signals are nearly fused and sum up to a large amplitude, the decrease in the rate constant of decline can be determined easily at the end of the compound response. Since the slowing is progressive, at a constant stimulation

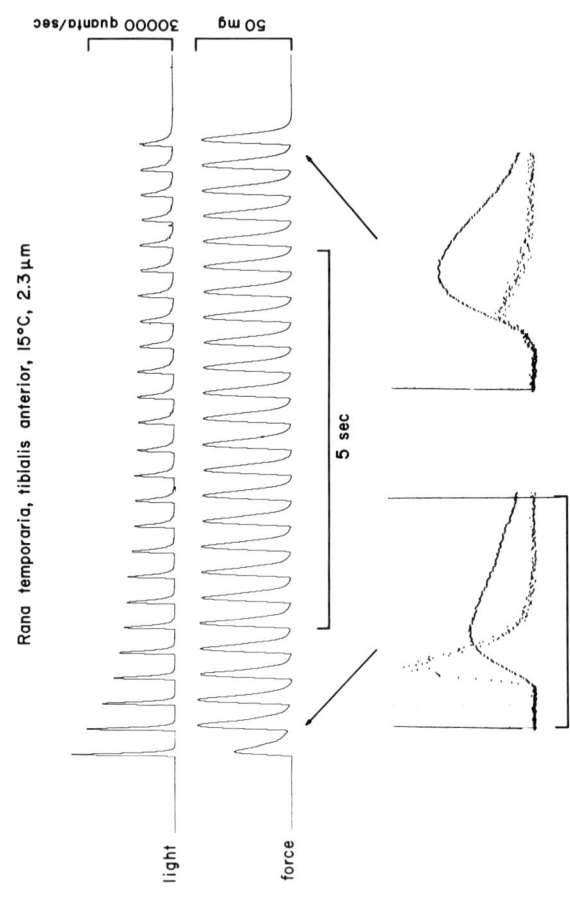

Fig. 4. Rana temporaria, tibialis anterior, 15° C, 2.3 μm. Staircase phenomena in response to repetitive stimulation. C.A.T. averaged records of ten trains of 25 twitches at 3 Hz. Upper panel shows chart record of all light and force responses. Lower panels show higher speed oscilloscope records of the first and last responses in the series.

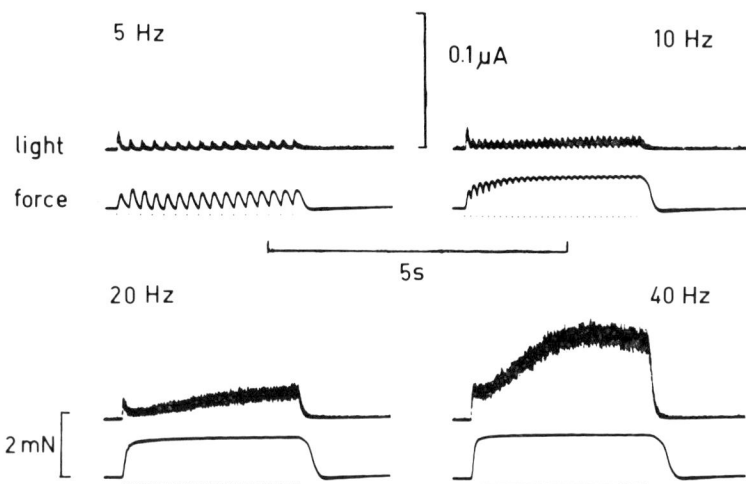

Fig. 5. Rana pipiens, semitendinosus, 2.3 μm, 5° C.
The aequorin signal in response to tetanic stimulation of
various frequencies. All panels are from the same fiber;
3 sec-tetani were separated by periods of at least 5 min, so
that the fiber was rested for each tetanus. Calibrations
are the same in all panels.

frequency it should be proportional to the duration of
stimulation. In fact, at 50 Hz we found that it takes about
0.5 sec of stimulation for the rate constant of decline to
fall from the 50/sec of a single aequorin response to 30/sec.
After 1.5 sec it has further dropped to 25/sec.

At high stimulation frequencies a third process becomes
obvious as a late fall in amplitude (Fig. 5, 40 Hz). This
"fade" is more conspicuous when a fiber is bathed in Ca-free
solution, which suggests that it might be due to loss of
intracellular Ca during prolonged stimulation.

Thus, the aequorin signal in response to tetanic
stimulation can be described as resulting from the summation
of signals that progressively decrease in amplitude and
increase in duration. Since the relative contribution of
these processes to the tetanic aequorin signal very much
depends on frequency and duration of stimulation and on
temperature, the luminescent signals accompanying tetani are
not so stereotyped as the aequorin response to a single
stimulus.

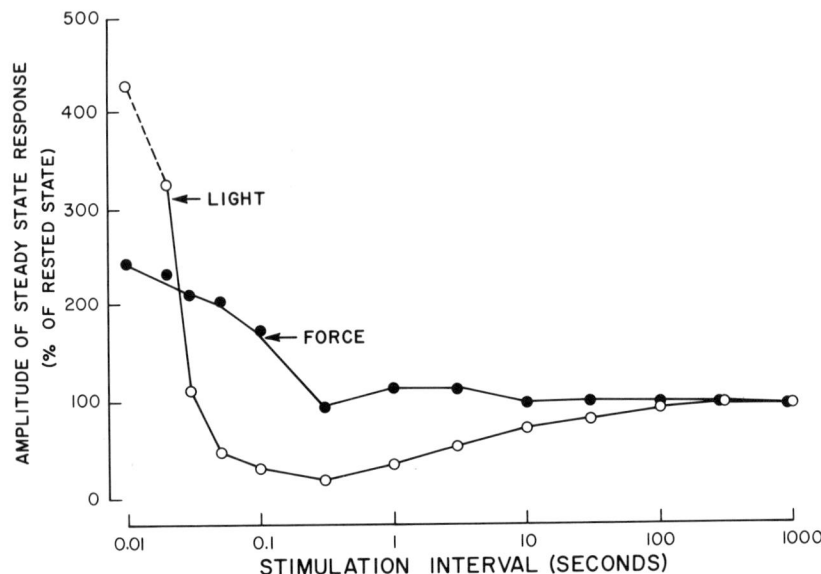

Fig. 6. Rana temporaria, semitendinosus, 15° C. Steady state peak values of aequorin (open circles) and mechanical (filled circles) responses attained during trains of stimuli as a function of the stimulus interval in the train. (At the shortest interval the aequorin response did not reach a steady state, so the peak intensity was plotted.)

The mechanical responses to low-frequency stimulation are much more variable than the luminescent responses. While the peak light falls, peak force may rise, stay constant, or fall. In Fig. 4, the descending staircase of light is accompanied by only a single-step rise in force. In other fibers it was accompanied by an ascending mechanical staircase with more than 60 steps (see Taylor *et al.* 1975, Fig. 2). On the contrary, the mechanical responses to high-frequency stimulation are much less variable than the luminescent responses. The large fluctuations in light intensity caused by the three progressive processes described above are usually not accompanied by changes of force (Fig. 5, 40 Hz). Apparently they reflect changes in intracellular calcium level occurring while the calcium binding sites of troponin are saturated. A convenient way of comparing the differences in the dependence of mechanical and luminescent responses on stimulation frequency is to stimulate a fiber with decreasing

stimulation interval and plot the steady state peak values
attained with reference to the rested-state peaks. This is
done in Fig. 6 for a fiber held at 15° C.

An interpretation of the aequorin signals during
repetitive stimulation can be given on the basis of the
earlier discussion of the aequorin response to a single
shock. Accordingly, the descending staircase indicates a
progressive decrease in the amount of calcium released in
response to successive stimuli, and the progressive slowing
of decline indicates a simultaneous slowing of the
sequestration of calcium by the sarcoplasmic reticulum.

These two processes could have a common origin in the
progressive displacement of calcium from sites of release
(terminal cisternae) to sites of uptake (longitudinal
elements) in the sarcoplasmic reticulum (Winegrad, 1968).
The slow return of sequestered calcium from the longitudinal
elements to the terminal cisternae (half-time 28 sec at 4° C,
Winegrad, 1970) could explain why an aequorin response
occurring 100 sec after a single rested-state twitch is
significantly reduced in amplitude (Fig. 6). On the other
hand, the same slow return after an activation should create
an increased level of reticular calcium at the uptake site.
This would account for a slowed sequestration of calcium,
because the pumping rate is inversely related to the amount
of reticular calcium (Makinose & Hasselbach, 1965).

In contrast to a single shock, tetanic stimulation should
activate a fiber long enough for an equilibration of the free
calcium in the sarcoplasm with calcium binding sites to occur.
In this condition there should be no considerable gradients
of calcium within a sarcomere, so that the aequorin signal
then monitors the calcium concentration at the calcium binding
site of troponin. The light intensity in a nearly fused
tetanus could therefore be taken as an indicator of the
calcium concentration giving nearly full activation. This
intensity turns out to be much lower than the peak of a
rested state response from the same fiber (Fig. 5, 10 Hz;
Fig. 6, 0.2 sec interval), a result which is understood in
light of the previously discussed situation, that in a single
twitch the aequorin response is not representative for the
calcium concentration at the site of the contractile
proteins. A comparison between amplitude of the aequorin
responses obtained in twitch and tetanus therefore seems of
little relevance. The large increase in intensity of the
aequorin signal observed when the stimulus interval is
decreased below 0.2 sec is probably due to the fact that a
given increment in the amount of released calcium will raise

the concentration of free calcium much more once the calcium binding sites on troponin have been saturated. In addition, the exponential increase of light intensity with calcium concentration (Allen *et al.* 1977) exaggerates the impression of changes in free $[Ca^{2+}]$ that actually take place.

IV. THE AEQUORIN SIGNAL DURING POST-TETANIC ACTIVATION

The two progressive changes occurring in the aequorin signal during repetitive stimulation persist for some time after termination of stimulation. Fig. 7 (left-hand panel) shows that an aequorin response to a single shock given 8 sec after a tetanus is much reduced in amplitude in comparison to a pre-tetanic test response. The decline of the post-tetanic single aequorin response is slowed, as is that of the conditioning tetanic response itself. Enhancement of contractile force in muscle can be reevaluated in terms of the data from such aequorin experiments. In Fig. 7 one can see post-tetanic potentiation of the duration of the twitch. In other experiments the post-tetanic twitch also showed increased force while the post-tetanic aequorin response was reduced and prolonged.

Although we have stressed in the beginning that the overall consumption of aequorin in a twitch is very little, there remains the possibility of a temporary local depletion of aequorin in the vicinity of the terminal cisternae. Such a reduction would considerably contribute to the descending staircase during successive contractions. However, a significant local depletion of aequorin can be ruled out by calculating the time needed to replenish presumably depleted zones around the sites of calcium release, using the diffusion coefficient determined in the experiment of Fig. 3. Even if a sphere around the terminal cisternae containing 1% of a sarcomere were totally depleted of aequorin, it would take only 7 msec for 96% repletion. The reduction of the peak aequorin signal of a nonrested-state response, however, persists for many seconds, as can best be seen in post-tetanic activation (Fig. 7, left-hand panel). If the conditioning tetanus is followed by a second tetanus instead of a twitch, the amplitude of the aequorin response in the second tetanus can even be higher than in the first. This, again, does not suggest aequorin depletion.

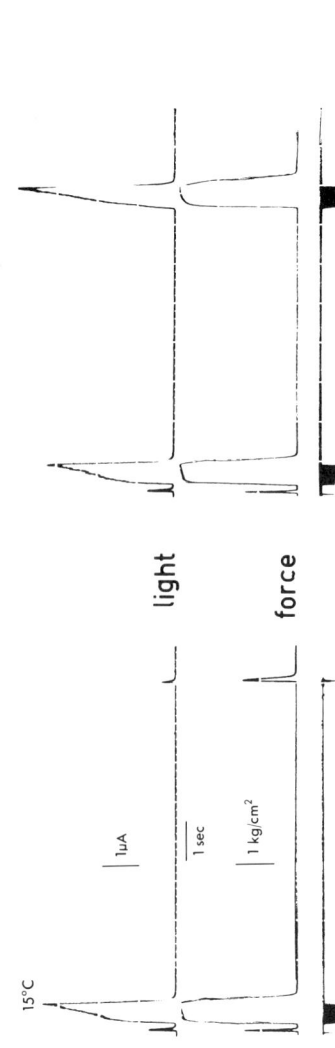

Fig. 7. *Xenopus laevis, iliofibularis, 2.5 μm. Post-tetanic aequorin signals. In both panels the first twitch is a rested-state contraction, which is followed by a 50-Hz tetanus. In the left-hand panel this is followed by a single twitch; in the right-hand panel by another tetanus. The two records were taken from the same fiber at a 5-min interval.*

These and similar observations (Fig. 5) suggest that the force developed in a twitch is more related to the length of time that the calcium concentration near the regulatory proteins remains elevated than to the actual amount of calcium liberated.

REFERENCES

1. Allen, D.G., Blinks, J.R. & Prendergast, F.G. (1977). Aequorin luminescence: Relation of light emission to calcium concentration--A calcium-independent component. *Science, 195*, 996-998.
2. Ashley, C.C., Moisescu, D.G. & Rose, R.M. (1974). Aequorin-light and tension responses from bundles of myofibrils following a sudden change in free calcium. *J. Physiol. 241*, 104-106.
3. Ashley, C.C. & Ridgway, E.B. (1970). On the relationships between membrane potential, calcium transient and tension in single barnacle muscle fibres. *J. Physiol. 209*, 105-130.
4. Blinks, J.R. & Koch-Weser, J. (1961). Analysis of the effects of changes in rate and rhythm upon myocardial contractility. *J. Pharmac. exp. Ther. 134*, 373-389.
5. Blinks, J.R., Prendergast, F.G. & Allen, D.G. (1976). Photoproteins as biological calcium indicators. *Pharmac. Rev. 28*, 1-93.
6. Hastings, J.W., Mitchell, G., Mattingly, P.H., Blinks, J.R. & van Leeuwen, M. (1969). Response of aequorin bioluminescence to rapid changes in calcium concentration. *Nature, Lond. 222*, 1047-1050.
7. Makinose, J. & Hasselbach, W. (1965). Der Einfluss von Oxalat auf den Calcium-Transport isolierter Vesikel des sarkoplasmatischen Reticulum. *Biochem. Z. 343*, 360-382.
8. Rüdel, R., Taylor, S.R. & Blinks, J.R. (1978). Calcium transients in isolated amphibian skeletal muscle fibers: detection with aequorin. *J. gen. Physiol.* submitted for publication.
9. Shimomura, O., Johnson, F.H. & Saiga, Y. (1962). Extraction, purification, and properties of aequorin, a bioluminescent protein from the luminous hydromedusan. *Aequorea, J. Cell. Comp. Physiol. 59*, 223-239.
10. Taylor, S.R., Rüdel, R. & Blinks, J.R. (1975). Calcium transients in amphibian muscle. *Fedn. Proc. 34*, 1379-1381.

11. Weber, A. & Herz, R. (1963). The binding of calcium to actomyosin systems in relation to their biological activity. *J. biol. Chem. 238*, 599-605.

12. Winegrad, S. (1968). Intracellular calcium movements of frog skeletal muscle during recovery from tetanus. *J. gen. Physiol. 51*, 65-83.

13. Winegrad, S. (1970). The intracellular site of calcium activation of contraction in frog skeletal muscle. *J. gen. Physiol. 55*, 77-88.

V

E-C Coupling in Heart Muscle

THE EFFECT OF EXTRACELLULAR POTASSIUM
ON THE EXCITATION CONTRACTION COUPLING
IN FROG HEART

Michèle Ildefonse
Michel Roche
Oger Rougier

Laboratoire de Physiologie des Eléments Excitables
Université Claude Bernard
F 69621, Villeurbanne, France

In 1972, Kavaler, Hyman & Lefkowitz showed that in
mammalian ventricular muscle a small increase in the
extracellular potassium concentration can induce an increase
or a decrease of the twitch contraction associated with an
action potential. We thought that it would be interesting
to reinvestigate this problem in more detail in measuring
the contraction in voltage clamp experiments.

We used small trabeculae of frog auricle which were put
in a double sucrose gap system as described by Rougier,
Vassort & Stämpfli (1968). The contraction was measured
together with the electrical events (action potential or
ionic currents) on the test node of the preparation by means
of a 5734 RCA transducer as originally done by Vassort &
Rougier (1972).

In the first experiments, we tried to reproduce Kavaler's
findings in frog atrial muscle. In most of the preparations
(Fig. 1A), when the extracellular potassium is increased
from 2.5 to 5 mM, we observe together with a depolarization
of nearly 15 mV a slight reduction in the amplitude and
duration of the action potential and an important decrease

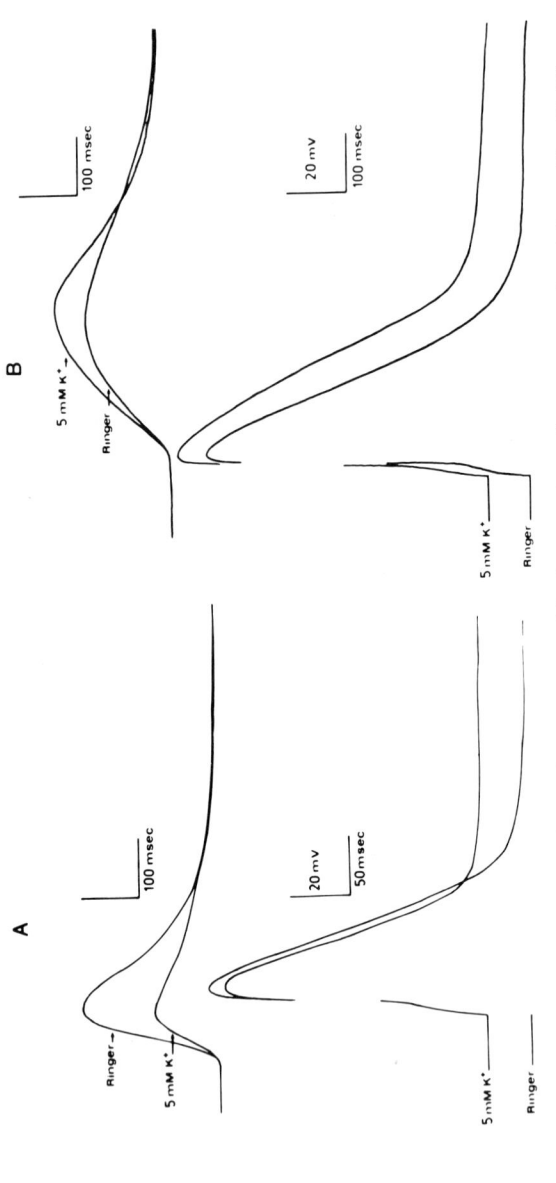

Fig. 1. Effect of increasing the extracellular potassium concentration (5 mM K) on the resting and action potentials (lower traces) and on the contraction (upper traces) for two preparations. In A, the contraction records have been shifted to the left for clarity.

in the amplitude of the contraction. In a few preparations
(Fig. 1B), the same increase in extracellular potassium
concentration gives a comparable depolarization, but the
action potential increases in amplitude and duration and
the contraction is bigger.

A straightforward explanation of the potassium alterations
in the action potential duration and amplitude can be given
by looking at the current voltage relation for the background
current. The most frequent case is illustrated in Fig. 2A,
where, after a crossing over, the outward current is greater
in high potassium solution, a situation which can explain
the shortening of the action potential (see Fig. 1A). In a
few cases (Fig. 2B), the current voltage relation in high
potassium solution crosses back the current voltage relation
in Ringer. In these conditions, the outward current,
weaker in high potassium, can induce a lengthening of the
action potential.

As shown by Vassort & Rougier (1972) and by Léoty &
Raymond (1972), the contraction of frog atrial trabeculae
is composed of two phases: a phasic contraction that can
be associated with the slow inward current, and a tonic
contraction dependent on the membrane potential and which
seems to be correlated with a sodium-calcium exchange
mechanism (Vassort, 1973). These two phases, which are
present during an action potential, can only be separated in
voltage clamp experiments. So, if at first sight it seems
possible to correlate the inotropic effects of high
potassium solution with the modifications of the action
potentials, we need voltage clamp experiments to decide
whether the tonic tension or the phasic tension are affected,
and in what manner.

When the preparation is depolarized by 85 mV, a potential
where the slow inward current and phasic tension are
maximum (Fig. 3, upper traces), we observe that an increase
in extracellular potassium induces immediately an increase
in the outward background current, a decrease in the slow
inward current and a decrease in the amplitude of
contraction. If we do the same experiment at 150 mV
depolarization, a potential where the tension is only tonic
(Fig. 3, lower traces), an increase in extracellular
potassium gives an increase in the outward background
current and an increase in the amplitude of contraction.

These observations are also illustrated in Fig. 4,
which shows the current-voltage relations for the background
current and for the slow inward current and the tension

voltage relation. It must be remarked that (1) the outward
background current is greater in 5 mM potassium for all the
potentials where the slow inward current is flowing; (2) the
slow inward current is decreased and its reversal potential
is 15 mV more negative in 5 mM potassium; (3) the phasic
tension is weaker and the tonic tension is greater in 5 mM
potassium. Thus the two components of tension are affected,
but in the opposite direction, by increasing the extracellular
potassium concentration.

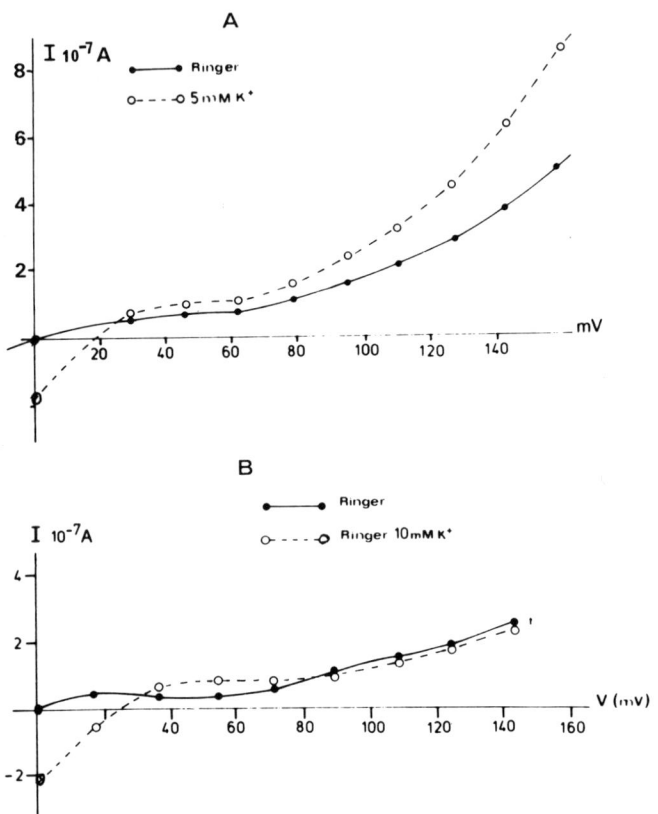

Fig. 2. *Current voltage relations of the background
current in different external potassium concentrations in
two different preparations. The current is measured at the
end of a pulse of 100 msec.*

The question which now arises is: are the modifications
in the two phases of tension the consequence of a single
mechanism, or are they due to two independent ones?

The increase in tonic tension and the modifications of
the slow inward current can be considered as good indications
that there is an increase in the intracellular concentration

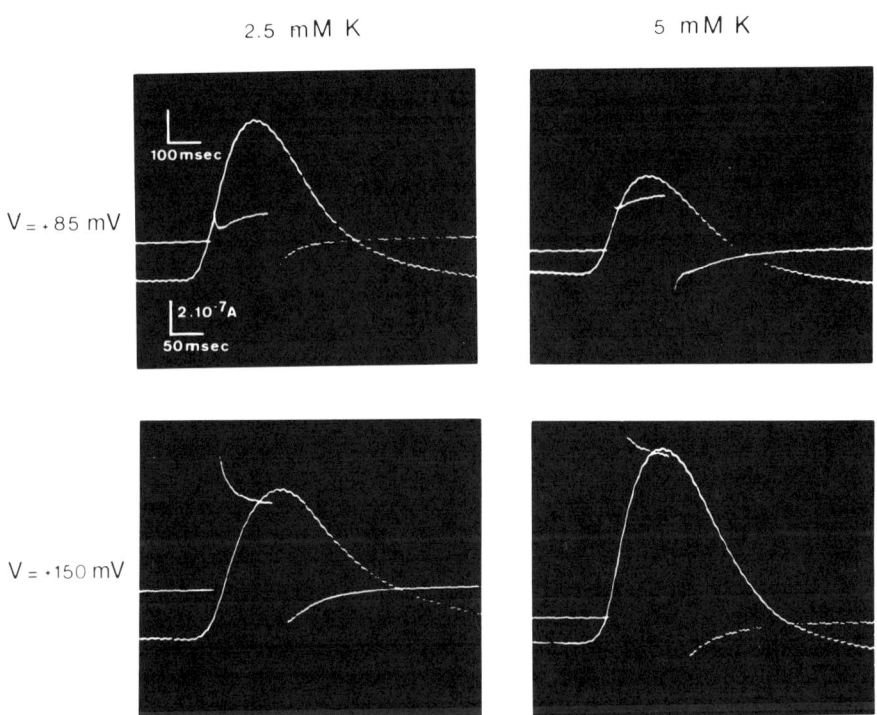

2.5 mM K   5 mM K

V = ·85 mV

100 msec

2.10⁻⁷A
50 msec

V = ·150 mV

*Fig. 3. Effect of increasing the extracellular potassium
concentration (5 mM K) on the membrane currents and on the
contraction obtained for two pulse values.   The
experiment was performed in Ringer with tetrodotoxin
(5 × 10⁻⁷ g/ml).*

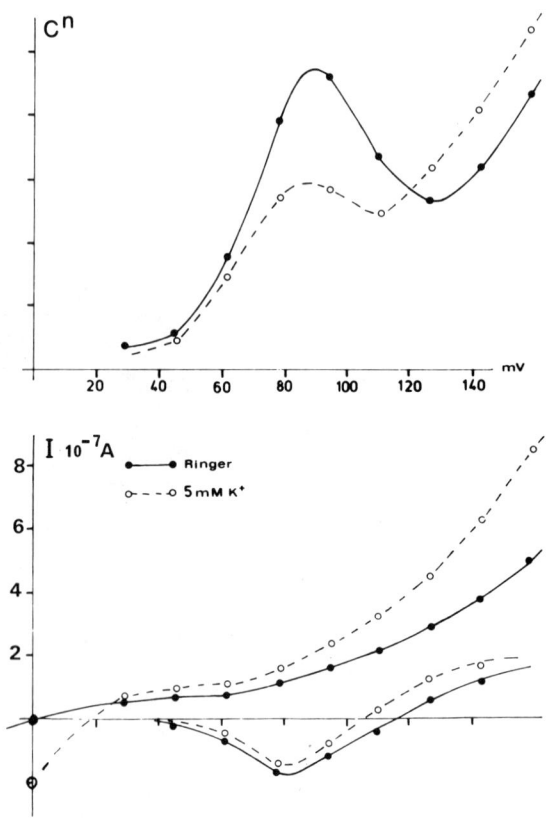

Fig. 4. *Effect of increasing the extracellular potassium concentration (5 mM K) on the current voltage (lower curves) and tension voltage (upper curves) relations. The slow inward current is measured as the difference between the maximum and the minimum of current. The background outward current is measured at the end of the pulse. Duration of depolarization 100 msec; same experiment as in Fig. 3.*

of calcium ions in high potassium solutions. These
observations are important because they discard a possible
intervention of the sodium-calcium exchange mechanism due
to a stimulation of the sodium pump--such a stimulation would
in fact decrease the intracellular calcium concentration.
A possible hypothesis would be the existence of a mechanism
of potassium-calcium exchange. Such a mechanism has been
postulated by Harris & Morad (1971) and Morad & Goldman
(1973) in heart muscle and hypothetically envisaged by
Baker (1972) in squid axon.

In order to test whether the movements of potassium ions
are related in some manner with modifications of the
intracellular concentration of calcium ions (measured as
modifications of the tonic tension), we carried out voltage
clamp contracture experiments under different experimental
conditions.

Fig. 5 shows the results obtained during a depolarization
induced contracture. The membrane is depolarized step by step
to a potential of 150 mV where it is maintained; the tension

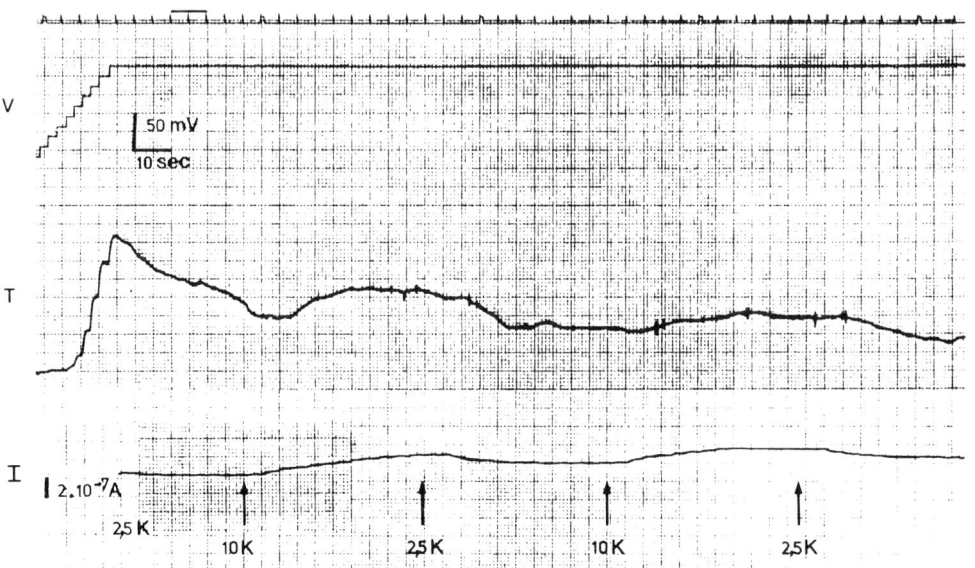

*Fig. 5. Effect of increasing the extracellular potassium
concentration (10 mM K) on contracture obtained by a large
depolarization.*

increases to a maximum value and then decreases slowly despite the maintaining of depolarization; the current reaches a steady state. If the potassium concentration is increased (from 2.5 mM to 10 mM), an outward current develops and the tension increases. This effect is reversible and can be repeated.

Fig. 6 shows an experiment with sodium free induced contractures obtained at different holding potentials.

In panel *A*, the membrane is first depolarized by 120 mV (*HP* = +120 mV); this induces a contracture which relaxes spontaneously to a steady level (this is not shown in the figure but is similar to the one described in Fig. 5). Then we turn to a sodium free solution and we obtain a contracture which relaxes slowly. If, during this phase of relaxation, the potassium concentration is increased (from 2.5 mM to 40 mM), an outward current develops and the tension increases simultaneously, without any time lag. This effect is reversible when returning to a low potassium concentration and can be repeated. When we finally return to Ringer solution, the tension relaxes completely and very rapidly.

In panel *B*, the same protocol has been followed, but starting from a holding potential of +70 mV. This potential was chosen because it nearly corresponds to the value for which the current voltage relations in 2.5 and 40 mM potassium cross each other. In these conditions, the increase in extracellular potassium concentration changes neither the current nor the tension. Returning to Ringer solution induces a rapid and complete relaxation like in the preceding experiment.

In panel *C*, the membrane potential was held at +50 mV. When the 40 mM potassium solution is turned on, we observe an inward current and a small decrease in tension. This effect is reversible when returning to the 2.5 mM potassium solution. The preparation relaxes completely and rapidly when we return to Ringer solution.

The preceding experiments have shown that the movement of potassium ions is linked with variations in tension; they develop synchronically and with the same time course. If we consider, as previously mentioned, that the level of tension is the traduction of the internal concentration of calcium ions, we can imagine that an outward movement of potassium is associated with an inward movement of calcium and that an inward movement of potassium is associated with an outward movement of calcium.

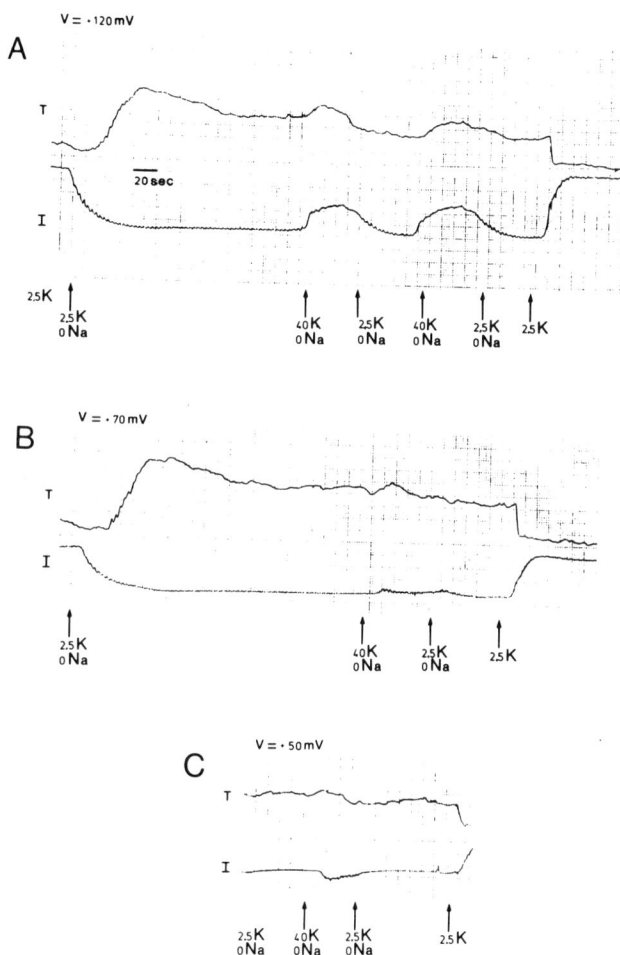

Fig. 6. Effect of increasing the extracellular
potassium concentration (40 mM K) on contracture obtained
in a "sodium free" solution (sucrose substitute) for three
values of imposed potential. The inward current observed
during the passage to the "sodium free" solution is due to
an outward movement of chloride ions.

DISCUSSION

McNaughton: I would like to make the point that your
results may be compatible with that we already know about
sodium potassium exchange. The conventional explanation of
the negative inotropic effect of potassium is that potassium
stimulates the sodium pump, and thus increases the sodium
gradient, which therefore stimulates the sodium-calcium
exchange that is supposed to run off the sodium gradient.
That explanation could accommodate your results as well,
if it is supposed that an induced outward current of
potassium leads to an accumulation of potassium in the
extracellular space outside the membrane and a consequent
stimulation of the potassium-sodium exchange pump, and the
sequence of events that I previously noted. The decrease in
tension that you observed when the potassium current is
inward would have a likewise inverse explanation. Do you
have any experiments?

Rougier: No, but if we had an effect of the sodium
pump, we would attempt to have some kind of time lag in
order to obtain something. What is sure is that when we
change from 2.5 to 10 mM potassium solution, at 150 mV or
more, we immediately observe an increase in outward current
and an increase in contraction. A stimulation of Na pump
would give exactly the opposite, i.e., a decrease in
contraction.

Reuter: How did you determine the reversal potential
of the slow inward current?

Rougier: That is a very important question. It depends
whether we are working on frog atrial or frog ventricular
muscle, because the delayed rectification is much greater
in atrium. But we always work with short pulses, no longer
than 100 msec, and too short to activate any delayed outward
current even for very high depolarizations. We decided to
measure the slow inward current as the difference between
the end of the current and its maximum value.

REFERENCES

1. Baker, P.F. (1972). Transport and metabolism of
    calcium ions in nerve. *Prog. Biophys. molec. Biol.*
    24, 177-223.

2. Harris, J.E. & Morad, M. (1971). The effect of Ca-K interaction on the plateau of the cardiac action potential. *The Physiologist 14*, 159.

3. Kavaler, F., Hyman, P.M. & Lefkowitz, R.B. (1972). Positive and negative inotropic effects of elevated extracellular potassium level on mammalian ventricular muscle. *J. gen. Physiol. 60*, 351-365.

4. Léoty, C. & Raymond, G. (1972). Mechanical activity and ionic currents in frog atrial trabeculae. *Pflügers Arch. ges. Physiol. 334*, 114-128.

5. Morad, M. & Goldman, Y. (1973). Excitation-contraction coupling in heart muscle: Membrane control of development of tension. *Progr. Biophys. molec. Biol. 27*, 259-308.

6. Rougier, O., Vassort, G. & Stämpfli, R. (1968). Voltage clamp experiments on frog atrial heart muscle fibres with the sucrose gap technique. *Pflügers Arch. ges. Physiol., 301*, 91-108.

7. Vassort, G. (1973). Influence of sodium ions on the regulation of frog myocardial contractility. *Pflügers Arch. ges. Physiol. 339*, 225-240.

8. Vassort, G. & Rougier, O. (1972). Membrane potential and slow inward current dependence of frog cardiac mechanical activity. *Pflügers Arch. ges. Physiol. 331*, 191-203.

A Ca-TRANSPORT SYSTEM FOR ACTIVATION OF TENSION
IN FROG VENTRICULAR MUSCLE

M. Morad
T. Klitzner

Department of Physiology
University of Pennsylvania
Philadelphia, Pennsylvania

The source of activator Ca^{2+} and the mechanism by which
calcium is made available to the myofilaments of the cardiac
cell has been the subject of considerable interest and
investigation since Ringer (1883) first showed that in the
absence of external Ca^{2+} frog heart failed to contract. In
the past 20 years new technical developments have made it
possible to identify subcellular structures such as the
sarcoplasmic reticulum which are capable of storage and
release of Ca^{2+} (Fawcett & McNutt, 1969; Hasselbach, 1964;
Endo, 1977). However, in cardiac cells, which are much
smaller in diameter (5-20 μm) and develop tension much more
slowly than the skeletal muscle fibers, the possibility
remains viable that activator calcium is transported across
the cell membrane during the action potential. Such a
possibility was highly strengthened when Weidmann (1959)
showed that rapid injection of a bolus of Ca^{2+} during the
plateau of cardiac action potential was capable of altering
the contraction of that same beat. A number of
investigators have recently proposed fairly detailed models
in which both extracellular and intracellular calcium
contribute to the development of contraction in mammalian
myocardium (Morad & Goldman, 1973; Bassingthwaighte & Reuter,
1972).

    In this report we will deal primarily with the mechanism
by which extracellular Ca^{2+} contributes to the development

285

and maintenance of force in cardiac muscle. In order to minimize the contribution of internal Ca^{2+} stores to development of tension, we have chosen to study the frog ventricular muscle which is known to have no t-tubular systems and much less sarcoplasmic reticulum than mammalian myocardium or fast skeletal muscle fibers (Page & Niedergerke, 1972; Peachey, 1965; Sommers & Johnson, 1969). We shall first review the evidence that development of tension in frog ventricle, unlike mammalian ventricular or frog skeletal muscle, is under direct control of membrane potential, and further, that the sarcoplasmic reticulum and mitochondria serve primarily as Ca^{2+} uptake systems and do not recirculate the sequestered Ca^{2+} in significant amounts under physiological conditions. In the second part of this report we shall present evidence that a K^+-linked calcium transport system serves as the primary mechanism by which Ca^{2+} moves across the cell membrane.

MEMBRANE CONTROL OF DEVELOPMENT OF TENSION

Using a single sucrose gap voltage clamp technique (Morad & Orkand, 1971; Goldman & Morad, 1977) it was possible to determine the relation between membrane potential and development of tension in frog ventricular strips (diameter 300-500 μm). Fig. 1 illustrates the relation between the duration of an action potential or a voltage clamp step and the time course of development of tension. The top left panel shows that premature termination of the action potential abruptly terminates contraction. The relation between duration of depolarization and the maintenance of tension is more rigorously investigated using voltage clamp steps of various durations. The results for a series of such clamp steps to +30 mV are plotted in the lower panel. Note that the relation between the duration of depolarization and maintenance of tension is linear. The initial 80-100 msec of depolarization fail to generate any measureable force. This observation is quite different from that of the mammalian heart where over 50-70% of tension is developed in the initial 100 msec of depolarization (Morad & Trautwein, 1968). These results suggest that development of tension in frog ventricle is under direct and continuous control of the membrane potential.

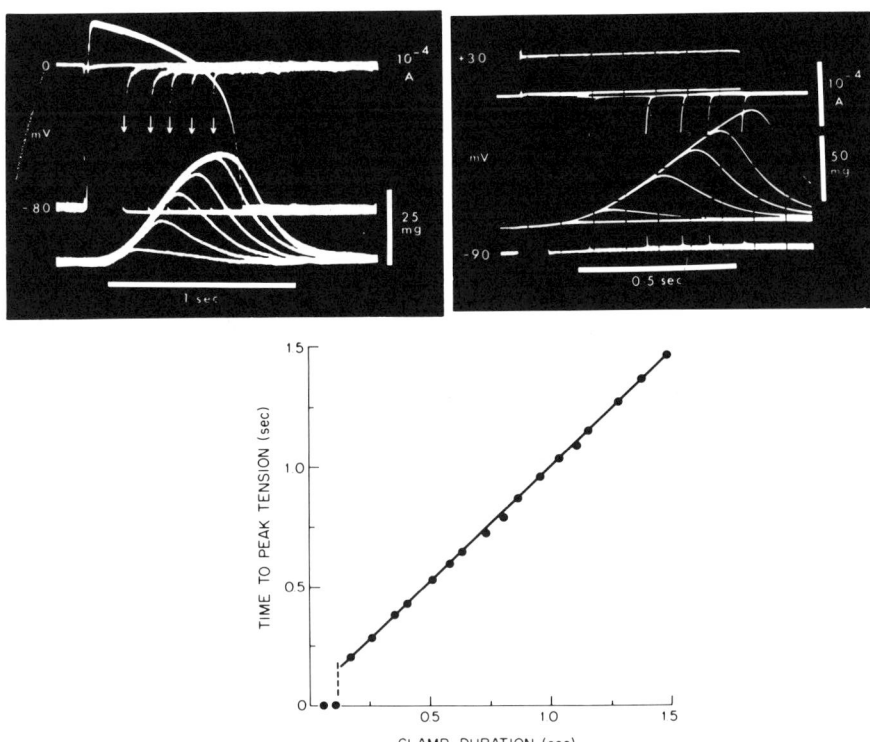

Fig. 1. Contractile response elicited by depolarization of varying duration. Top left, superimposed traces of shortened action potentials and accompanying contractions. Top right, superimposed traces of clamp steps of various durations. Bottom, correlation of duration of depolarization with time to peak of tension for voltage clamp step to +30 mV.

Fig. 2 shows the effect of clamp steps to different membrane potentials on the development and maintenance tension. It can be clearly seen that tension always develops monotonically upon depolarization and is maintained for the duration of the clamp pulse. It is perhaps of interest to note that such clamp steps are capable of generating contractions several fold larger than those accompanying the action potential. Termination of the clamp

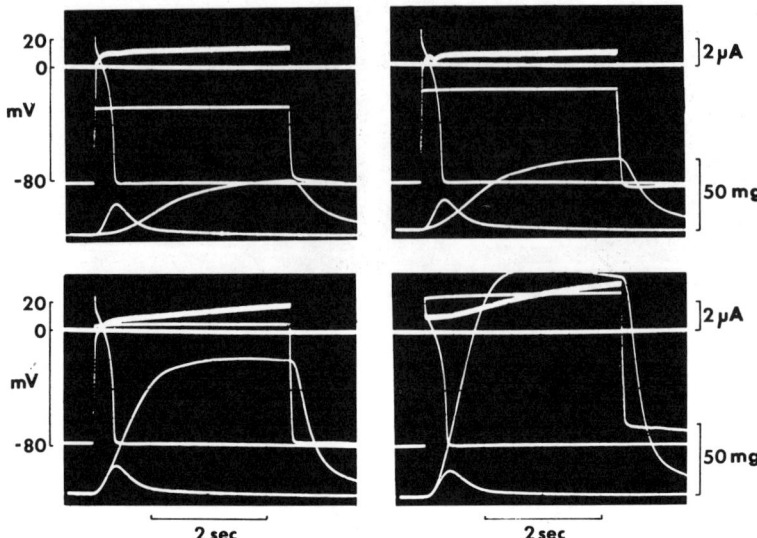

Fig. 2. Each panel shows an action potential and accompanying twitch superimposed with a voltage clamp step and accompanying tension and current. Counterclockwise from upper left clamp potentials are -30 mV, -20 mV, +5 mV, +30 mV. Note the variation in the post clamp afterpotential which reflects the extracellular K^+ accumulation during each clamp step.

step always institutes relaxation. The qualitative relation between the duration of depolarization and maintenance of tension is maintained even when considerably longer clamp steps are used.

ARE THERE PHYSIOLOGICALLY SIGNIFICANT TRIGGERABLE INTERNAL CALCIUM STORES IN FROG VENTRICULAR MUSCLE?

The presence of functionally triggerable internal calcium stores which contribute to the development of tension has been considered by numerous investigators in both mammalian and frog ventricular muscle (Morad & Goldman, 1973; Fabiato & Fabiato, 1972; Endo, 1972; Constantin, 1977). Evidence from morphological, physiological, and pharmacological investigations suggest that an internal calcium store which may be released during contraction of frog ventricle is either absent or physiologically insignificant when compared to the internal stores of mammalian heart (Antoni, Jacob & Kaufmann, 1969; Morad & Goldman, 1973; Morad & Orkand, 1971; Fabiato & Fabiato, 1975; Bassingthwaighte & Reuter, 1972; Endo, 1977; Page & Niedergerke, 1972; Sommer & Johnson, 1969; Winegrad, 1971).

In order to investigate further the role of internal stores in frog ventricle we carried out the following series of experiments. The rationale for these experiments is based on the assumption that if there is a significant and active releasable internal calcium store in the frog heart, then it should be possible to alter its content physiologically. The alterations in the contents of such Ca^{2+} stores would then be expected to influence the strength of the subsequent contractions. Such procedures have been previously used to identify the capacity and the kinetics of the internal calcium stores in mammalian hearts (Antoni *et al.* 1969; Wood, Heppner & Weidmann, 1969; Morad & Goldman, 1973).

Fig. 3 shows that potentiation of tension caused by interposing a large depolarizing pulse is not reflected in the subsequent beats in frog ventricular strips. In contrast,

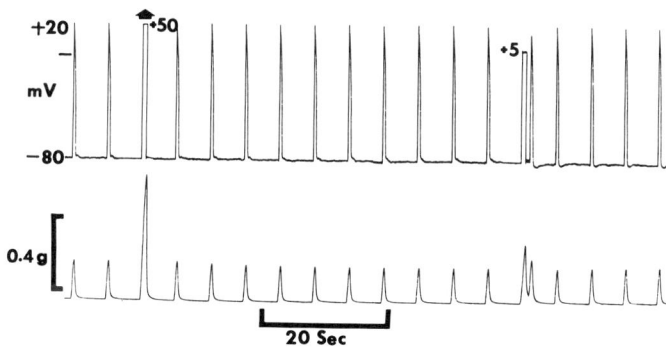

Fig. 3. A large voltage clamp step is interposed in a series of normal twitches. Note that although the voltage clamp step produces tension many fold greater than the normal twitch, tension accompanying the action potential immediately following the clamp step is unaffected. Later clamp step to +5 indicates that post clamp potentiation may be seen only if the post clamp interval allows insufficient time for complete relaxation.

the same experiment in cat ventricular muscle produces a large
potentiation of subsequent beats. This altered contractile
state decays exponentially in a beat dependent manner
(Morad & Goldman, 1973) consistent with the view that cat
ventricular cells possess significant releasable internal
stores. Although the results of Fig. 3 strongly argue
against the presence of significant releasable internal
calcium stores in frog ventricle, it still may be argued
that these stores exist but have such a large capacity that
the effect of calcium sequestered after a single depolarizing
clamp step may be insignificant. In the experiment
illustrated in Fig. 4, an attempt is made to overload any
internal stores with a long clamp step so that the
preparation develops maintained tension. Under these
conditions, we tested for the effect of internal stores on
the development of tension. Fig. 4 shows that a long
depolarizing clamp step to around -30 mV produces tension
which is maintained for 40-50 sec. Short test clamp pulses
to +30 mV superimposed on such a conditioning clamp pulse
generate large twitch-like tensions. Comparison of the
amplitude of the first and the fifth such test pulses
readily indicates that no beat dependent potentiation in
developed tension has occurred. These results indicate
that even in the presence of significantly increased
myoplasmic concentrations of Ca^{2+} the internal stores fail
to participate in the recirculation of the activator calcium.

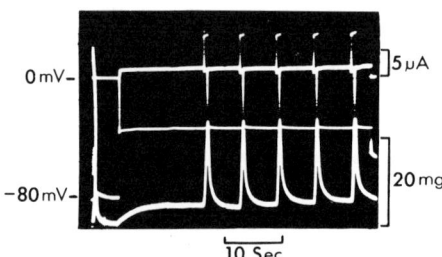

Fig. 4. Membrane is clamped to -30 mV for 10 sec and
steady tension develops. A series of 1 sec clamps to +30
is superimposed on the holding potential of -30 mV. Each
short clamp elicits the same tension.

THE SLOW INWARD CURRENT AS A POSSIBLE SOURCE OF THE ACTIVATOR
CALCIUM

The observations that there is no significant tension
history in frog heart and that tension responds rapidly to
variations in $[Ca]_o$ (Weidmann, 1958; Kavaler & Anderson,
1975) suggest strongly that the primary source of activator
calcium is the extracellular pool of calcium. Included in
this pool are extratrabecular Ca^{2+}, paracellular Ca^{2+}, and
Ca^{2+} bound to the membrane. Since the source of activator
calcium is primarily extracellular, it is reasonable to ask
how calcium is transported across the cell membrane.

Voltage clamp experiments once again provide some
insight into the mechanism of calcium transport. Although
the slow inward current (the "calcium current") in heart
muscle has been the subject of considerable criticism, it
must still be considered as a possible mechanism for the
transport of activator Ca^{2+}.

In the frog heart there are a few problems with the
calcium current as a candidate for providing the activator
calcium.

1. The first problem is that the slow inward current
when observed in a TTX-treated preparation seems to be almost
completely inactivated within 100-150 msec, a period during
which little or no tension develops in frog ventricular
muscle (see Fig. 1). Fig. 5 illustrates the results from

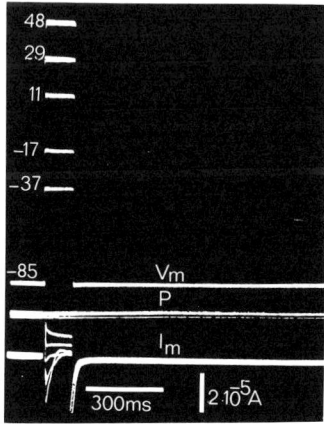

*Fig. 5. Clamp steps of 100 msec duration to various
potentials. Note that no tension develops independent of the
size or direction of the accompanying membrane current.*

six superimposed clamps of 100 msec in duration. Note that little or no tension is generated by these depolarizations whether the initial membrane current is inward or outward.

    2. The second problem with the calcium current hypothesis is that if the activator Ca^{2+} is primarily provided by the secondary inward current mechanism, then it should be possible to suppress tension as the clamped potential approaches the calcium equilibrium potential (E_{Ca}). E_{Ca} may be roughly calculated by assuming that the cytoplasmic calcium concentration in the frog heart is about 5×10^{-8} or 10^{-7} M at the onset of contraction (Winegrad, 1971). In the experiment illustrated in Fig. 6 the external calcium

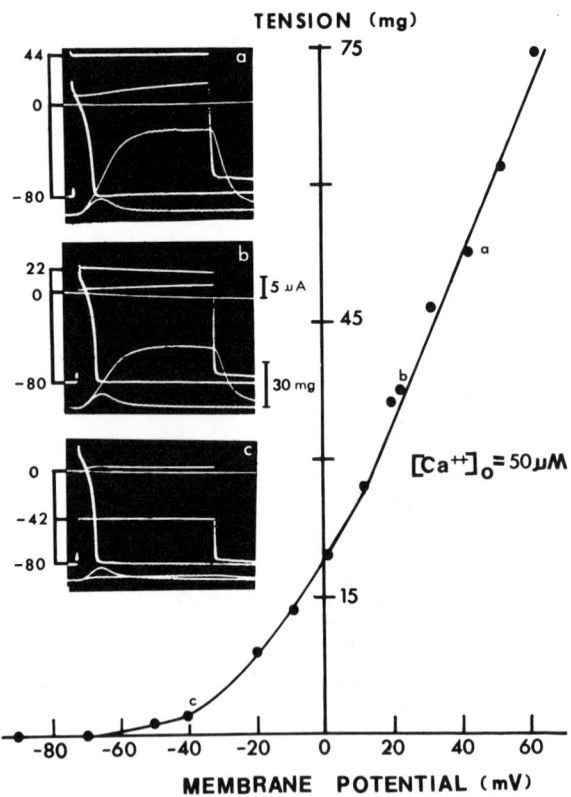

Fig. 6. *Tension voltage relation in 50 µm Ca^{2+}. E_{Ca} is at most 29 log 50 $\times$ 10^{-6}/5 $\times$ 10^{-8} = 87 mV. The calcium current hypothesis predicts that tension should begin to decrease well below E_{Ca}. Experiment done with chopped current clamping technique. Clamp duration = 3 sec.*

concentration has been reduced to 50 μM so that E_{Ca} is at most +70 to 87 mV. Note that developed tension is not reduced as the clamped potential approaches the assumed E_{Ca}. The results of this experiment are consistent with the findings of Morad & Orkand (1971). In the experiment of Fig. 6 the extracellular resistance is functionally eliminated by the use of a chopped current pulse voltage clamp procedure (Goldman & Morad, 1977). In some experiments the membrane potential was clamped to very positive levels (+100 to +150) and although the developed tension plateaued, no decrease in developed tension could be recorded. Results of this type provide strong evidence against the translocation of Ca^{2+} across the membrane through a slow inward current channel.

3. A third type of evidence against the slow inward current system as a mechanism by which activator calcium is made available to the myofilaments comes from the use of calcium antagonists such as Mn^{2+}. Mn^{2+} has been shown to block the slow inward current in cardiac muscle (Vassort & Rougier, 1972; Ochi, 1970). In frog ventricular muscle addition of 0.2 mM Mn^{2+} to a strip bathed in Ringers (1.0 mM Ca^{2+}) produces a suppression of voltage-tension relation and blocks the secondary inward current. However, the voltage-tension in the presence of 1 mM Mn^{2+} plus 1 mM Ca^{2+} is equivalent to the voltage-tension relation in 0.2 mM Ca^{2+} in the absence of Mn^{2+} despite the fact that the slow inward current is suppressed in the presence of Mn^{2+}. Such studies suggest that the concentration of Mn^{2+} necessary to block the secondary inward current is smaller than the concentration of this agent needed to suppress the developed tension.

The three different types of evidence summarized above suggest that the secondary inward current cannot serve as a major Ca^{2+} transport system for activation of tension in frog ventricular muscle.

Ni^{2+} AS A SPECIFIC CALCIUM ANTAGONIST

Divalent cations such as Mn^{2+}, Mg^{2+}, and Co^{2+} uniformally suppress the voltage-tension relation, block the slow inward current, and shorten the action potential. Ni^{2+} behaves like other divalent cations except that it prolongs the action potential in a manner similar to the removal of calcium from the bathing solution (Harris & Morad, 1976). Since Ni^{2+} does not substitute for calcium on the troponine-tropomyosin site (Fischman & Swan, 1967), we used Ni^{2+} as an ionic probe to investigate the calcium transport mechanism in the frog heart.

Fig. 7 shows simultaneous recording of action potential and contraction from a frog ventricular muscle as it is exposed to a solution containing $NiCl_2$ (0.5 mM). Ni^{2+} suppresses the plateau potential and increases the action potential duration. Simultaneous with these events the overshoot potential is increased by 20 to 25 mV.

Comparison of the effect of the extracellular sodium concentration on the overshoot potential in the presence and absence of Ni^{2+} suggests that the sodium "selectivity" of the membrane is markedly increased in the presence of Ni^{2+}. In Fig. 8 a comparison of the overshoot potential to the log of $[Na]_o$ shows a slope of 18 mV/decade in ventricular

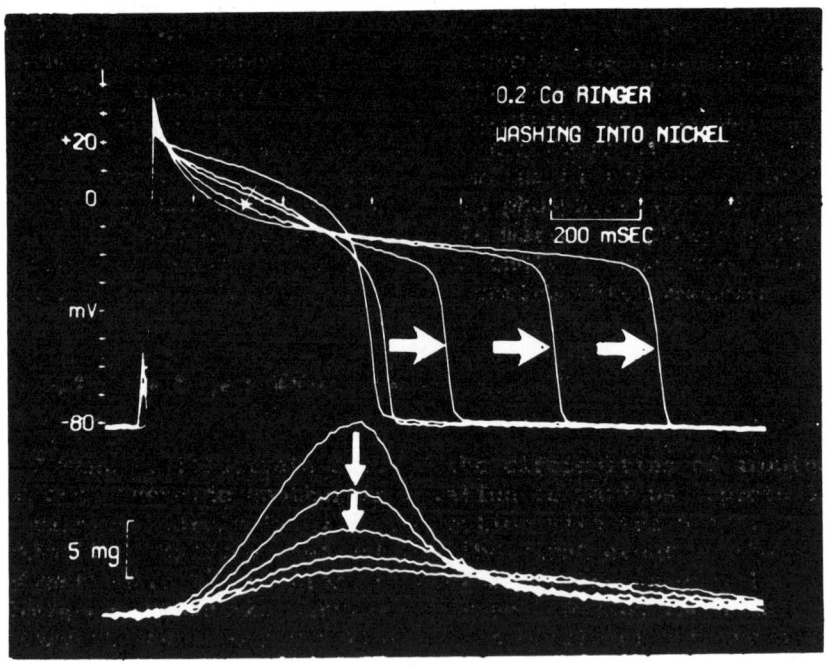

Fig. 7. Response of ventricular muscle upon exposure to 1.0 mM Ni^{2+}. Note prolongation of action potential, decrease in plateau potential and suppression of tension. Traces have been superimposed by an on-line computer data acquisition system.

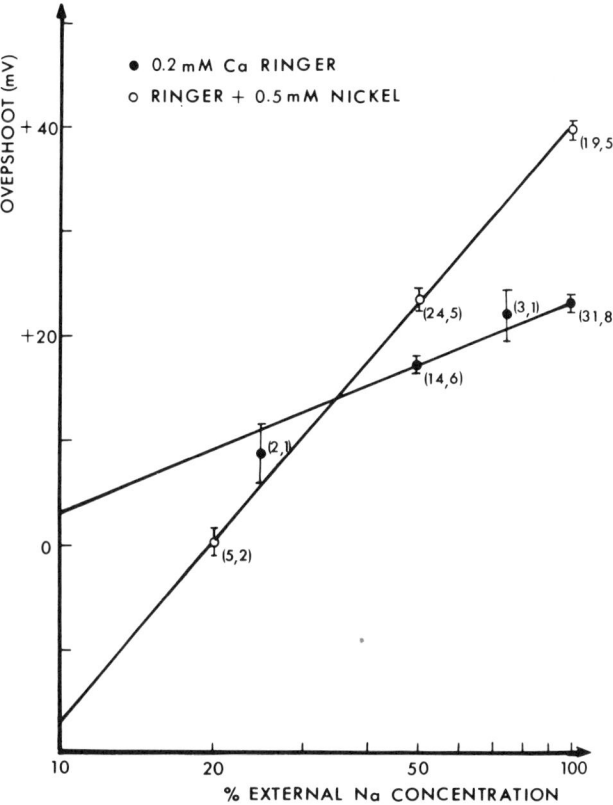

Fig. 8. Maximum overshoot of frog ventricular muscle plotted vs. log [Na]$_o$. Slope changes from 20 mV/decade in control situation (closed circles) to 58 mV/decade in Ni^{2+} (open circles). Numbers in parentheses indicate the number of punctures and different preparations averaged to give each point.

preparations bathed in normal Ringer (see also Niedergerke & Orkand, 1966). Addition of Ni^{2+} increases the Na-dependence of the overshoot potential to about 58 mV/decade. This increased Na-"selectivity" of the membrane is confirmed when the effect of TTX on the cardiac action potential is studied in the presence and absence of Ni^{2+}. In control preparation, the effect of TTX (10^{-7} to 5×10^{-6} M) is confined to blocking the upstroke of the action potential,

leaving the duration of the action potential and the level
of the plateau potential unaffected when the physiological
node is cathodally depolarized in the sucrose gap chamber.
In the presence of Ni^{2+}, however, both the upstroke and the
plateau of the action potential are blocked by TTX (10^{-7} to
10^{-6} M), and cathodal depolarizations fail to produce any
regenerative activity in the ventricular strip. Ni^{2+} also
blocks the secondary inward current and suppresses the
voltage tension relation. Fig. 9 shows the effect of 1 mM
Ni^{2+} on the voltage tension relation of a ventricular strip
bathed in normal Ringer. Developed tension is suppressed
uniformly at all potentials investigated. The suppressive
effect of Ni^{2+} on tension can be easily reversed either by
removal of Ni^{2+} or addition of higher concentrations of Ca^{2+}
to the bathing solution. The effect of Ni^{2+} on the
tension-voltage relation is quite similar to the effect of
other Ca^{2+} antagonists such as Mn^{2+} and Co^{2+}.

   Ni^{2+} also exposes a tension-suppressing effect of TTX.
In the presence of Ni^{2+}, TTX, which normally has no effect
on developed tension, strongly suppressed the voltage

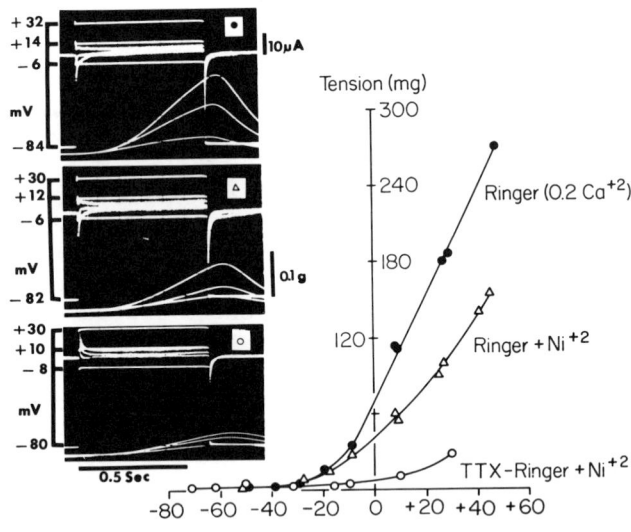

Fig. 9. Effect of Ni^{2+} (1.0 mM) and TTX (10^{-6} M) on the
tension voltage relation of frog ventricular muscle.

tension relation (Fig. 9). In this state the developed
tension is much less sensitive to elevation of $[Ca]_o$.

The tension-suppressing effect of TTX in the presence of
Ni^{2+} is accompanied by a marked decrease in the
time-dependent outward membrane current. Fig. 10 shows
that the steady I-V relation (current measured at the end
of a 600 msec clamp pulse) is N-shaped in a TTX treated
ventricular strip. Addition of Ni^{2+} is only observed in the
presence of TTX. In normal Ringer, TTX or Ni^{2+} alone have
little or no effect on the steady state current-voltage
relation.

In order to identify the ionic current component which is
responsible for the decrease in outward membrane current
in the presence of Ni^{2+} and TTX, we measured the post clamp
afterpotential in order to quantify the contribution of K^+ flux
to the net membrane current. The accuracy of such a technique
has been evaluated and checked against the response of a

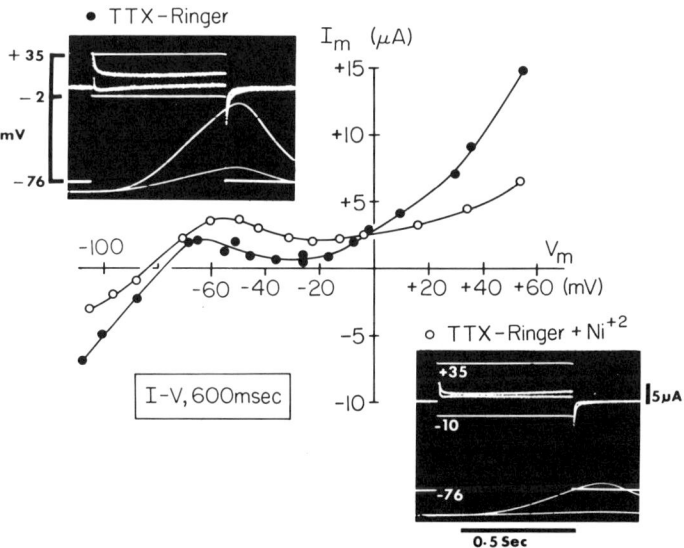

*Fig. 10. Comparison of the steady state (600 msec)
current-voltage relation in a TTX treated preparation before
and after the addition of 1.0 mM Ni^{2+}. Note that addition
of Ni^{2+} strongly suppresses the outward currents in the
positive range of membrane potentials.*

K⁺-selective microelectrode placed in the extracellular
space of the muscle (Cleemann & Morad, 1976; Cleemann & Morad,
this volume). These results suggest that the post clamp
afterpotential represents fairly accurately the magnitude
of the net K^+ efflux. Fig. 11 compares the membrane current,
afterpotential (ΔV), and the developed tension during a
depolarizing clamp pulse in a frog ventricular strip bathed
in normal Ringer, in Ringer plus Ni^{2+} and in Ringer plus
Ni^{2+} and TTX. It is clear from experiments such as the one
illustrated here that the decrease in time dependent outward
current which occurs upon addition of TTX to the Ni^{2+}-treated
preparation is accompanied by a decrease in the magnitude of
the post clamp afterpotential. The decrease in the
afterpotential in the presence of TTX and Ni^{2+} was seen at
all clamped potentials examined. Comparison of the change
in the outward membrane current (ΔI_m) or the change in the
extracellular K^+ concentration (ΔK_{acc}) with the change in
developed tension (ΔT) yields a linear relation (Fig. 12).
K_{acc} has been estimated from the afterpotential using the
Nernst equation as described by Cleemann & Morad (1978).
The results from such experiments suggest a direct relation

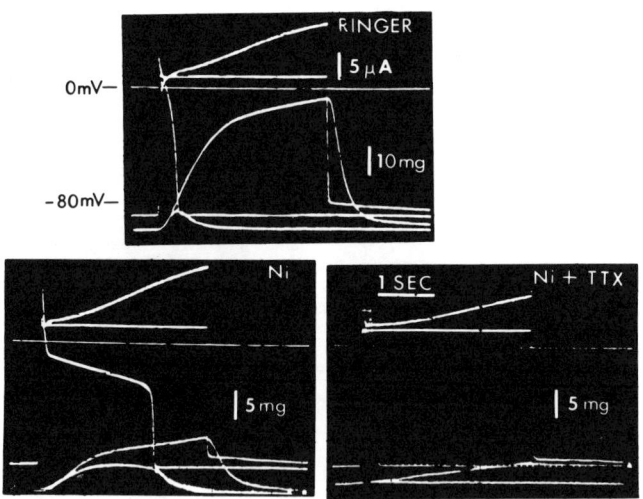

Fig. 11. Comparison of 3 sec voltage clamps to +10 mV
in Ringer (0.2 mM Ca^{2+}), Ringer + Ni^{2+} (1.0 mM), and
Ringer + Ni^{2+} + TTX (10^{-6} M). Note the suppression of
tension, outward current and afterpotential upon addition
of TTX and Ni^{2+}.

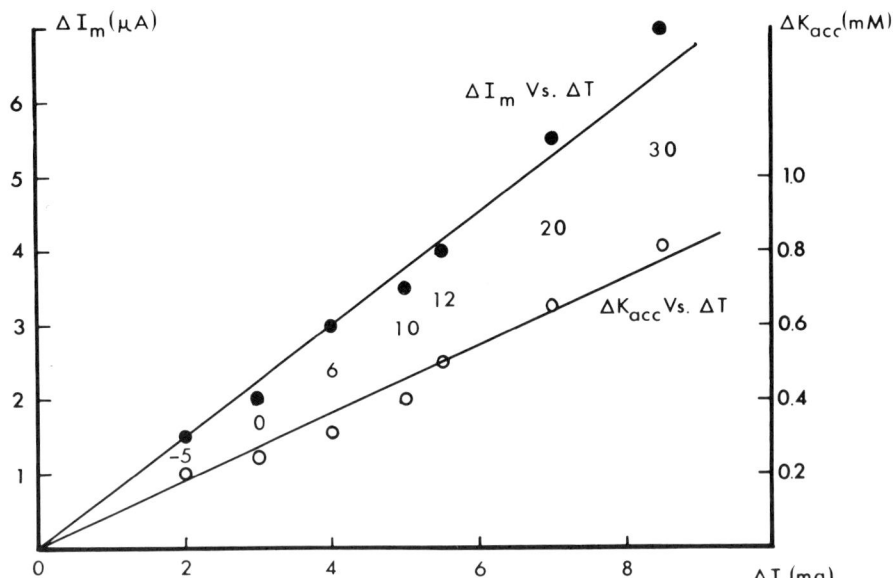

Fig. 12. Decrease in final outward current (ΔI_m) and decrease in K^+ accumulation (ΔK_{acc}) plotted vs. decreases in tension (ΔT), brought about by addition of TTX to the Ni^{2+} treated preparation. Note that the decrease in tension brought about by TTX in the presence of Ni^{2+} is linearly related to the decrease in both I_m and K_{acc}. Numbers represent membrane potential at which each measurement was made.

between K^+ efflux and the development of tension. Making a few assumptions as to the size of K^+ accumulation space and the relation between developed tension and the myoplasmic Ca^{2+} we can estimate a transport ratio of K^+ to Ca^{2+} of 4-6 to 1.

A preliminary schematic of a model which is consistent with results presented above is shown in Fig. 13. The schematic membrane contains a Na^+ channel which can be blocked by TTX and an inwardly rectifying K^+ channel, the characteristics of which have already been presented by Dr. Cleemann (in this volume). An indifferent channel which can carry Na^+, Ca^{2+}, and K^+ is in some cases in contact with a counter transport carrier. This channel-carrier system has a K^+ selective site on the myoplasmic side and a Ca^{2+}/Na^+ site on the extracellular side. The effect of Ni^{2+}

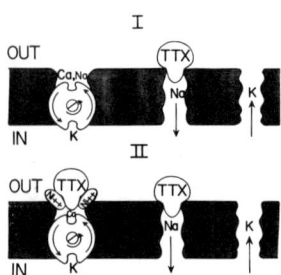

Fig. 13. A schematic representation of the proposed
K^+-linked Ca^{2+} transport mechanism. In the absence of Ni^{2+}
(top panel) TTX blocks only the Na^+ channel. In the
presence of Ni^{2+}, TTX also blocks access to a Ca^{2+}-K^+
transport carrier (bottom panel).

in this model is brought about by altering the selectivity
filter of the indifferent channel while at the same time Ni^{2+}
competes for the Ca^{2+}-selective site on the extracellular
side. The indifferent channel under normal conditions
cannot be blocked by TTX, but in the presence of Ni^{2+}, the
mouth of the channel may be sufficiently altered as to allow
TTX binding. Such a mechanism provides a possible
explanation for the observation that while TTX has no effect
on developed tension or outward K^+ current in normal Ringer
solution, in the presence of Ni^{2+}, it suppresses both the
outward K^+ current and development of tension.

This model is essentially similar to that proposed by
Morad & Orkand (1971) to provide for the uphill movement
of Ca^{2+}. The model is consistent with the observations
that increasing the $[Ca]_o$ shortens the action potential
and increases the rate of development of tension. Removal
of Ca^{2+}, on the other hand, slows down the turnover of the
K^+-Ca^{2+} transport system and provides for suppression of
tension and prolongation of the action potential. The
model also explains the negative inotropic effect of
increasing the $[K]_o$.

SUMMARY AND CONCLUSIONS

The experiments described above suggest that in frog ventricular muscle tension develops under direct control of the membrane potential. Extracellular calcium seems to be the primary source of the activator calcium. Internal releasable stores of calcium are either absent or contribute insignificantly to the development of tension. The secondary inward current does not serve as the primary calcium transport systems in frog ventricular muscle. Experiments using Ni^{2+} as a calcium antagonist suggest that a K^{+}-linked calcium transport system serves as the primary mechanism to transport the activator calcium across the cell membrane.

DISCUSSION

Winegrad: Is there any reason why the calcium couldn't be coming through the sodium channel and the TTX blocks that channel so that when you combined nickel plus TTX inhibition of contraction is really nickel on one channel and TTX on the sodium plus calcium channel?

Morad: If Ca^{2+} were in part being transported through the Na^{+} channel, then one would expect a decrease in developed tension when a ventricular strip is exposed to TTX. This effect has not been seen either in normal Ringer or in the presence of any other divalent Ca^{2+}-antagonists except Ni^{2+}. We further have a problem in explaining the decrease in the outward current simultaneous with decrease in tension in the presence of Ni^{2+} and TTX. Also, if a significant fraction of tension in Ni^{2+} were being generated by Ca^{2+} flowing through the fast Na^{+} channel then we should see some indication of reversal of the Ca^{2+} current as E_{Ca} is approached. In the presence or absence of Ni^{2+} we have found no indication of such a reversal.

Winegrad: You show that after the first 100 msec of depolarization the rate of rise of tension increases with the degree of depolarization. If you have calcium coming in from the outside, that would suggest that the rate of calcium entry increases as you depolarize. Yet you show that the duration of the latent period is unchanged even when the latent period is followed by differences in the rate of rise of tension. That would argue against the latent period's being due to calcium increasing up to a threshold level.

Morad: I think that the range of potentials that I showed you was rather narrow. One was at +2 and the other was at +20. The difference really isn't that great. The latent period is 100 msec at 0.2 calcium, about 30 msec at 1 mM calcium, and is even smaller in higher calcium concentrations, suggesting that the myoplasmic calcium concentration in part determines the latent period. We don't understand all the complexities of the latent period, but so far, what I have found is consistent with the idea that the calcium entering modulates the myoplasmic [Ca] and alters the pCa^{2+} vs. tension relation. This effect can explain many of the rate-rhythym inotropic effects in this preparation.

Reuter: How do you explain the discrepancy between your results and the results which were obtained 100 years ago regarding *treppe*?

Morad: Staircase is a very complex phenomenon. When I control the membrane potential, that is, if I clamp always to the same holding potential, and I always depolarize for the same duration in a range of 300 to 600 msec clamp pulses, there isn't much staircase. But there often is a very small increase in baseline tension. Staircase is a complex phenomenon in which a combination of potassium accumulation, the shortening or prolongation of the action potential, and the internal calcium concentration play a role in determining the degree of staircase in frog ventricular preparations.

Carmeliet: How much of the reduction in the potassium outward current in the presence of nickel is due to the change in the time dependent current and how much is due to the change in background current? In neurofibers, nickel is known to reduce the resting potassium permeability to a large extent, so there is no more depolarization when you increase the potassium.

Morad: These are concentrations of 0.5 mM or less, and in such concentrations the background potassium current or at least the current involved with the voltage step, that is the initial current, doesn't change much. Perhaps at a higher nickel concentration we would see it. But most of the effect seems to be due to the time dependent current. We also find that in the presence of nickel the membrane often hyperpolarizes. In this regard I should also point out that before we began measuring afterpotentials in the presence of Ni^{2+} we checked to make sure that the resting

membrane's response to changes in $[K]_o$ was in fact
"Nernstan." Some indication of P_K/P_{Na} at rest may be
obtained from the shape of this curve at low $[K]_o$. Although
this is not a terribly sophisticated measure of P_K the
results indicate that the P_K/P_{Na} ratio at rest is not altered
by Ni^{2+}.

  Rougier: May I make some comment about the action of
slow inward current and contraction in frog ventricle?
We looked at the action of acetylcholine in frog atrium and
ventricle on background current and on slow inward current.
We measured the contraction and the current in a double
sucrose gap voltage clamp experiment. We have pulses no
longer than 150 msec. We began in Ringer solution and
we obtained tension together with development of slow
inward current. If we gave acetylcholine we obtained a
very rapid increase in the outward background current
together with a modification of the time course of the slow
inward current, the inactivation time constant being
more rapid under the action of acetylcholine. We also
observed a decrease in contraction. We interpret this as
follows: first, when the slow inward current is modified,
there is a modification of the contraction in frog
ventricle, and second, the first modification observed
on the slow inward current in the frog under the action of
acetylcholine is an increase in the rate of the inactivation.

  Morad: In 1967 I did a series of voltage clamp
experiments on acetylcholine in the frog ventricle. I was
puzzled to find that the time course of the secondary
inward current decreased, while the developed tension
increased. However, more rigorous experiments, although
confirming the decrease in the secondary inward current,
failed to show any change in developed tension in the
presence of acetylcholine. I found that the background
current, rather than the time dependent outward current, had
increased. Perhaps the difference in our results can be
attributed to American vs. French frogs!

REFERENCES

1. Antoni, H., Jacob, R. & Kaufmann, R. (1969). Mechanical
    response of the frog's and mammalian myocardium to
    modifications of the action potential duration by constant
    current pulses. *Pflügers Arch. 306*, 33-57.

2. Bassingthwaighte, J.N. & Reuter, H. (1972). Calcium movements excitation-contraction coupling in cardiac cells. In: *Electrical Phenomena in the Heart*, ed. deMello, W.C., 353-395.

3. Cleemann, L. & Morad, M. (1976). Extracellular potassium accumulation and inward-going potassium rectification in voltage clamped ventricular muscle. *Science 191*, 90-92.

4. Cleemann, L. & Morad, M. (1978). Extracellular potassium accumulation in voltage-clamped frog ventricular muscle. *J. Physiol.* (in press).

5. Constantin, L.L. (1977). Activation in striated muscle. *Handbook of Physiology, Section 1: The Nervous System (1)*, 215-259.

6. Endo, M. (1972). Ca-release from the sarcoplasmic reticulum in skinned muscle fibers. *J. Physiol. Soc. Jap. 34*, 88.

7. Endo, M. (1977). Calcium release from the sarcoplasmic reticulum. *Physiol. Rev. 57*, No. 1, 71-108.

8. Fabiato, A. & Fabiato, F. (1972). Excitation-contraction coupling of isolated cardiac fibers with disrupted or closed sarcolemmas. Calcium-dependent cyclic and tonic contractions. *Circ. Res.* 293-307.

9. Fabiato, A. & Fabiato, F. (1975). Contractions induced by a calcium-triggered release of Ca from the sarcoplasmic reticulum of single skinned cardiac cells. *J. Physiol. 249*, 469-495.

10. Fawcett, D.W. & McNutt, N.S. (1969). The ultrastructure of the cat myocardium. *J. cell Biol. 42*, No. 1, 1-45.

11. Fischman, D. & Swan, R. (1967). Nickel substitution in excitation-contraction coupling of skeletal muscle. *J. gen. Physiol. 50*, 1709-1728.

12. Goldman, Y. & Morad, M. (1977). Measurement of transmembrane potential and current in cardiac muscle: A new voltage clamp method. *J. Physiol. 268*, 613-654.

13. Harris, J.E. & Morad, M. (1976). A $Ca^{+2}-K^{+}$ coupling mechanism for the plateau of the cardiac action potential. *J. gen. Physiol.* (in press).

14. Hasselbach, W. (1964). Relaxing factor and the relaxation of muscle. *Prog. Biophys. Biophys. Chem. 14*, 169-222.

15. Kavaler, F. & Anderson, T.W. (1975). Rapid alterations in extracellular ionic composition during voltage clamp in frog ventricular muscle. *Fedn. Proc. 34*, No. 3.

16. Morad, M. & Goldman, Y. (1973). Excitation-contraction coupling in heart muscle: Membrane control of development of tension. *Prog. in Biophys. and mol. Biol. 27*, 257-312.

17. Morad, M. & Orkand, R.K. (1971). Excitation-contraction coupling in frog ventricle: Evidence from voltage clamp studies. *J. Physiol. 219*, 167-189.

18. Morad, M. & Trautwein, W. (1968). The effect of the duration of the action potential on contraction in the mammalian heart muscle. *Pflügers Arch. 299*, 66-82.

19. Niedergerke, R. & Orkand, R.K. (1966). The dependence of the action potential of the frog's heart on the external and intracellular sodium concentration. *J. Physiol. 184*, 312-334.

20. Ochi, R. (1970). The slow inward current and the action of manganese ions in guinea pig's myocardium. *Pflugers Arch. 316*, 81-94.

21. Page, S.G. & Niedergerke, R. (1972). Structures of physiologic interest in the frog heart ventricle. *J. cell. Sci. 2*, 179-203.

22. Peachey, L.D. (1965). The sarcoplasmic reticulum and transverse tubules of the frog's sartorius. *J. cell Biol. 25*, No. 3, Part 2, 209-231.

23. Sommer, J.R. & Johnson, E.A. (1969). Cardiac muscle: A comparative ultrastructural study with special reference to frog and chicken hearts. *A. Zellforsch 98*, 437-468.

24. Vassort, G. & Rougier, O. (1972). Membrane potential and slow inward current dependence of frog cardiac mechanical activity. *Pflügers Arch. ges Physiol. 331*, 191-203.

25. Weidmann, S. (1958). Effect of increasing calcium concentration during a singel beat. *Experimental 15*, 128.

26. Weidmann, S. (1959). Effect of increasing the calcium concentration during a single heart-beat. *Experientia 15*, 128.

27. Winegrad, S. (1971). Studies of cardiac muscle with a high permeability to calcium produced by treatment with ethylenediaminetetraacetic acid. *J. gen. Physiol. 58*, 71-93.

28. Wood, E.H., Heppner, R.L. & Weidmann, S. (1969). Inotropic effects of electric currents. *Circulation Res. 24*, 409-445.

E-C COUPLING STUDIES ON SKINNED CARDIAC FIBERS

M. Endo
T. Kitazawa

Department of Pharmacology
Tohoku University School of Medicine
Seiryo-machi, Sendai 980, Japan

I. INTRODUCTION

Skinned skeletal muscle fibers, introduced by Prof.
Natori more than twenty years ago (Natori, 1954), proved to
be a very useful preparation for the study of the
contractile proteins and the sarcoplasmic reticulum (SR).
In skinned fibers these subcellular structures are kept in a
more or less physiological state. Removal of the
sarcolemma makes it possible to alter the ionic environment
of the myoplasm easily. In cardiac muscle, however, skinned
fibers are rather difficult to prepare, because cardiac cells
are much smaller than those of skeletal muscle. Fabiato &
Fabiato (1972, 1973) have described a good technique for
making skinned fibers of single cardiac cells. Winegrad
(1971) using a different technique found that chelating
agents such as EDTA (ethylenediamine-tetraacetic acid) raised
the permeability of the surface membrane of cardiac cells
to various substances. In this case it is not necessary to
isolate single cells before skinning. Some other approaches
to make multicellular preparations of cardiac muscle
equivalent to skinned fibers also have been reported
(see Fabiato & Fabiato, 1977).
    Our approach is also to skin cardiac fibers chemically,
using saponin, plant-origin glycosides. Ohtsuki suggested
to us that, since saponin produced holes in the membrane by
combining with cholesterol molecules (Ohtsuki, Palade &
Jamieson, in press), and since the content of cholesterol
in the SR membrane was much less than that of the surface

membrane (Martonosi, 1968; Waku, Uda & Nakazawa, 1971), saponin might specifically act on the surface membrane without affecting the SR. We demonstrated that this was in fact the case in skeletal muscle fibers (Endo, 1976a). Using similar procedures we have been able to chemically skin cardiac fibers and maintain a functional SR (Endo, 1976b; Kitazawa, 1977).

II. MATERIALS AND METHODS

Papillary muscles of guinea-pig right ventricles were used. A small bundle with a diameter of about 100 μm was dissected and was connected to a strain gauge transducer, for measurement of isometric tension. The preparation was suspended in a trough, through which solutions could be perfused rapidly. The experiments were performed at 25° C.

Normal external solution contained 150 mM NaCl, 2 mM KCl, 2 mM Ca-methanesulfonate, 5 mM HEPES (N-2-hydroxyethyl-piperazine-N'-2-ethanesulfonic acid) neutralized with Tris-(hydroxymethyl)-aminomethane (pH 7.4) and 5.6 mM glucose. *High potassium solution* was obtained by replacing all of NaCl in the normal solution with K-methanesulfonate. The composition of *normal relaxing solution* was 110 mM K-methanesulfonate, 4 mM Mg-methanesulfonate, 4 mM ATP-Na$_2$, 2 mM EGTA (ethylene-glycol-bis-(β-aminoethylether)-N-N'-tetraacetic acid), 10 mM PIPES (piperazine-N-N'-bis[2-ethanesulfonic acid]), brought to pH 6.8 with KOH. The concentration of EGTA was altered when necessary. In *activating solutions*, 10 mM total EGTA was used and a specified amount of Ca-methanesulfonate was added. Free Ca ion concentration was calculated by assuming an apparent association constant of 5×10^5 M^{-1} for CaEGTA, (Ogawa, 1968). In all of the above alterations from the normal relaxing solution, ionic strength was kept constant by adjusting the concentration of K-methanesulfonate.

Saponin obtained from Merck & Co., Inc. was dissolved in a relaxing solution shortly before each use. Caffeine was dissolved in a suitable solution.

To examine the effect of saponin, intact and skinned skeletal muscle fibers of iliofibularis of *Xenopus laevis* were also used at 2° C in solutions with compositions slightly different from those described above (Endo & Nakajima, 1973).

III. RESULTS

A. *The Effect of Saponin on Skeletal Muscle Fibers*

An intact single skeletal muscle fiber could not maintain tension but eventually was kept relaxed in a K solution (an activating solution) containing 3×10^{-6} M free Ca, which was enough to produce 70–80% maximal tension if directly applied to the contractile system (Endo, 1972). With the Ca concentration kept constant, however, when 50 µg/ml saponin was added to the solution, the fiber quickly developed tension, which could be abolished by replacing Ca in the solution with EGTA. Reapplication of Ca again produced a similar magnitude of tension. Thus the treatment with saponin appears to have made the surface membrane of the fiber freely permeable to substances such as Ca buffer. Even after this treatment, however, the SR seems to be still functioning; it is capable of accumulating Ca, which could be released by applying caffeine or by "depolarizing" the SR by replacing methanesulfonate in the solution with Cl (Endo & Nakajima, 1973).

The effects of saponin were examined more closely with mechanically skinned skeletal muscle fibers. Responses of the contractile system were examined by activating the skinned fibers with 3×10^{-6} M free Ca. During the Ca treatment, the duration of which was fixed at 60 sec, the SR accumulated Ca which could be released by application of 25 mM caffeine. Pairs of such a Ca stimulation and a subsequent caffeine stimulation were given once in every 5 min, and during intervals various saponin treatments were made. As shown in Fig. 1, both Ca and caffeine responses gradually declined in size without any saponin treatments, indicating deterioration of the fibers. Fig. 1 also shows that whereas 50 µg/ml of saponin did not affect the time course of the deterioration, the glycosides with a concentration higher than 150 µg/ml specifically reduced the caffeine responses progressively. From this type of experiment, it could be concluded that the skeletal muscle SR is unaffected by a 50 µg/ml treatment of saponin for at least 30 min. Contractile system can tolerate higher concentrations, 500 µg/ml saponin. Concentrations as small as 5 µg/ml of saponin were found to strongly affect the surface membrane of skeletal muscle.

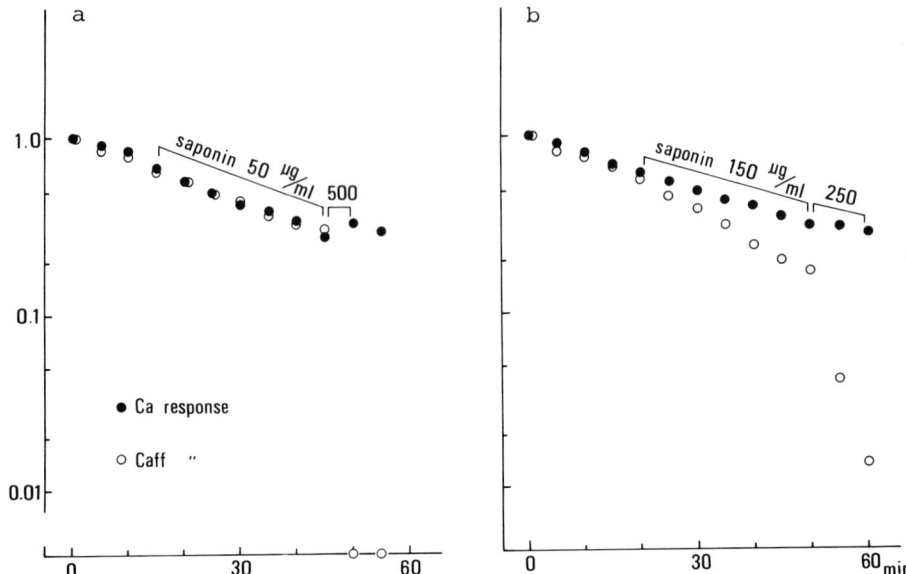

Fig. 1. The effect of saponin on the SR of skinned fibers of skeletal muscle of Xenopus laevis. *Peak tension produced by 3 × 10^{-6} M free Ca during 60 sec application (filled circles) and area under the subsequent caffeine contractures (open circles) were plotted in values relative to the first respective responses of each fiber. (a) Fiber 50924b, (b) Fiber 50926a.*

B. Responses of Intact Fiber Bundles of Papillary Muscle

Since we dissected small bundles from papillary muscle, which we felt have no guarantee for preserved functions, we examined the contractile response of these preparations before saponin treatment. As seen in Fig. 2, they twitched normally, and show the normal dependence on $[Ca]_o$. On the other hand, K contractures were very large compared with twitch tension and were insensitive to alteration of $[Ca]_o$.

The effect of caffeine on these bundles was similar to those already reported (Suzuki, 1962; Blinks, Olson, Jewell & Bravený, 1972; Chapman & Miller, 1974; Jundt, Porzig, Reuter & Stucki, 1975). Caffeine (25 mM) either failed to produce contractures in normal solution (Fig. 3a), or caused only small and very brief transient contractures. However, twitch tension was enormously potentiated by caffeine (Fig. 3d). If Na in the external solution was replaced by

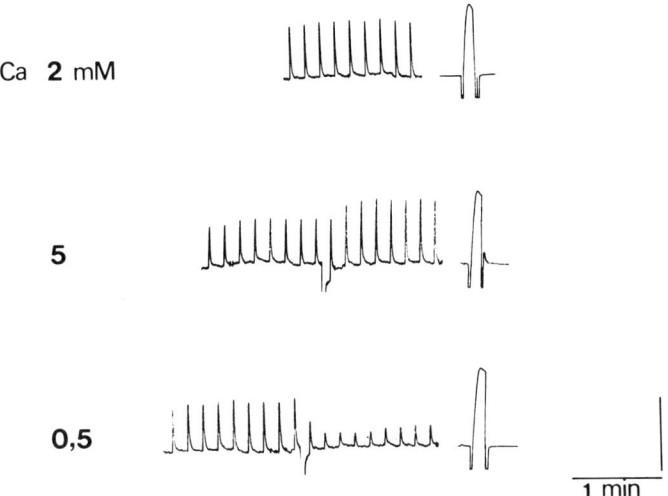

Ca **2** mM

5

0,5

1 min

Fig. 2. Twitches and K-contractures of a small bundle of guinea-pig papillary muscle and their dependence on external Ca. In the middle and lower panels, Ca concentration in the external solution was changed from 2 mM to 5 or 0.5 mM, respectively, at the time indicated by artifacts. Although twitch tension was altered quickly, K-contracture was not significantly affected by the change in Ca concentration. Bundle 761223. Tension calibration : 1 mg for twitch, 20 mg for K-contracture.

either Tris, Li, or K, caffeine produced large contractures which "inactivated" with time (Fig. 3b, c). Similar caffeine contracture could be produced even in the presence of 4 mM EGTA (Fig. 3e, f), indicating that the contracture was due to Ca release from some internal store(s). A dose-response relation for caffeine contractures in intact cardiac bundles is shown in Fig. 4. A high concentration of caffeine appears to be effective in producing tension in cardiac SR, provided that external Na is absent. Fig. 4 (B,C) also shows the "inactivation" of the caffeine effect in cardiac muscle. Such inactivations are not apparent in amphibian skeletal muscle.

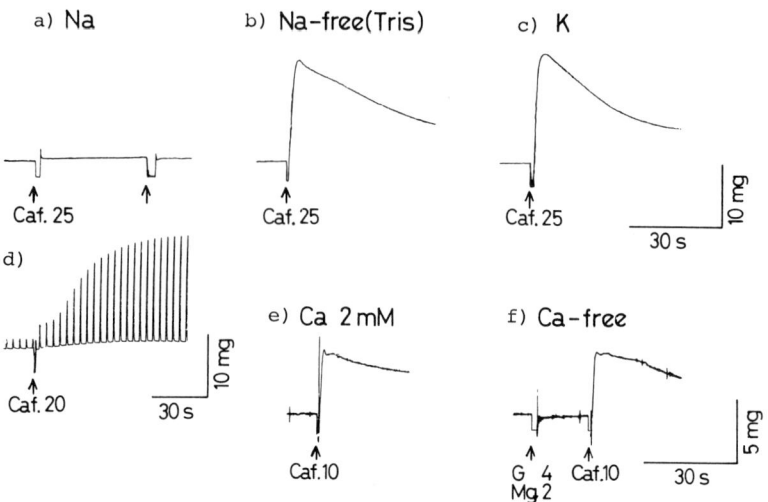

Fig. 3. The effect of caffeine on small bundles of intact guinea-pig papillary muscle. (a, b, c) The effect of 25 mM caffeine on an unstimulated bundle in solutions of various main cations. (c) Caffeine was applied after the K-contracture subsided. (d) The twitch potentiating effect of 20 mM caffeine. (e, f) The absence of effect of external Ca on contracture produced by 10 mM caffeine. (e) Control. (f) At G4 Mg2, 2 mM Ca in normal external solution was replaced by 2 mM Mg, and at the same time, 4 mM EGTA was added. (a–d) Bundle 770216. (e, f) Bundle 770208.

C. The Effect of Saponin on Cell Membrane of Cardiac Muscle

After the intact fiber responses were examined, each cardiac bundle was immersed in a relaxing solution and treated with saponin. Fig. 5 shows the time course of the effect of 50 µg/ml saponin. As is seen, after 5 min of the saponin treatment, tension could be produced on application of 10^{-4} M Ca, but the rate of development of tension was rather slow. With further saponin treatments, the time course of tension development became faster, until maximum effect was obtained with a total about 30 min of the treatment at this saponin concentration. Thus, after saponin treatments, contractile responses could be controlled by altering free Ca ion concentrations in the external solution. If either ATP or Mg was removed at this stage, the bundle rapidly developed a rigor tension, which was

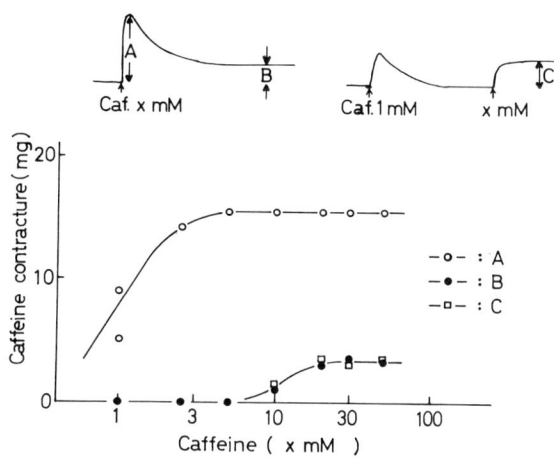

Fig. 4.   Dose response curves of caffeine contracture of
a small bundle of intact guinea-pig papillary muscle.   As
shown at top, tensions at the peak (A), at the steady state
(B), and at the steady state of caffeine contracture after
pretreatment with 1 mM caffeine were plotted.   Bundle 770202.

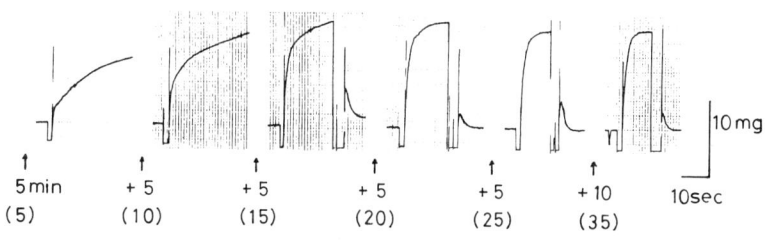

Fig. 5.   Time course of tension development of a small
bundle of guinea-pig papillary muscle by 10^{-4} M Ca after
each saponin treatment.   Between each contraction, the bundle
was treated with 50 µg/ml saponin for 5 or 10 min as
indicated by the arrows.   The numbers in parentheses indicate
total duration of saponin treatment by the time.   Bundle
770428.

abolished when ATP or Mg was reapplied. The surface membrane
of the saponin-treated fibers thus was sufficiently skinned
to allow substances such as Ca, Mg, EGTA, and ATP to pass
freely. From the results of Fig. 5, we adopted 30 min
treatment with 50 μg/ml saponin as a standard procedure of
chemical skinning.

The response to the 10^{-4} M Ca after saponin treatment was
usually greater than K contracture of the same bundle before
the saponin treatment. Potassium solution containing 25 mM
caffeine usually produced the largest possible contracture,
larger than K alone, and this response was approximately
the same as the maximum calcium tension obtained after
saponin treatment (Fig. 6). These results suggest that
probably all fibers in the bundle were skinned by saponin.

D. *The pCa-Tension Relation of Saponin-Treated Cardiac Fibers*

Fig. 7 shows an example of the pCa-tension relation of
skinned fiber bundle of guinea-pig papillary muscle. The
relation is slightly less steep than that of skeletal muscle
(Endo, 1972), but essentially similar to those already
reported for skinned or glycerinated cardiac fibers (Winegrad,
1971; Fabiato & Fabiato, 1975a; Kitazawa, 1976).

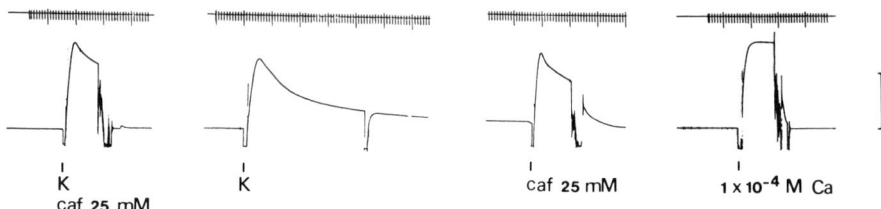

K
caf 25 mM   K   caf 25 mM   1 x 10⁻⁴ M Ca

*Fig. 6. Comparison of intact (first three) and skinned
(the fourth) fiber responses of the same bundle of papillary
muscle. Between the third and the fourth responses, the
bundle was treated with 500 μg/ml saponin for 5 min. Results
were the same after a treatment with 50 μg/ml saponin for
30 min. K, Caf 25 mM: normal external solution was replaced
by high K solution containing 25 mM caffeine. K: normal
external solution was replaced by high K solution.
Caf 25 mM: after the K contracture was subsided, 25 mM
caffeine was added to the high K solution. Bundle 761223.
Tension calibration: 20 mg. Time marker: 1 sec per small
division.*

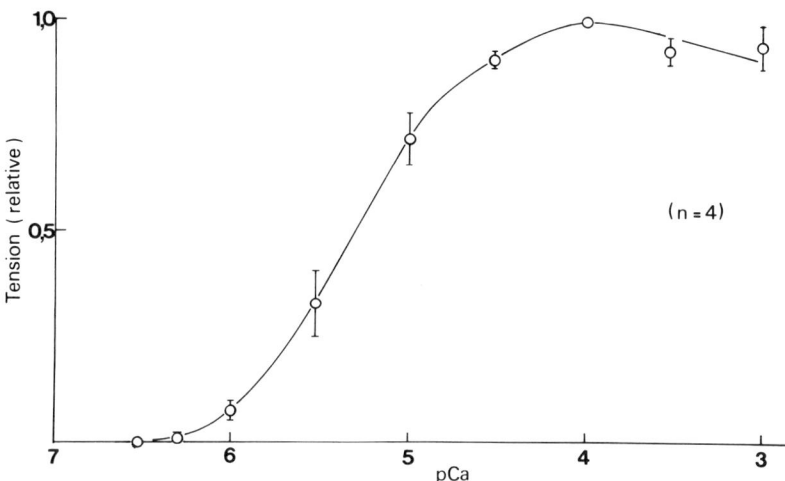

Fig. 7. The pCa-tension relation of saponin-treated guinea-pig papillary muscle. Average and S. E. of four bundles.

E. *The SR Responses of Saponin-Treated Cardiac Fibers*

The SR response of saponin-treated cardiac fibers is essentially similar to that of mechanically skinned skeletal fibers. If SR was loaded with Ca, an application of 30 mM caffeine in the presence of, e.g. 0.1 mM EGTA produced a transient contraction (Fig. 8, inset). After a single treatment with such a high concentration of caffeine, SR was depleted of Ca and no responses could be obtained on reapplication of caffeine. However, if the fiber was incubated with a solution containing 10^{-6} M free Ca, the SR accumulated Ca, and responses of the fiber to caffeine recovered gradually with an increase in duration of immersion in 10^{-6} M Ca (Fig. 8). Fig. 8 shows the time course of accumulation of Ca by the SR of saponin-treated cardiac fibers.

Fig. 9 shows the dose-response relation of caffeine in saponin-treated cardiac fibers. The saturating concentration of caffeine is about 30 mM, as compared to 5 mM for intact fibers (Fig. 4).

A few differences have been noted between the SR responses of saponin-treated guinea-pig papillary muscle and those of skinned fibers of amphibian skeletal muscle. One is the fact that the cadriac SR appears to be leakier than that of

amphibian skeletal muscle. As shown in Fig. 10, after the
SR was loaded to a fixed extent, fibers were immersed in a
solution containing various concentrations of EGTA. Calcium
content of the SR then was gradually lost with a rate that
was dependent on EGTA concentration. The rate of loss was
much more rapid in cardiac fibers. With 10 mM EGTA most of
the Ca is lost in 5 min, whereas most of the Ca is retained
in the SR of skeletal muscle. This leakiness of the SR of
cardiac fiber does not seem to be due to the saponin
treatment, since (1) an additional 20-min treatment with
saponin did not increase the leakiness of the SR of the
fiber already treated with saponin for 25 min, and (2) with
mechanically skinned cardiac fiber, similar leakiness was
observed (Fabiato, personal communication).

The second difference is the fact that Ca-induced Ca
release is easier to evoke in cardiac SR than in amphibian
skeletal muscle SR (see also Fabiato & Fabiato, 1972, 1973,

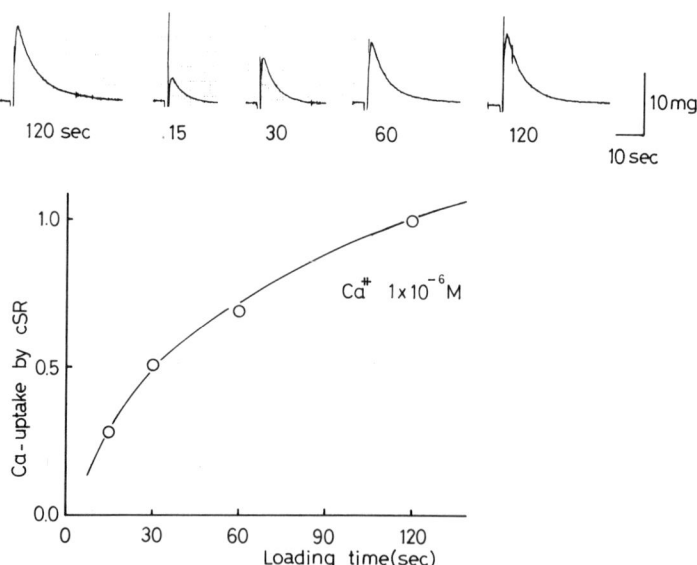

*Fig. 8. Time course of recovery of caffeine responses of
a saponin-treated guinea-pig papillary muscle showing that
of Ca uptake by the SR. 30 mM caffeine was applied in a
relaxing solution containing 0.15 mM EGTA. Responses of
the areas under caffeine were plotted in values relative to
that obtained with 120 sec loading with 10^{-6} M free Ca.
Bundle 770503.*

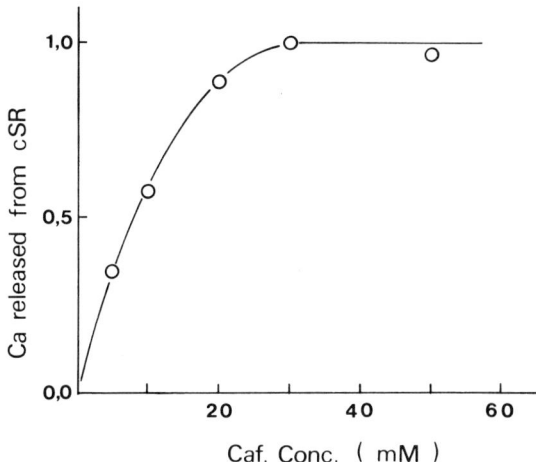

Caf. Conc. (mM)

Fig. 9. Dose-response curve of caffeine effect on a saponin-treated guinea-pig papillary muscle. The SR was loaded each time with 10^{-6} M Ca for 2 min. Various concentrations of caffeine were applied in a relaxing solution containing 0.1 mM EGTA. The area of resulting contraction was plotted in relative values. Bundle 761211.

1977, 1975*b*). Fig. 11 shows the time course of Ca uptake with various free Ca ion concentrations. With 3×10^{-6} M free Ca, cardiac SR takes up Ca rapidly, but after some time it appears to lose some of the Ca, which might be an indication of Ca-induced Ca release. With 10^{-5} M or higher concentrations of Ca, the initial rate of uptake of Ca by the cardiac SR seems to be fairly high, but the amount of Ca accumulated does not reach the high levels attained with lower concentrations of free Ca. This finding suggests that Ca-induced Ca release has occurred at the concentration of Ca higher than 10^{-5} M, so that the steady state uptake of Ca by the SR under this higher Ca concentration was much smaller than it was with lower Ca concentration where the Ca-induced Ca release mechanism did not operate. In similar experiments in skeletal muscle skinned fibers, the effect was observed only with 10^{-4} M or higher concentrations of free Ca (Endo, 1975).

In Fig. 12, Ca-induced Ca release was more directly examined. The cardiac SR was loaded to a fixed extent and then exposed to various free Ca ion concentrations for a short time. The amount of Ca remaining in the SR after

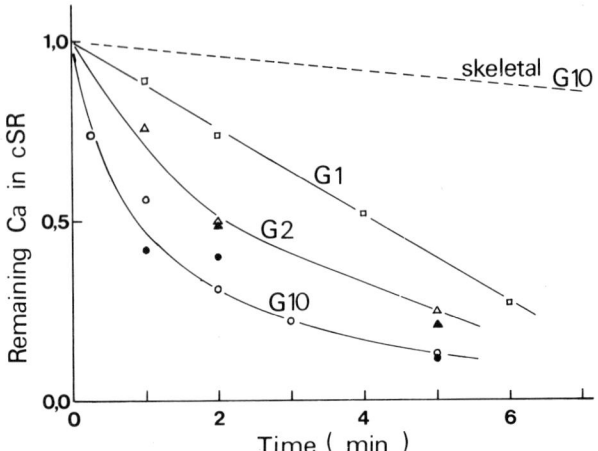

Fig. 10.   The time course of leak of Ca from the SR of a
saponin-treated guinea-pig papillary muscle.   The SR was
loaded first with 10^{-6} M Ca for 2 min, and then immersed in
relaxing solutions containing 1 mM (G1), 2 mM (G2), and 10 mM
(G10) EGTA.   The amount of Ca remaining in the SR was plotted
against time of immersion.   Open and filled symbols indicate
leak in the presence and absence of ATP, respectively.   Bundle
761204.   In comparison, the results with a mechanically
skinned fiber of Xenopus skeletal muscle were drawn with a
dashed line.

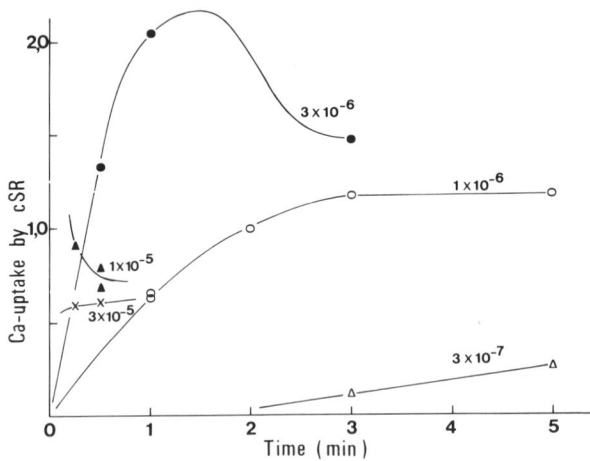

Fig. 11.   Time course of Ca uptake by the SR of
saponin-treated guinea-pig papillary muscle under various free
Ca levels (M) given in the figure.   Experiments similar to
those in Fig. 8.   Bundle 760928.

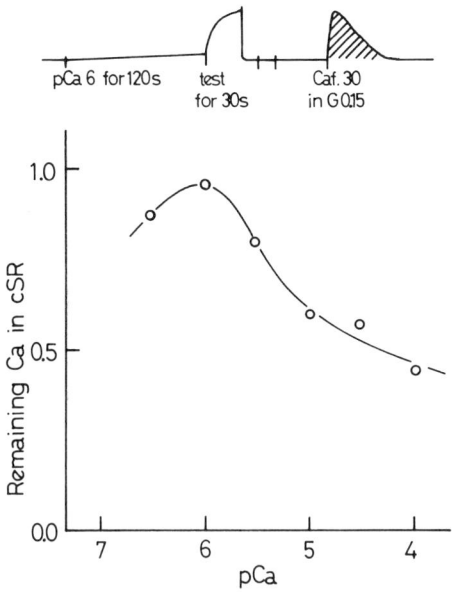

Fig. 12. *Dependence of Ca-induced Ca release in saponin-treated guinea-pig papillary muscle on Ca ion concentration. Experiments done as at top of the figure. At "test," various pCa solutions were applied, and remaining Ca values, estimated by responses to 30 mM caffeine in a relaxing solution containing 0.15 mM EGTA, were plotted as values relative to those without "test" treatment. Bundle 770507.*

these treatments was estimated by applying a high concentration of caffeine. In good agreement with the results of Fig. 11, the Ca remaining in the SR was smaller after treatment with Ca, concentrations of which were higher than 3×10^{-6} M. These results suggest a Ca-induced Ca release during the high Ca treatments (Fig. 12). Again, similar experiments in skinned skeletal muscle fibers indicated Ca release only with Ca concentrations higher than 3×10^{-5} – 10^{-4} M (Endo, 1975).

One of the expressions of Ca-induced Ca release is spontaneously repeated contractions (Fabiato & Fabiato, 1972; Endo, Tanaka & Ogawa, 1970). We could obtain conspicuous cyclic contractions in our saponin-treated cardiac bundles if we reduced the EGTA concentration below 50 μM (Fig. 13). In the case of frog skinned skeletal muscle fibers, only

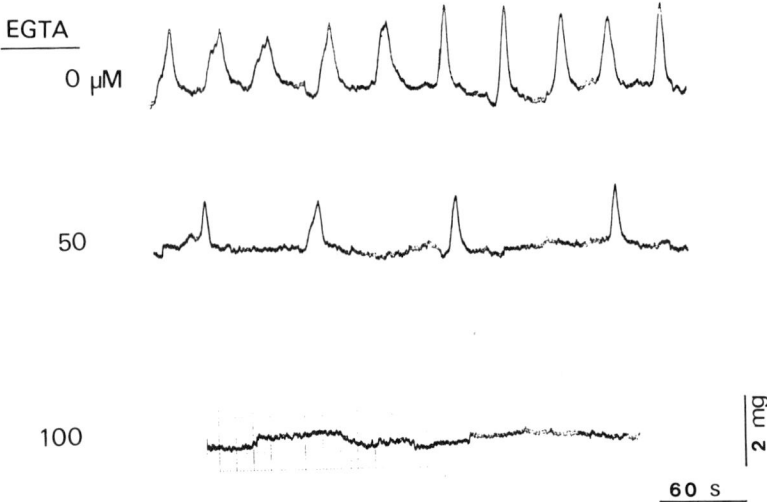

Fig. 13. Spontaneous cyclic contractions of a saponin-treated guinea-pig papillary muscle and their inhibition by EGTA. The bundle was immersed in relaxing solutions containing EGTA concentrations indicated in the figure. Bundle 761124.

small and sluggish responses were observed with contractions below 25 μM EGTA (Endo *et al.* 1970). This result agrees with the greater difficulty of evoking Ca-induced Ca release in this tissue. The cyclic contractions of saponin-treated cardiac fibers just as those of skeletal fibers were facilitated by caffeine, and were suppressed by procaine or free Mg (Endo, 1977).

A further difference might be in the ionic permeability of the SR membrane. In skinned skeletal muscle fibers, the replacement of methanesulfonate with Cl caused a Ca release from the SR due to "depolarization" of the SR (Endo & Nakajima, 1973). However, Fabiato & Fabiato (1977) have shown that the effect of Cl was very weak in cardiac SR. In agreement with this, the Ca release due to replacement of methanesulfonate with Cl was minimal, if present at all (Fig. 14). This result, however, does not indicate that "depolarization" is ineffective in causing Ca release in cardiac SR, since as is shown in Fig. 14, if K is replaced by choline, a definite response was obtained. These results suggest that Cl permeability of the cardiac SR membrane is much lower than that of the skeletal muscle.

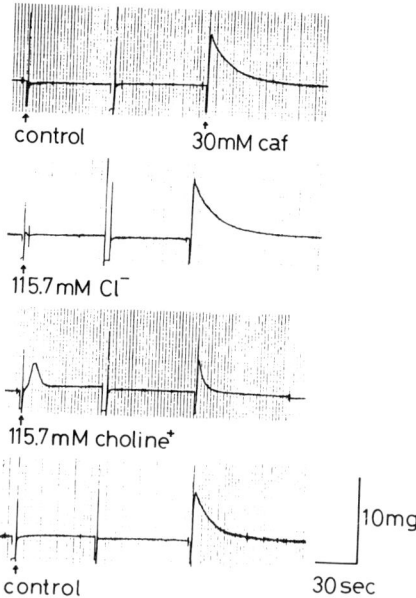

Fig. 14. "Depolarization"-induced Ca release in the SR
of a saponin-treated guinea-pig papillary muscle. Before
each run, the SR was loaded with 10^{-6} M Ca for 2 min. Control:
no change in ionic composition. Cl and choline replaced
methanesulfonate and K, respectively, in the presence of
0.1 mM EGTA. Remaining Ca was measured by 30 mM caffeine in
the presence of 0.15 mM EGTA. Bundle 770428.

IV. SUMMARY AND CONCLUSIONS

It was clearly demonstrated that by the use of saponin,
cardiac muscle fibers were chemically skinned within a short
time. Since this does not require any particularly skillful
technique such as that required for mechanically skinned
cardiac cells (Fabiato & Fabiato, 1972, 1973), the method
could be used widely for studies of cardiac functions. The
only precaution necessary is to have the bundles as thin as
possible, since inadequate diffusion makes the results
unreliable.

Saponin-treated fibers have the following advantages
over glycerinated fibers. (1) In glycerinated fibers,
denaturation or removal of some of the contractile proteins
could occur, whereas it is much less likely in saponin-treated

fibers.   (2) When functions of the SR are to be studied,
glycerinated fibers usually cannot be used, while
saponin-treated fibers were shown to retain good SR
responses.   (3) Since skinning with saponin can be
accomplished easily within a short time, intact and skinned
fiber responses can be compared rather quickly by the use
of saponin.
        Saponin-treated fibers are different in the following
respects from Winegrad's fibers treated with chelating
agents.   While the former have holes in the membrane so
large that ferritin can easily pass through (Ohtsuki et al.
in press), the surface membrane of the latter appears to
remain impermeable to large molecules such as protein
(Winegrad, 1971).   While high permeability produced by
chelating agents could be reversed by a high concentration
of calcium (Winegrad, 1971), holes made by saponin have never
sealed again.   These different characteristics could either
be an advantage or a disadvantage depending on the purpose
of an experiment.
        In our thin bundles of guinea-pig papillary muscle,
caffeine contractures appeared to be more conspicuous than
has been previously reported.   This result could be explained
by the fact that calcium release in the intact cardiac
fibers by caffeine appears to be "inactivated" with time
(Fig. 3, 4) unlike that in amphibian skeletal muscle.   In
thick bundles, therefore, caffeine contractures are expected
to be smaller in proportion to those in thin bundles, since
(1) surface fibers might already be "inactivated" when the
caffeine reaches fibers in the core of bundles, and (2) a
slow rise in caffeine concentration around central fibers
might proceed to "inactivate" contractures of the central
fibers.   The mechanism of "inactivation" is not clear at
present.   Similar "inactivation" does not seem to occur in
skinned preparations.
        Another prominent feature of caffeine contracture of
cardiac muscle is a strong inhibitory effect exerted by Na,
ion (Suzuki, 1962; Chapman & Miller, 1974; Jundt et al. 1975),
which was never seen in skeletal muscle.   This inhibitory
effect was attributed by Jundt et al. (1975) to the presence
of a Na-Ca exchange system in cardiac muscle membrane.
However, it is rather difficult to explain our results in
this way, because the Na-Ca exchange system must have had an
extremely high rate of transport in order, for example, to
have abolished the contractile response completely (Fig. 3a).
Na ions did not affect the caffeine-induced calcium release
in saponin-treated cardiac fibers.   These results all raise

a very puzzling question as to how extracellular Na ions can inhibit the effect of caffeine to cause Ca release that most probably is exerted directly on the SR.

If the dose-response curve of caffeine in intact cardiac fibers (Fig. 4) and that in saponin-treated fibers are directly comparable, it follows that the amount of Ca present in the internal store of cardiac fibers under the condition of Fig. 4 is more than enough to fully activate the contractile system, since with 5 mM caffeine, which releases only a fraction of Ca in the SR (Fig. 9) the maximal contracture could be produced. This argument would not hold if the condition of Fig. 9 somehow made the SR much less sensitive to caffeine than in intact fibers, which is conceivable though not likely.

A few differences between the cardiac SR and the amphibian skeletal muscle SR were found or confirmed and further extended. The leakiness of the cardiac SR might be related to its often-assumed property that the Ca content of the cardiac SR changes rather quickly in response to changes in experimental conditions (Morad & Goldman, 1973).

The cardiac SR appears to have a low permeability to Cl ion, unlike the skeletal muscle SR. This is not particularly puzzling, since it is known that anion permeability of the surface membrane is quite different in cardiac muscle than in skeletal muscle (Horowicz, 1964).

The Ca-induced Ca release can be evoked in the SR of guinea-pig papillary muscle with Ca concentrations more than one order of magnitude lower than in amphibian skeletal muscle SR. This was first noted by Fabiato & Fabiato (1972, 1973, 1975*b*) qualitatively. Quantitatively, they reported that much lower concentrations of Ca than reported here could induce Ca release in the rat cardiac SR (Fabiato & Fabiato, 1977). Whether this difference is explained simply by species difference is not clear at present. Although it is fairly certain that the Ca-induced Ca release mechanism of cardiac muscle gives the basis for after-contractions (Endo, 1977), it is not yet certain whether it also plays an important role in physiological excitation-contraction coupling. Further studies, including the determination of the extent of physiological loading of the cardiac SR, are urgently needed.

DISCUSSION

Jahromi: In the lobster abdominal extensor muscles, we were not able to produce any potassium contracture, but after we pretreated the muscle fibers with caffeine solutions and washed out caffeine we could produce potassium contractures. This is consistent with your finding that caffeine and potassium produce large tensions.   I wonder whether you could comment on the mechanism?   The next part of my question is that in both phasic and tonic muscle fibers, we have seen caffeine contracture which lasted for more than 30 minutes. Therefore the notion that the twitch fibers are not capable of holding tension as it has been seen in the potassium contracture could not be applicable to the lobster case. Nevertheless, under the dissecting microscope we could see that the fibers were clearly deteriorating.   I wonder if you have seen this process of deterioration in your preparations. Or if this deterioration in any way has affected your tension measurement.

Endo: It is well-known that in frog skeletal muscle fibers a small concentration of caffeine does not produce a contracture by itself but enhances the twitch and potassium contracture.   In the case of frog skeletal muscle, my interpretation is that since caffeine enhances calcium-induced calcium release, if a certain amount of calcium is released from the SR by any means, say, with potassium depolarization or electrical stimulation, the raised free calcium concentration might then cause further calcium release from the SR under the influence of caffeine.   In the absence of caffeine, the secondary release hardly occurs.   A similar thing might happen in lobster muscle.   For the second problem, in living muscle, a high concentration of caffeine certainly causes irreversible changes in skeletal muscle probably because of long-lasting and strong contractions. In skinned fibers this could be easily avoided by using a suitable concentration of EGTA when caffeine is applied. However, as far as minor deterioration is concerned, we usually detect it in amphibian skinned skeletal muscle fibers, not only after the caffeine application, but also when we apply a high concentration of calcium, especially when, as a result, large tension was maintained for a long time. Sarcomeres are then no longer uniform nor well-aligned so that optical diffraction patterns due to sarcomeres completely disappear.   However, with these fibers we still can get nice tension responses usually.

Amvari: The internal calcium that you mentioned in your caffeine-EGTA treated preparations. Can you give me some explanation or biochemical explanation of existence of this calcium? Is the stored calcium bound to a protein?

Endo: The SR can accumulate calcium from a medium containing low free calcium, utilizing energy of ATP. This can be demonstrated by isolating the SR from muscle homogenate as microsome fraction, and then measuring its calcium accumulation in the presence of ATP. When I mentioned the internal calcium store, I had the SR in mind since it is well-known that the SR is present in cardiac muscle as well. What is generally thought is that calcium is transported across the sarcoplasmic reticulum membrane by a calcium pump mechanism, and most of the calcium transported is bound to some intraluminal calcium-binding sites, probably to a calcium-binding protein, calsequestrin. But a part of the calcium is in free form.

REFERENCES

1. Blinks, J.R., Olson, C.B., Jewell, B.R. & Braveny, P. (1972). Influence of caffeine and other methylxanthines on mechanical properties of isolated mammalian heart muscle. Evidence for a dual mechanism of action. *Circulation Res. 30*, 367–392.
2. Chapman, R.A. & Miller, D.J. (1974). Effects of caffeine on contraction of frog heart. *J. Physiol. 242*, 589–613.
3. Endo, M. (1972). Stretch-induced increase in activation of skinned muscle fibres by calcium. *Nature, Lond. 237*, 211–213.
4. Endo, M. (1975). Conditions required for calcium-induced release of calcium from sarcoplasmic reticulum. *Proc. Jap. Acad. 51*, 467–472.
5. Endo, M. (1976a). Effects of some detergents on surface-membrane and sarcoplasmic-reticulum of skeletal muscle. *Folia pharmac. jap. 72*, 9P–10P.
6. Endo, M. (1976b). *J. physiol. Soc. Jap. 38*, 157. (in Japanese).
7. Endo, M. (1977). Calcium release from the sarcoplasmic reticulum. *Physiol. Rev. 57*, 71–108.
8. Endo, M. & Nakajima, Y. (1973). Release of calcium induced by depolarization of sarcoplasmic-reticulum membrane. *Nature, Lond. 246*, 216–218.

9. Endo, M., Tanaka, M. & Ogawa, Y. (1970). Calcium induced release of calcium from the sarcoplasmic reticulum of skinned skeletal muscle fibres. *Nature, Lond. 228*, 34-36.

10. Fabiato, A. & Fabiato, F. (1972). Excitation-contraction coupling of isolated cardiac fibers with disrupted or closed sarcolemmas. Calcium-dependent cyclic and tonic contractions. *Circulation Res. 31*, 293-307.

11. Fabiato, A. & Fabiato, F. (1973). Activation of skinned cardiac cells. Subcellular effects of cardioactive drugs. *Eur. J. Cardiol. 1*, 143-155.

12. Fabiato, A. & Fabiato, F. (1975a). Effects of magnesium on contractile activation of skinned cardiac cells. *J. Physiol. 249*, 497-517.

13. Fabiato, A. & Fabiato, F. (1975b). Contractions induced by a calcium-triggered release of calcium from sarcoplasmic reticulum of single skinned cardiac cells. *J. Physiol. 249*, 469-495.

14. Fabiato, A. & Fabiato, F. (1977). Calcium release from the sarcoplasmic reticulum. *Circulation Res. 40*, 119-129.

15. Horowicz, P. (1964). Effects of anions on excitable cells. *Pharmac. Rev. 16*, 193-221.

16. Jundt, H., Porzig, H., Reuter, H. & Stucki, J.W. (1975). Effect of substances releasing intracellular calcium ions on sodium-dependent calcium efflux from guinea-pig auricles. *J. Physiol. 246*, 229-253.

17. Kitazawa, T. (1976). Physiological significance of Ca uptake by mitochondria in heart in comparison with that by cardiac sarcoplasmic reticulum. *J. Biochem., Tokyo 80*, 1129-1147.

18. Kitazawa, T. (1977). Ca uptake and release of sarcoplasmic reticulum in mammalian cardiac skinned fibers. *Jap. J. Pharmac. 27*, 155P.

19. Martonosi, A. (1968). Sarcoplasmic reticulum. V. The structure of sarcoplasmic reticulum membranes. *Biochim. biophys. Acta 150*, 694-704.

20. Morad, M. & Goldman, Y. (1973). Excitation-contraction coupling in heart muscle: Membrane control of development of tension. *Progr. Biophys. mol. Biol. 27*, 257-313.

21. Natori, R. (1954). The role of myofibrils, sarcoplasma and sarcolemma in muscle contraction. *Jikei. med. J. 1*, 18-28.

22. Ogawa, Y. (1968). Apparent binding constant of glycoletherdiaminetetraacetic acid for calcium at neutral pH. *J. Biochem. 64*, 255-257.

23. Ohtsuki, I., Palade, G.E. & Jamieson, D.J. *J. microsc. cell. Physiol.* (in press).

24. Suzuki, K. (1962). Studies on the mechanism of the excitation-contraction coupling in cardiac muscle, with special reference to the caffeine contracture. *Jap. J. Physiol. 12*, 186-199.

25. Waku, K., Uda, Y. & Nakazawa, Y. (1971). Lipid composition in rabbit sarcoplasmic reticulum and occurrence of alkyl ether phospholipids. *J. Biochem., Tokyo 69*, 483-491.

26. Winegrad, S. (1971). Studies of cardiac muscle within a high permeability to calcium produced by treatment with ethylene diaminetetraacetic acid. *J. gen. Physiol. 58*, 71-93.

MEMBRANE CONTROL OF CARDIAC CONTRACTILE SYSTEMS

Saul Winegrad
George B. McClellan

Department of Physiology
University of Pennsylvania
Philadelphia, Pennsylvania

Most of the discussion during this symposium has been concerned with modulation of contractility by variation in the amount of calcium made available to the contractile proteins from changes in either the action potential or the sarcoplasmic reticulum. Modification of the properties of the contractile proteins themselves, however, might be important in altering the contractility of cardiac muscle. Our interest in this possibility was stimulated by the observation that the ATPase activity of fragmented cardiac muscle decreases when the cell fragments are concentrated, but as soon as the fragments are diluted, the decline in ATPase activity stops. There is no subsequent recovery to the original level. This experiment was interpreted as indicating that the fragmentation of the surface membrane had initiated the production of some substance which when concentrated had a negative effect on the contractile proteins.

To study this phenomenon, it was desirable to have a preparation in which the properties of the contractile proteins and their response to calcium could be directly tested without losing a significant part of the cellular metabolic activity. The preparation that we ultimately used in these experiments is one in which the surface membrane has been made highly permeable to small ions and molecules by a period of exposure to the chelator EGTA (Winegrad, 1971).

Supported by USPHS grants HL 16010 and HL 15835.

The gross morphology of surface membrane remains unchanged as
far as can be determined by conventional electron microscopy,
but the membrane is very leaky to small ions and molecules
up to about 1000 Daltons. The membrane remains a permeability
barrier for most, if not all, of the cellular proteins,

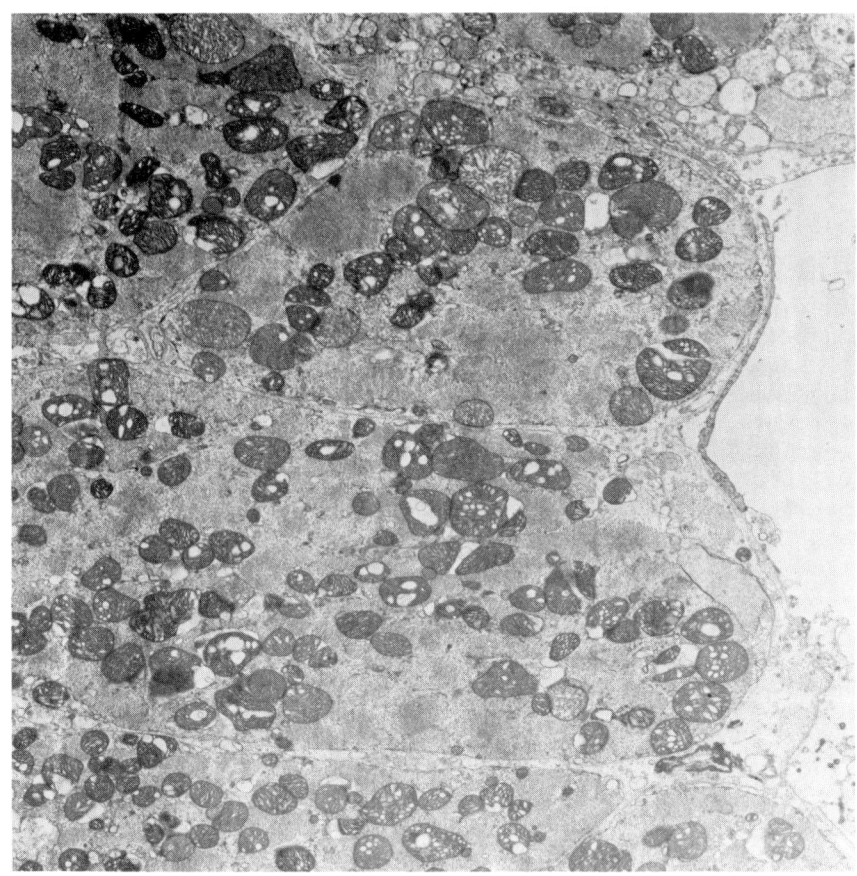

*Fig. 1. Electron micrograph of rat cardiac muscle that
has been treated with EGTA. Note the normal appearance of the
sarcolemma, myofibrils, and sarcoplasmic reticulum. (Electron
micrograph courtesy of Dr. Thomas F. Robinson.)*

including soluble ones. One of the ways in which this was
demonstrated was by assaying the activity of the soluble
protein, creatine phosphokinase, in the tissue. A bundle of
hyperpermeable rat ventricular cells was put into rigor by
replacing the usual 5 mM ATP in the bathing solution with
5 mM ADP. After tension had reached a new steady level,
15 mM creatine phosphate without any creatine phosphokinase
was added to the bathing solution. The muscle promptly
relaxed, indicating that ATP had been formed from the ADP and
creatine phosphate in the absence of added creatine
phosphokinase, and demonstrating the presence of a high level
of creatine phosphokinase in the hyperpermeable bundle.
Analagous experiments have demonstrated that other soluble
proteins such as pyruvate kinase are retained in this
preparation. With this preparation it has been possible to
examine some of the mechanisms for the control of the
properties of the contractile proteins.

REGULATION OF CA CONCENTRATION REQUIRED FOR ACTIVATION

   Calcium sensitivity (the concentration of calcium
necessary to activate the contractile proteins and produce
maximum force) is shown with a plot of the relative force
against the negative logarithm of the calcium concentration
(pCa). For the hyperpermeable cardiac fiber the well-known
sigmoidally shaped relation exists (Winegrad, 1971; Fabiato
& Fabiato, 1975). If theophylline, a phosphodiesterase
inhibitor, is added to the bathing medium, the curve shifts
to the left (lower concentrations of Ca) by a factor of
about 3 to 6, indicating that in the presence of the drug
as little as one-sixth of the original concentration of
calcium is necessary to produce a given amount of activation.
The change in Ca sensitivity is completely reversible with
the removal of theophylline.
   Treatment of the tissue with a nonionic detergent like
Brij or Triton X100, which inactivated cellular membranes
including the sarcolemma and the sarcoplasmic reticulum,
produces the same shift of the tension-pCa curve. Since
there is 3 mM EGTA present in all of the experiments to
minimize any influence of the sarcoplasmic reticulum on
sarcoplasmic Ca concentration, it is unlikely that the
reticulum is important in the changes observed. Furthermore,
the Ca sensitivity of the mechanically skinned cardiac cell,

in which the surface membrane is absent but the sarcoplasmic
reticulum is present and functioning, is insensitive to
detergents (Fabiato & Fabiato, 1975). Consequently, the
effect of the detergent on the hyperpermeable fiber must be
on the sarcolemma rather than the intracellular membrane.
As expected, the addition of theophylline to the solution
bathing a hyperpermeable bundle that has already been treated
with a detergent causes little or no change in Ca sensitivity.
Theophylline requires the integrity of the sarcolemma for its
effect. Treatment with a phosphodiesterase inhibitor or a
detergent results in the same Ca sensitivity as total removal
of the surface membrane.

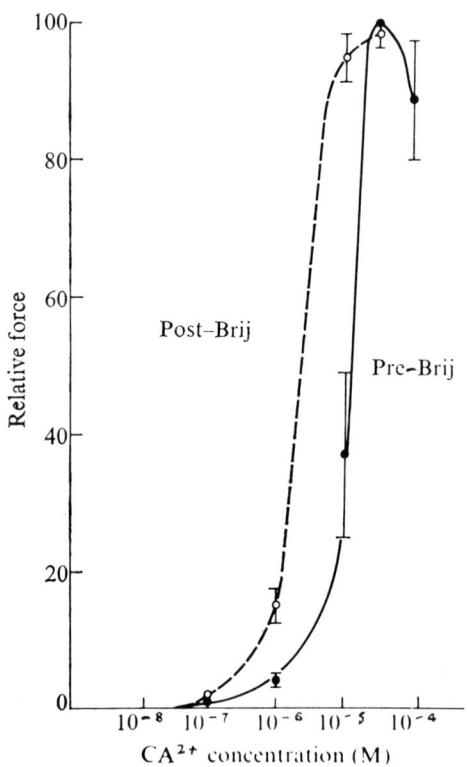

*Fig. 2. Relation between tension and concentration of
calcium ions in the EGTA-treated rat ventricular bundle
before and after treatment with 0.5% Brij 58 for 40 min.
The treatment with the detergent decreases the concentration
of Ca^{2+} required to activate the contractile system. (From
McClellan & Winegrad, 1977.)*

The marked response to inhibition of phosphodiesterase activity indicated that cyclic nucleotides should be involved, but addition of the cyclic nucleotides themselves in concentrations up to 10^{-5} M had no effect without an accompanying phosphodiesterase inhibitor. In the presence of 5 mM theophylline, however, 10^{-7} M cGMP caused a further increase in the calcium sensitivity. In young rats, which had a relatively low Ca sensitivity initially, adding cAMP to a bathing solution that already contained theophylline decreased the extent to which Ca sensitivity had been

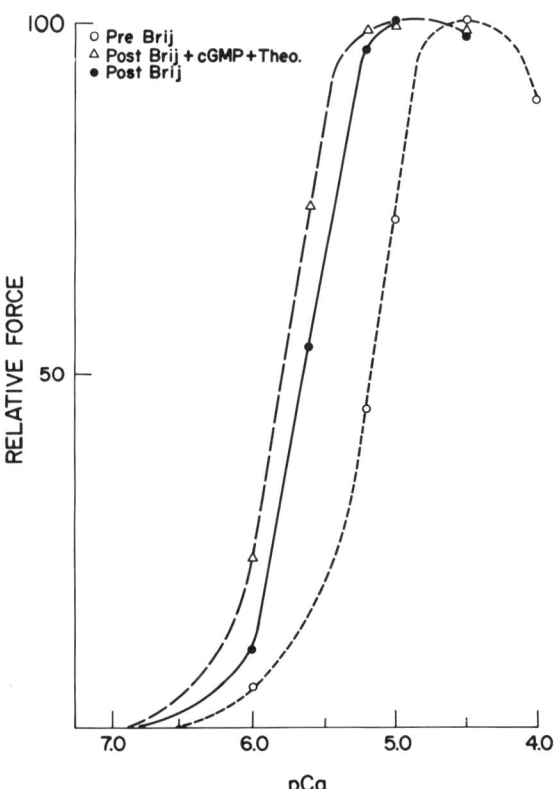

Fig. 3. *The effect of* 10^{-7} *M cGMP on the concentration of Ca required to activate EGTA-treated rat ventricular muscle. The addition of cGMP plus a phosphodiesterase inhibitor increases the* Ca^{2+} *sensitivity of the system to an even greater extent than treatment with Brij.*

increased by theophylline. In may older rats that initially
have higher calcium sensitivities, cAMP plus theophylline
caused a net decrease in Ca sensitivity. In these tissues
maximum force is generated in pCa of 5.0 and about 75% in
pCa of 5.2, but when cyclic AMP and theophylline have been
added, the force in pCa of 5.2 drops by about 65% to about
25% of maximum even though the value for maximum
Ca-activated force is unchanged (Fig. 4). Cyclic AMP and
cyclic GMP therefore respectively decrease and increase
calcium sensitivity.

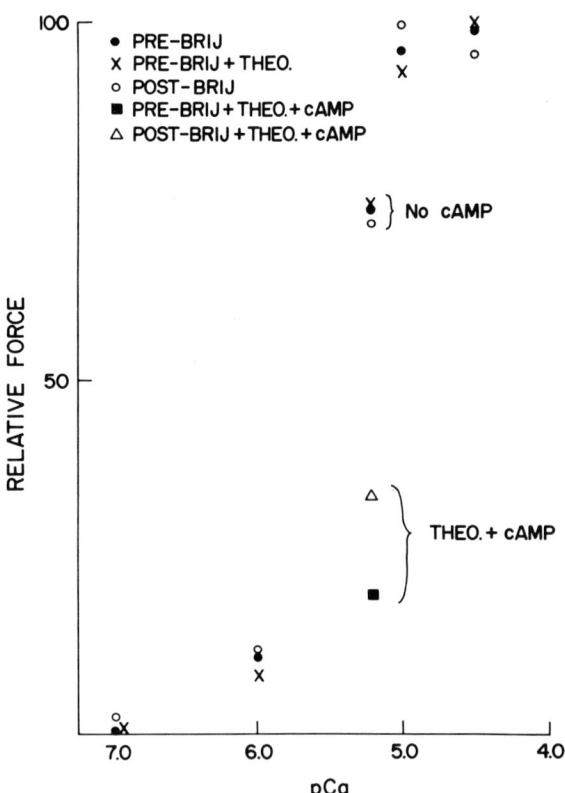

*Fig. 4. The effect of cAMP with a phosphodiesterase
inhibitor on the concentration of Ca^{2+} required for activation
of EGTA-treated rat ventricular bundles that showed an
initially high sensitivity to the ions. The cAMP decreases
the Ca sensitivity as shown by the relative force production
at pCa of 5.2.*

Additional evidence in favor of a phosphorylation reaction in the regulation of Ca sensitivity comes from experiments in which the hyperpermeable bundle was deprived of a phosphate donor. H.H. Weber showed many years ago that cytosine triphosphate (CTP) was about 85% as effective as ATP in the cyclical interactions between actomyosin. CTP, however, will not support phosphorylation because it is a poor phosphate donor, and it cannot be converted to a cyclic nucleotide with demonstrable physiological effect. The substitution of CTP for ATP therefore should produce dephosphorylation of the contractile proteins. Experimentally, this substitution raises the Ca sensitivity of the contractile system to the same high level as theophylline or treatment with detergents (Fig. 5).

Phospholipids in the surface membrane appear to be important for the modulation of Ca sensitivity. Their addition to the bathing medium reverses the effect of detergent on Ca sensitivity. Both the original lower Ca sensitivity and response to theophylline are restored. As regards regulation of Ca sensitivity, detergent appears merely to inactivate nucleotide cyclases in the membrane, a change that can be reversed by phospholipids in the immediate environment of the sarcolemma.

In summary, these experiments strongly suggest that calcium sensitivity can be varied as a function of the state of phosphorylation of the contractile proteins. They indicate that the nonphosphorylated form has a high sensitivity that can be decreased by cAMP-regulated phosphorylation. The cAMP-regulated decrease in sensitivity can be reversed, not only by the removal of cyclic AMP and dephosphorylation but also by the addition of cyclic GMP even in the presence of cyclic AMP. It is not clear from these studies, however, whether the cGMP directly effects the contractile proteins or a cGMP-dependent protein kinase.

REGULATION OF FORCE GENERATION

The relation between force and calcium concentration is a function of the concentration of magnesium ATP, the substrate for the contractile proteins. When the concentration of MgATP is reduced below 200 µM, the maximum Ca-activated force is increased, and the concentration of Ca required for activation is decreased. At 10-20 µM

MgATP, maximum force is increased by 50-100% and a
significant force is developed by the bundle in the lowest
achievable Ca concentration (about 10^{-9} M). Bremel & Weber
(1972), who have observed an analogous response to low MgATP
by isolated skeletal muscle contractile proteins, have
proposed an intriguing model to explain this phenomenon.
It attributes the force in very low Ca to the release of
troponin-tropomyosin inhibition by the formation of rigor
links that occurs in the low MgATP.

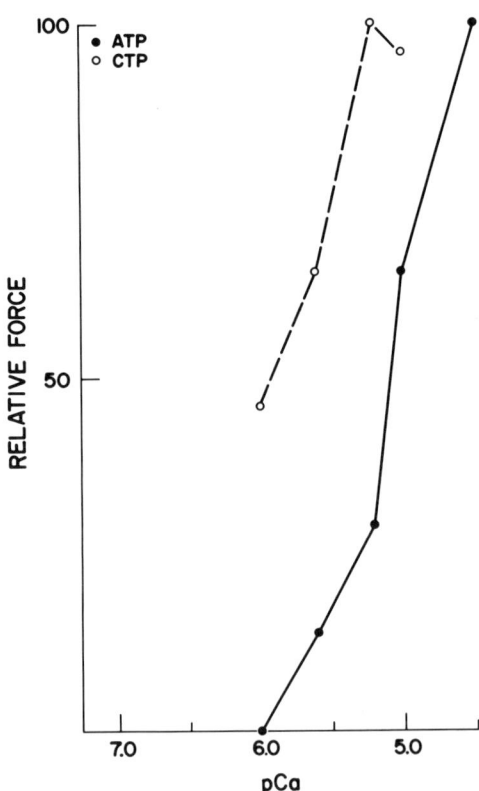

Fig. 5.   The effect on Ca^{2+} sensitivity of the
EGTA-treated rat ventricular bundle of replacing all of the
ATP with CTP, which does not act as a phosphate donor in
phosphorylation reactions of contractile proteins.   In the
presence of the CTP, where presumably the contractile
proteins are not phosphorylated, the Ca sensitivity is high,
as in the presence of cGMP.

The concentration of MgATP can also influence the mechanical response of cardiac muscle to Ca in another way, and this is illustrated in Fig. 6. The reproducible generation of force in pCa of 5 is shown in four consecutive runs, after which the MgATP concentration is lowered to 20 μM and the Ca concentration to 1 μM. Tension is not

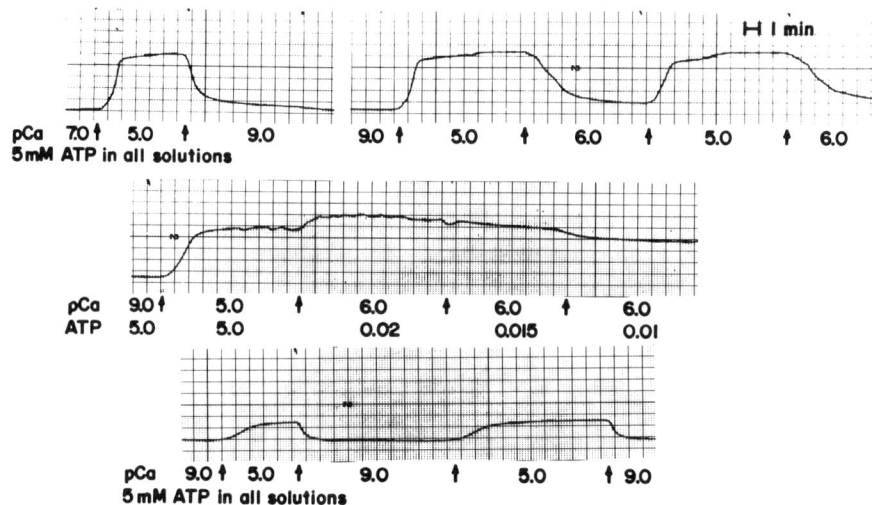

Fig. 6. *Force record produced by a rat ventricular trabecula that had been functionally skinned by an overnight soak at 4° C in a solution containing 10 mM EGTA, 4 mm MgAc$_2$, 5 mM ATP, and 120 mM K propionate. The force developed in 5 mM ATP during maximal activation is very reproducible in each of the first four exposures, but when the ATP concentration is lowered to 0.01-0.02 mM, the force declines. After return to 5 mM ATP, the force in optimal pCa is again stable, but it has been markedly reduced. The shape of the tension-Ca relation is the same after the exposure to low ATP as it was before; a uniform relative decrease in force at each concentration of Ca has occurred. In all solutions the concentration of free Mg^{2+} ions was held constant at 2 mM after taking into consideration its interactions with ATP, EGTA, and creatine phosphate. The MgATP concentration was controlled by varying ATP concentration; in view of the large excess of Mg ions, it was assumed that all ATP was in the form of MgATP. (From McClellan & Winegrad, 1977.)*

maintained in this solution. The decline in the ability to
generate force persists in normal relaxing solution containing
5 mM ATP although Ca-activated force has become stable.
The necessary condition for the change in the contractile
response to Ca is exposure to the combination of low ATP and
high calcium. Some ATP must be present, as removal of all
ATP produces only a period of rigor that is terminated with
the restoration of ATP. The progressive development of an
inhibition of contractility continues for a few minutes
after the ATP and Ca concentrations have been restored to
their normal values in relaxing solution, and it is several
minutes before a new steady state has been achieved after
the ATP concentration has been raised. Apparently some
active substance has been synthesized as a result of the
combination of low ATP and elevated Ca, and this substance
is critical in activating the reactions that depress
contractility. As the rate at which contractility declines
is accelerated by a phosphodiesterase inhibitor, a
phosphorylation reaction may be involved; and the conditions
that favor the inhibition resemble those which stimulate
guanylatecyclase activity.

The depression in contractility is partially reversible.
When a tissue is exposed first to a normal ATP concentration
and a very high calcium concentration, a large force develops.
A decrease in the ATP concentration causes a further increase
in tension due to the activating effect of rigor links as
described above, but then force begins to fall. A decrease
in Ca concentration at the same low ATP concentration
produces a decline in force that is greater than would be
expected in low concentration of ATP, but then force
gradually begins to increase, and much of the decline is
reversed. This two-phase response appears to involve first
a decline in tension from the drop in calcium and then the
increase in tension due to the removal of the inhibition
that had been produced by the low ATP itself.

In order to evaluate the role of the surface membrane in
the inhibition of contractility, the sarcolemma was
inactivated with detergent. This resulted in a marked
change in the ATP requirement for the inhibition. Now
inhibition could be produced in 2-3 mM of ATP, which is
very close to the physiological range, instead of 20 μM or
less that was required before the exposure to detergent.

The addition of a microsomal fraction to the bathing
medium protected the hyperpermeable bundle from the
depression of contractility produced by the combination of

low ATP and elevated Ca. Since the membrane of the
hyperpermeable fibers will not allow the entry of vesicles
into the sarcoplasm, the protective effect of the microsomal
vesicles must be exerted by the synthesis of a relatively
small active molecule that can permeate the cell membrane
of the EGTA-treated fibers (Fig. 7).

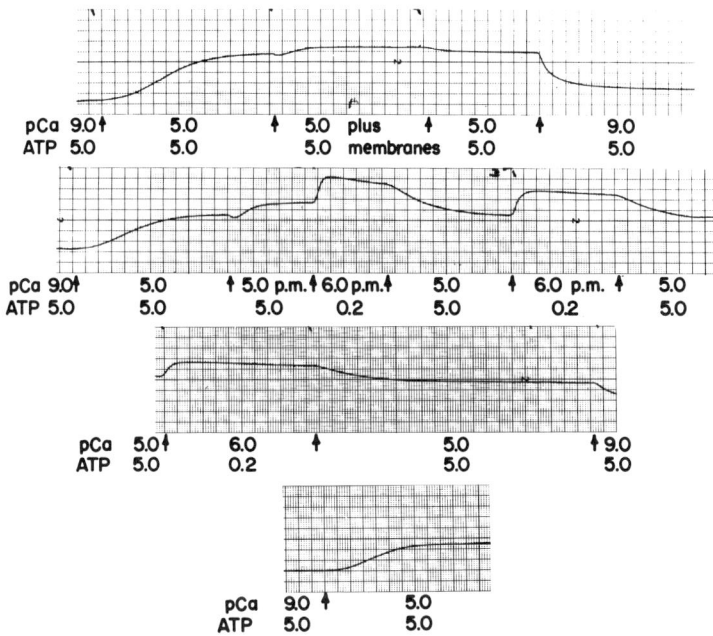

*Fig. 7. The presence of isolated microsomal vesicles
in the bathing medium protects the EGTA-skinned rat
ventricular cells from the negative inotropic effect of an
exposure to the combination of low ATP and high Ca. Force
in optimal Ca and 5 mM ATP is reproducible. The exposure
of the tissue to pCa 5.0 and 20 μM ATP does not inhibit
the subsequent ability of the tissue to generate force on
pCa of 5.0 when vesicles are present in the bathing solution.
(From McClellan & Winegrad, 1977.)*

The phenomenon is sensitive to lipids. The addition of a chloroform-methanol extract of rat ventricle to the bathing medium reproduced the effect of lowering ATP. Tension development at threshold concentration of calcium was unaltered, but a further increase of concentration caused an inhibition of force production. Apparently whether Ca produces or inhibits force may depend upon the state of the membrane, and the relevant reactions are sensitive to the concentration of ATP. The membrane therefore can detect the concentration of sarcoplasmic ATP and exert some important control over contractility, the nature of which depends on the ATP concentration.

A model of the mechanism of the inhibition of contractility in the presence of reduced ATP concentrations is illustrated in Fig. 8. Inside the cell there is a reaction that can inhibit the contraction system. This reaction requires Ca and a cyclic nucleotide, and it may involve phosphorylation of the contractile system at a critical site. A small diffusible molecule that is synthesized by the membrane at a rate determined by the MgATP concentration regulates the reaction. There are no data as yet concerning the nature of the reversal of the inhibition as this reaction appears to be either defective or inactive in the hyperpermeable cells. Since the

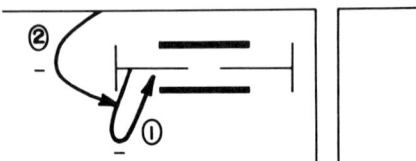

REACTION I: INHIBITS CONTRACTILITY   cGMP
REACTION 2: CONTROLS REACTION I   ? cAMP

Fig. 8. A schematic model of one mechanism for the regulation of the contractility of the contractile proteins in response to a decrease in the cytoplasmic concentration of MgATP. An intracellular reaction involving cGMP inhibits the contractile system, and this reaction in turn is regulated by a small molecule produced by another reaction that occurs in the membrane.

regulation of the contractility by this mechanism resides in the membrane, the system can be sensitive to both intracellular and extracellular factors.

FUNCTIONS OF THE MEMBRANE-CONTROLLED SYSTEM

The intriguing possibility about regulation of contractility in response to ATP concentration is that the work of the cardiac cell can be adjusted to the energy supply, a mechanism which would obviously be of great utility to the cell. As the ATP concentration drops, so does the work done by the cell. Control of the system, however, is exercised at the appropriate location to respond to override signals from the outside of the cell.

The alteration in calcium sensitivity could be important in setting diastolic tone, in regulating the degree of coupling between generation of force and other calcium sensitive reactions, in modifying the inotropic state of the heart, and even in activating the contractile system. Simply by shifting the calcium concentration required for activation to and from the existing Ca contraction, the contractile system could be activated or inactivated with no change in Ca concentration. The lack of information about the kinetics of the involved reactions prevents any statement about whether these control systems might be operative within a single contractile cycle or require several cycles.

DISCUSSION

Bassange: Could part of the depressing effect on contractility in your experiments be explained by changes in ATP concentration, since it has been shown recently by Schrader and by Bourne and his coworkers that adenosine acts as a very strong calcium antagonist, and you have such large changes in ATP concentrations in your experiments?

Winegrad: I think we can rule that out quite promptly, because we have 15 mM creatine phosphate plus creatine phosphokinase present in all solutions as a phosphate regenerating system, and this is approximately twice as much as is necessary to maintain the ATP phosphorylated. I don't think we have a significant, uncontrolled change in ATP concentration.

Morad: Do you want to comment on how your system would control the catecholamine or digitalis action?

Winegrad: I wouldn't dare take on digitalis, especially since we have limited data.   Catecholamines are probably important in the regulation of both Ca sensitivity and the force production of the contractile proteins, the former almost certainly by a membrane-controlled phosphorylation of the inhibitor subunit of troponin.   As regards force, we know that the surface membrane sets the ATP level at which contractility is totally inhibited.   I believe that catecholamines are important in determining the set point. Let me give you an example of how this might operate in an intact organism.   As a result of a decrease in blood flow to a certain portion of the heart, the ATP concentration in myocardial cells in that region drops.   The available energy is used not for generating force, but for maintaining cell viability, that is, supporting such vital reactions as ion pumps and protein syntheses.   If, however, the mass of the heart involved is so large that without its contraction function the organism is threatened, catecholamines are released and the ATP concentration required for inhibition is lowered.   The entire heart begins to beat again, and it continues to beat until either circulation is improved or the cells die.   That is actually what cardiac surgeons have observed in bypass surgery for coronary artery disease.

REFERENCES

1.   Baba, N., Kim, S. & Farrell, E. (1976).   Histochemistry of creatine phosphokinase. *J. molec. cell. Cardiol.* *8*, 599.
2.   Bremel, R. & Weber, A. (1972).   Cooperation within actin filaments in vertebrate skeletal muscle. *Nature 238*, 97.
3.   Fabiato, A. & Fabiato, F. (1975).   Contraction induced by a calcium triggered release of calcium from the sarcoplasmic reticulum of single skinned cardiac cells. *J. Physiol. 249*, 469.
4.   McClellan, G.B. & Winegrad, S. (1977).   Membrane control of cardiac contractility. *Nature, 268*, 261.
5.   Winegrad, S. (1971).   Studies of cardiac muscle with a high permeability to calcium produced by treatment with EDTA. *J. gen. Physiol. 58*, 71.

VI

Mechanisms of Drug
Action in Cardiac Muscle

DIGITALIS: INOTROPIC AND ARRHYTHMOGENIC EFFECTS
ON MEMBRANE CURRENTS IN CARDIAC PURKINJE FIBERS*

*R.W. Tsien, R. Weingart,** and R.S. Kass[+]*

Department of Physiology
Yale University School of Medicine
New Haven, Connecticut 06510

Questions about the mechanism of the action of digitalis
really began about 200 years ago with William Withering's
use of this drug in the treatment of congestive heart
failure (see Withering, 1785). Fig. 1 helps frame the
problem by giving a simplified view of the inotropic and
toxic responses to increasing doses of cardiotonic steroid.
The main difficulty with digitalis is to produce beneficial
actions, such as the increase in the strength of contraction
of the heart, without undesirable toxic effects, such as
contractures or ventricular arrhythmias. One of the first
clinical indications of ventricular arrhythmias is the
phenomenon of premature ventricular contractions, and these
occur at a lower dose than the contractures Dr. Morad
discussed. This article will address the question of how
the arrhythmias come about, before reviewing ongoing research
about the mechanism of the positive inotropic effect.

*This paper reviews work supported by grants HL 13306 and
AM 17433 from the U.S. Public Health Service. R. Weingart
was supported by the Swiss National Science Foundation.
R.W. Tsien is an Established Investigator of the American
Heart Association.*
    ***Present address: Physiological Institute, University of
Bern, Bern, Switzerland.*
    *+Present address: Department of Physiology, University of
Rochester School of Medicine and Dentistry, Rochester, New
York 14602.*

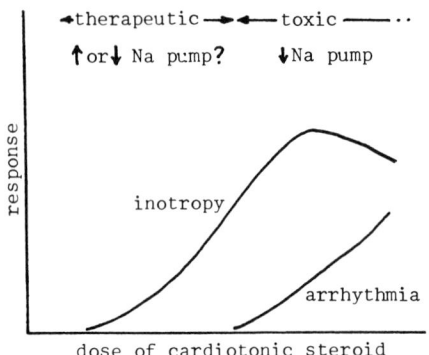

*Fig. 1. Schematic representation of positive inotropic
and arrhythmogenic effects of cardiac glycosides and aglycones.
Over the "therapeutic" range, increased strength is not
accompanied by detectable arrhythmia. Present controversy
concerns reports of stimulation or inhibition of sodium pump
activity and their possible involvement in the positive
inotropic action. Arrhythmias are observed over the "toxic"
range and are associated with inhibition of the Na pump.
The positive inotropic response reaches a maximum, and then
falls off with increasing doses.*

When people talk about beneficial and harmful effects of
digitalis, they often describe drug doses as "therapeutic" or
"toxic." These terms are somewhat confusing, because there
is a range of concentrations over which inotropy and
arrhythmias both increase. In this discussion the word
"therapeutic" describes a range of low cardiac glycoside
concentration which produces an appreciable positive inotropic
effect *without* causing significant arrhythmias (see Fig. 1).
The problem is that over this range of cardiac glycoside
concentrations, it is not clear whether the sodium pump is
stimulated or inhibited. In the range where toxic effects
such as premature ventricular contractions occur, it is
generally agreed that the effects of the cardiotonic steroids
are inhibitory and that the pump activity is decreased. The
safety margin between the onset of the therapeutic and
arrhythmogenic effects may be as small as a factor of 3 (see
Rhee, Dutta & Marks, 1976). Thus it is very important to
understand the basic mechanisms underlying these two effects.

We shall consider first the basis of spontaneous ventricular impulses under toxic conditions. Fig. 2 is from the work of Ferrier & Moe (1973), who were one of four different groups that independently published papers on the mechanism of the arrhythmic effects (see also Davis, 1973; Hogan, Wittenberg & Klocke, 1973; Rosen, Gelband, Merker & Hoffman, 1973). The figure shows transmembrane potential recordings from a dog Purkinje fiber which has been intoxicated with acetylstrophanthidin. Note that repetitive stimulation produces a beat by beat increase in the amount of diastolic depolarization. The last action potential was followed by an oscillatory afterpotential. The oscillatory afterpotentials (called "transient depolarizations" or TDs by Ferrier & Moe) were enhanced by increasing the rate of

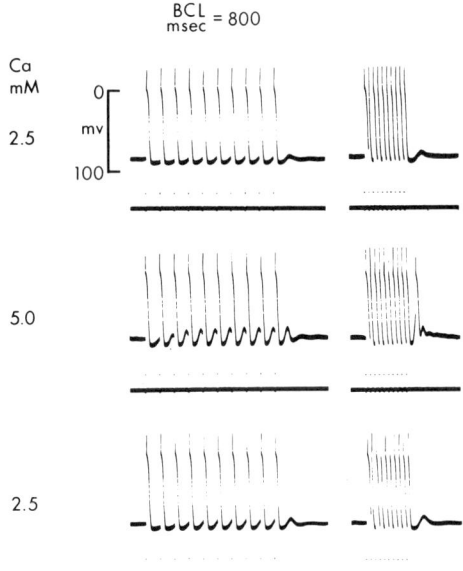

Fig. 2. Effect of increased calcium concentration on transient depolarizations produced by acetylstrophanthidin (1×10^{-7} g/ml) in dog Purkinje tissue. The top trace in each panel is an intracellular voltage recording, and the bottom trace is the stimulus pattern. Resting membrane potential at the beginning of all three records is 83 mV. Spikes were retouched. BCL = basic cycle length. From Ferrier & Moe (1973).

drive or by elevating the extracellular calcium concentration. In this experiment, the combination of rapid stimulation and 5 mM $[Ca]_o$ produces an afterdepolarization that exceeds the excitatory sodium threshold and produces a spontaneous impulse. If that impulse were connected to the rest of the heart in an intact animal, it would produce a premature ventricular contraction. The question is, then, what is the mechanism of this oscillatory afterpotential? Such potentials were first observed by Bozler (1943a,b), the same man who used the capillary electrometer to study the activity of smooth muscle and cardiac muscle (see paper by A.F. Huxley in this volume). Bozler observed these oscillatory afterpotentials by putting a turtle heart into very high calcium solutions. From the beginning then, some type of involvement of calcium ion was clear.

We carried out voltage clamp experiments in calf Purkinje fibers to investigate the basis of the TD. In Fig. 3 the top panel shows a series of action potentials followed by the application of the voltage clamp at the point of maximum repolarization. The accompanying membrane current changes in a smooth fashion toward a net inward value because of the shutting off of the pacemaker potassium current (see paper by R.H. Adrian in this volume). When we apply the cardiotonic steroid strophanthidin at a fairly high concentration, we obtain the same type of result as Ferrier and others observed, a beat by beat increase in the afterpotential. When the voltage clamp is applied at the new point of maximum diastolic repolarization, the current records look grossly different. Instead of a smoothly changing current record, there is a lot of jagged noise and a bump of inward current. This transient inward current was termed a "TI" by Lederer & Tsien (1976) because of its role in generating the TD, the transient depolarization. Fig. 3 shows that a similar TI is evoked following the "off" of a depolarizing clamp pulse. If we remove the drug, everything returns to the way it was originally. There is no increased diastolic depolarization, and smooth decay follows the action potentials and the voltage clamp.

The magnitude of the transient inward current in Fig. 3 is on the order of 10 nanoamps, and the time scale is very slow. These two considerations make the TI relatively easy to measure. It is also reassuring that the timing and magnitude of these currents are suitable to explain the increased diastolic depolarization, if the measured value of the membrane capacity is taken into account.

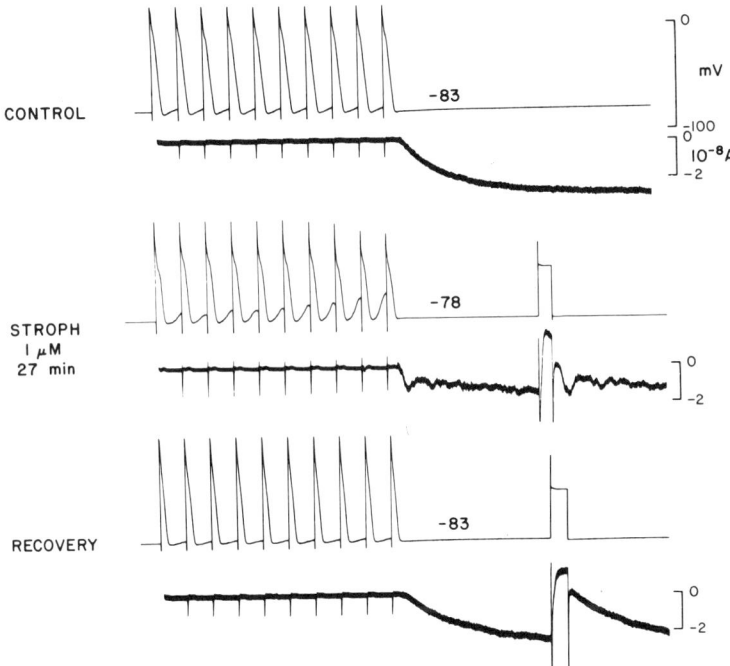

Fig. 3. Effect of 1 μM strophanthidin on electrical activity and membrane current over the pacemaker range of potentials. Each panel shows chart recording of membrane potential (above) and total membrane current (below). Top: a train of action potentials was stimulated by external shocks at 0.5 Hz after a rest period of 25 sec. Voltage clamp control was imposed following the tenth action potential at the point of maximum repolarization. Middle: same procedure after exposure to 1 μM strophanthidin for 27 min. Bottom: recovery after removal of drug. Preparation 127-2, 35° C, $Ca_O = 5.4$ mM. From Lederer & Tsien (1976).

Fig. 4 reintroduces the question of calcium involvement by showing that there is a contractile event which accompanies TIs or TDs. The top panel again shows the end of a series of action potentials obtained in the presence of strophanthidin. Each action potential is accompanied by a twitch, and each of the ensuing transient depolarizations is accompanied by a small contractile event, which was labeled aftercontraction by Reiter (1962). Clamping the potential at the maximum diastolic level produces another transient inward current, and an aftercontraction once again. Similarly, a depolarizing clamp pulse is followed by yet another transient inward current and yet another aftercontraction. Thus there is a definite correlation between these two events, although the time to peak of the current precedes the time to peak of the aftercontraction by about 100 msec. The aftercontraction is an indication of a transient increase in myoplasmic calcium.

Fig. 5 shows another correlation between changes in current and force. In this experiment, the oscillatory character of the TI is enhanced by heavy intoxication and a long depolarizing clamp pulse. The repolarizing step is followed by a series of clamped oscillations in current and force. The two signals have been photographically superimposed in the bottom panel by inverting the tension record and advancing it 80 msec. Note that the current and force records match very well.

What is the causal relation between current and force? Fig. 6 shows two possible hypotheses. The first hypothesis arises from a proposal by Ferrier & Moe (1973) who attributed the TD to an increase in the calcium permeability of the surface membrane. Calcium entry would lead to calcium-troponin interaction and direct activation of the aftercontraction. This is a sequential model, where the calcium influx is necessary in order to cause aftercontraction. Another less obvious possibility is that both the inward current and the aftercontraction may be common reflections of some underlying event, such as a transient or phasic release of calcium from an intracellular store like the sarcoplasmic reticulum. If this were correct, the current could be carried by an ion other than calcium--for example, sodium. The similarity between the transient inward current and the aftercontraction would arise because both signals were secondary to a single common event.

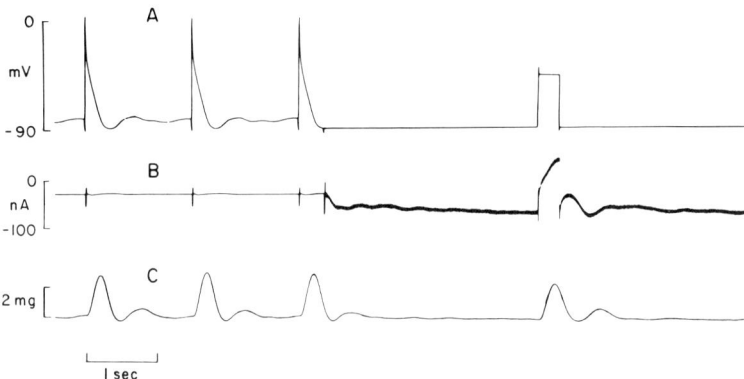

Fig. 4. Toxic effects of strophanthidin on membrane current and contractile activity. Chart records of membrane potential (A), membrane current (B), and force (C). Trace A displays the last three action potentials from a train of eleven responses. The action potentials were evoked by external shocks and were followed by transient depolarizations (TDs). Trace B shows the stimulus artifacts and the presence of a small amount of steady hyperpolarizing current. Trace C shows twitches accompanying the action potentials and aftercontractions in association with the TDs. Following the last action potential, the membrane was clamped at the maximum diastolic potential (-88 mV). This evoked a transient inward current (TI) on the current trace (B), as well as an aftercontraction (C). The right side of B shows a depolarizing clamp pulse from -88 to -44 mV for 300 msec. The depolarizing step gave rise to an inward current surge (B) and a twitch (C), and the repolarizing step evoked a TI (B) and an aftercontraction (C). Preparation R30-1; 1 µM strophanthidin; apparent cylindrical area, 0.009 cm². From Kass et al. (1978).

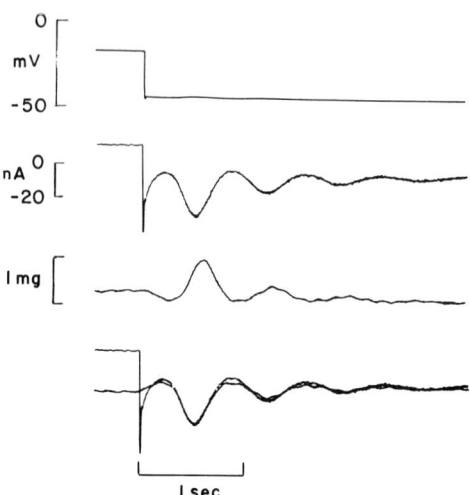

Fig. 5.   Temporal relationship between TIs and aftercontractions.   A repolarizing step to -44 mV terminated a 10 sec depolarizing pulse to -17 mV.   The repolarizing step evoked a series of inward current transient which were associated with discrete aftercontractions.   In the lower trace the TI and aftercontraction records are superimposed, the force record being inverted and advanced by 80 msec. Preparation R9-4; 1 μM strophanthidin; apparent cylindrical area 0.007 cm². From Kass et al. (1978).

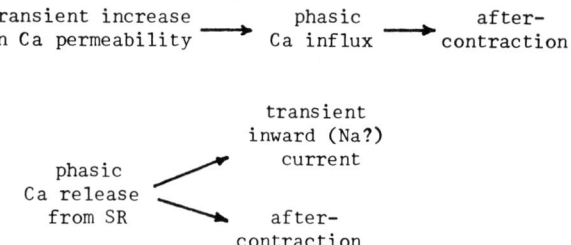

Fig. 6.   Two alternative hypotheses for the generation of transient inward current and aftercontraction.   From Kass et al. (1978).

VOLTAGE-DEPENDENT REVERSAL OF THE TI

Fig. 7 illustrates one type of experiment that helps distinguish between these hypotheses. The experiment shows how the magnitude and direction of the TI vary with the electrical driving force. The potential was stepped from a holding potential of -37 mV to +25 for 200 msec in order to induce the TI and aftercontraction. The membrane was then repolarized to various levels. Between -20 and +20 mV the current signal (B) and force signal (C) are influenced in rather different ways by varying the voltage level. As the repolarization level is displaced toward positive potentials, the aftercontraction grows somewhat larger and is followed by increasing amounts of tonic force. The associated current

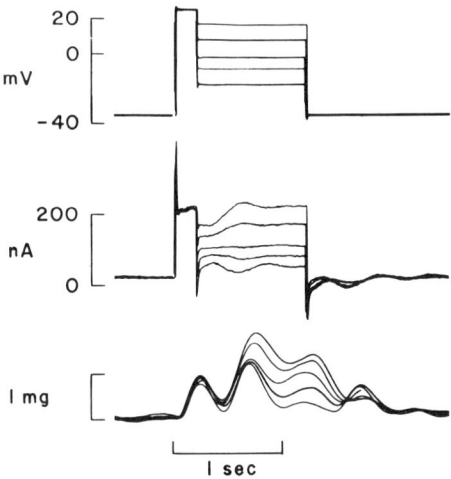

Fig. 7. Voltage-dependent reversal of the TI. Top: superimposed voltage traces from five trials during a single run. Step depolarization to +25 was followed by repolarizations to +17, +9, -1, -8, and -17, mV. Middle: associated records of membrane current (same vertical order as voltage traces). Bottom: associated contractile force (same vertical order as voltage traces). For further details, see text. Preparation R22-3; 1 µM strophanthidin; apparent cylindrical area, .009 cm^2. From Kass et al. (1978).

records show a different kind of voltage-dependence.
Transient current is inward at -17 mV, less inward at -8 mV,
and nearly flat at -1 mV. An outward transient was recorded
at +9 mV, and it becomes larger at +17 mV. Fig. 8 shows the
analysis of these records (solid circles) and other data
(open circles) from the same run. The amplitude of the
current transient follows a smooth curve which intersects the
abscissa at -2 mV. This intercept defines a reversal
potential, E_{rev}.

Voltage-dependent inversion of the TI was demonstrated
in a total of 38 runs in 15 preparations using the procedure
illustrated in Fig. 7. The reversal potential was determined
by graphical interpolation as in Fig. 8. An average value
of E_{rev} = -4.8 ± 1.2 mV (mean ± SEM, N = 38) was obtained by
pooling the results in the standard Tyrode solution.

These results rule out the idea that the TI is carried
by a calcium-specific channel. Since E_{rev} does not
correspond to the Nernst potential for any of the major ions,
one must consider the possibility of a rather nonselective
pathway. We have carried out other experiments to study the

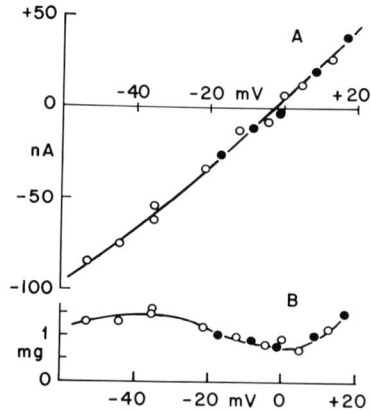

Fig. 8. Influence of repolarization potential (abscissa)
on the TI (A) and the aftercontraction (B). Transient
amplitudes were measured using results in Fig. 7 (solid
circles) and other records from the same experiments (open
circles) using sloping baselines drawn between troughs
flanking the peak in question. The smooth curves were drawn
by eye. The reversal potential, E_{rev} is given by the voltage
intercept in (A). From Kass et al. (1978).

influence of the ionic composition of the bathing solution on
E_{rev}. Replacement of external sodium with choline or Tris,
or substitution of NaCl by sucrose, causes a negative
displacement of the reversal potential. In nominally
"sodium-free" solutions, E_{rev} is displaced to about -35 mV.
On the other hand, E_{rev} was not changed by replacing
extracellular chloride with methylsulfate. The calcium and
potassium concentrations cannot be varied over a wide range
without causing unacceptably large variations in the degree
of toxicity; no significant shift in the reversal potential
was observed with limited variations in $[Ca]_o$ (1.8 to 10.8 mM)
or $[K]_o$ (1 to 8 mM).

These results are consistent with the idea that the
transient current reflects a rather nonselective increase in
membrane conductance to cations. There may be an analogy
with transmitter-induced conductances in post-synaptic
membranes such as the motor endplate, where the reversal
potential is largely determined by a combination of sodium
influx and potassium efflux. In the case of the transient
inward current, it seems likely that intracellular calcium
plays the role of a "transmitter" in causing a transient
conductance increase. We have no direct evidence about the
nature of the Ca-sensitive pathway which this hypothesis
requires. Purkinje fibers have a rather large,
time-independent "leak" conductance which contributes to the
normal pacemaker depolarization; it seems possible that
"background" or "leak" channels might be involved in carrying
the TI.

Our present thinking on the genesis of the TI may be
summarized by the following hypothesis:

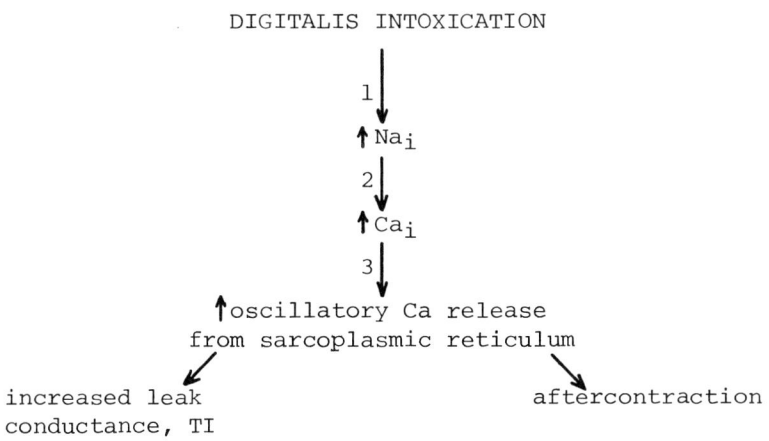

DIGITALIS INTOXICATION

1 ↓

↑ Na_i

2 ↓

↑ Ca_i

3 ↓

↑oscillatory Ca release
from sarcoplasmic reticulum

increased leak   aftercontraction
conductance, TI

In this scheme, step 1 is caused by the inhibition of the sodium pump which occurs during digitalis toxicity. Step 2 may be mediated by the Ca-Na exchange (Reuter & Seitz, 1968) or by Na-dependent release of Ca from the mitochondria (Carafoli, Tiozzo, Lugli, Crovetti & Kratzing, 1974). Step 3 occurs after the cells have reached a state of calcium overload and involves a Ca-dependent release of calcium from the sarcoplasmic reticulum of the type described by Fabiato & Fabiato (1975) and Endo (this volume). This sequence of events provides a framework for understanding the actions of agents which prevent transient depolarizations and subsequent arrhythmias. Manganese, verapamil, D600, and magnesium probably act by reducing calcium entry via the slow inward current, thus indirectly relieving the calcium overload. Local anesthetics may reduce transient depolarizations by interfering with calcium-dependent calcium release (see Endo (1977) for review). Fig. 9 shows the effect of tetracaine, a local anesthetic that has been widely used in studies of

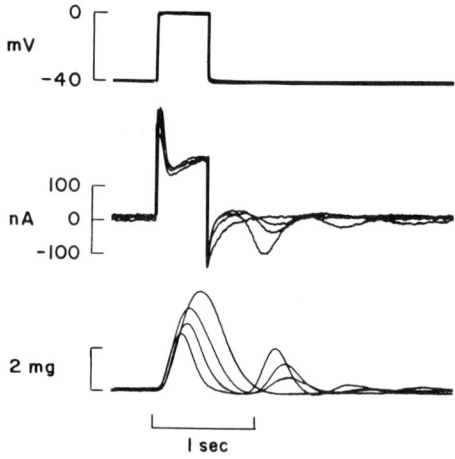

Fig. 9.   Tetracaine inhibits TIs and aftercontractions. Upper panel: depolarizing voltage clamp pulse. Middle panel: records of transient inward current before, and 3, 4, and 5 min after administration of 0.5 mM tetracaine. Lower panel: corresponding records of contractile force. Tetracaine not only reduces the TI and aftercontraction but also increased the time-to-peak of the twitch response. Preparation R38-4, 1 μM strophanthidin.

excitation-contraction coupling in skeletal muscle (see, for example, Almers & Best, 1977). Exposure to tetracaine causes a progressive inhibition of both the TI and the aftercontraction. The correlation between these effects is consistent with the idea that the local anesthetic is gradually reducing phasic calcium release from the sarcoplasmic reticulum. Although other explanations are also possible, it is exciting to think that some of the anti-arrhythmic action of local anesthetics may involve actions on SR as well as surface membrane.

THE POSITIVE INOTROPIC EFFECT OF DIGITALIS

The steps leading up to increased intracellular calcium in the scheme above were first proposed as a mechanism for increased contractility--the so-called "sodium lag" hypothesis (Baker, Blaustein, Hodgkin & Steinhardt, 1969; Glitsch, Reuter & Scholz, 1970; Langer & Serena, 1970). It seems likely that sodium pump inhibition and consequent increases in Ca_i contribute to the increase in contractile force that occurs at relatively high digitalis concentrations where arrhythmogenic effects are already clear (Fig. 1). But there is considerable controversy about the role of Na pump inhibition and increased $[Na]_i$ at low digitalis concentrations which produce increased contractile force without obvious arrhythmogenicity (see, for example, Tuttle, Wit & Farah, 1962; Gadsby, Niedergerke & Page, 1971; Rhee, Dutta & Marks, 1976; Godfraind & Ghysel-Burton, 1977; Blood and Noble, 1977a,b; Okita, 1977).

We have concentrated on a different aspect of the problem: the question of whether digitalis inotropy is associated with an increased entry of calcium via the slow inward current. Digitalis effects on the slow inward current have been suggested repeatedly (Katz, 1972; Fozzard, 1973) but have not been supported by recent voltage clamp experiments (Morad & Greenspan, 1973; Greenspan & Morad, 1975; McDonald, Nawrath & Trautwein, 1976). We now report conditions under which the slow inward current can be increased by a cardiotonic steroid.

Fig. 10 shows the influence of strophanthidin on the electrical and mechanical activity of a Purkinje fiber preparation. The records were obtained using a protocol which alternated between trains of stimulated action potentials and periods of voltage clamp in a stereotyped manner. This

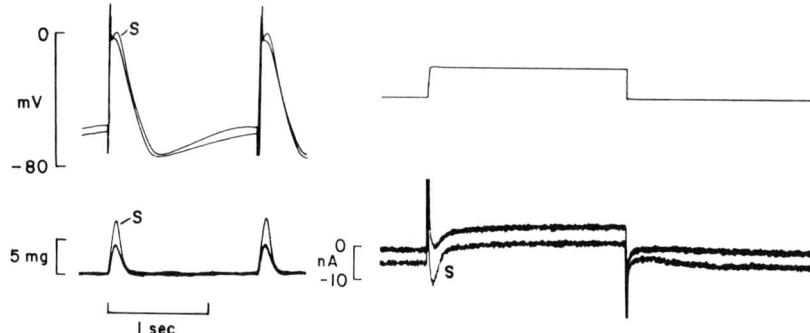

Fig. 10. Records of membrane potential, force and
membrane current in the absence and presence of strophanthidin
(1 μM, 3 min exposure). Action potentials (A) and twitches (B)
were elicited by external stimuli. Shock artifacts appear as
downward deflections preceding action potential upstrokes.
Traces in A and B show the last two responses in a train of
fourteen beats evoked during the application of a steady
hyperpolarizing current of -15 nA. The voltage clamp was
imposed at the point where trace A ends and held the membrane
potential at -38 mV for 2 sec before the depolarizing step
shown in C. The depolarizing clamp pulse to -21 mV was
associated with the current records in panel D. Records were
taken with a Brush 440 chart recorder and superimposed
photographically. The action potential overshoots were
attenuated by the recorder. The current signal was smoothed
by a low pass filter with an exponential time constant of
5 msec. Force traces which accompanied the voltage clamp
pulse were omitted for clarity. The step depolarization
produced a twitch amplitude of 0.06 mg in the absence of drug,
and 0.11 mg after exposure to strophanthidin for 3 min.
Preparation R13-1. The composition of the modified Tyrode
solution used in this and other experiments was as follows
(in mM): 150 Na, 4 K, 155.8 Cl, 5.4 Ca, 0.5 Mg, 10
Trismaleate (pH 7.2-7.4), 10 glucose. The temperature was
near 36-37° C and was held within ±0.2° C during a given
experiment. From Weingart, Kass & Tsien (1978).

procedure allowed ongoing comparison of changes in action
potential configuration, twitch height, and membrane current.
Fig. 10c shows a standard voltage clamp pulse from about -40
to -20 mV. The holding potential of -40 mV required very
little holding current and was used to eliminate interference
from the rapid excitatory sodium. Panel D superimposes the
associated current traces in the absence of drug and after
exposure to 1 µM strophanthidin for 3 min (s). The downward
surge of current following the step depolarization reflects
the activation and inactivation of the slow inward current.
Strophanthidin increases the magnitude of the slow inward
current as well as displacing the steady current levels in
the inward direction. This conclusion is supported by the
superimposed action potentials in panel A, which were
recorded within a few seconds of the current records.
Strophanthidin evokes a marked secondary depolarization during
the plateau. This change in action potential configuration
provides an additional indication of enhanced slow inward
current and is associated with a clear positive inotropic
effect (B). The experiment in Fig. 10 was done with a rather
high concentration of strophanthidin in order to obtain a
rapid and dramatic effect. But in preliminary experiments, we
have seen qualitatively similar increases in the height of
the plateau, increases in the height of the twitch, and
increases in the inward peak of the slow inward current, with
strophanthidin concentrations ranging down to 50 nM, that is,
close to a threshold dose. Thus, the increase in slow inward
current might be a general phenomenon which extends over a
broad range of cardiotonic steroid concentrations.

This result should be of interest, because as Dr. Reuter
will discuss in this symposium, catecholamines also increase
calcium currents. This is one of the principle mechanisms of
action of adrenaline on the heart. Because of the apparent
similarity of action on slow inward current, we tried to
compare the influence of strophanthidin and adrenaline on
contractions in the same preparation. Fig. 11 shows the
result we obtained. The lower right-hand panel shows a
contractile restitution curve. It is obtained by applying a
train of stimuli at a constant rate, followed by an extra
stimulus after a variable interval. The size of the last
contraction was plotted (ordinate) as a function of the
interval (abscissa). After the absolute refractory period of
the tissue, there is a progressive and fairly exponential
recovery of the ability to contract. Exposure to 1 µM
epinephrine produces a dramatic positive inotropic effect,
but no dramatic change in the time course of restitution.

The left panel shows another control curve and results from a run in the presence of strophanthidin. After exposure to the cardiotonic steroid, the recovery of tension is no longer smooth and monotonic: the size of the contraction increases, decreases, and then increases again as the interval is lengthened. Thus there is an obvious and important difference between the actions of catecholamines and digitalis.

This oscillation in the repriming time course led us to wonder whether there might be a correlate of that phenomenon in the slow inward current, and Fig. 12 shows that this is the case. Two voltage pulses lasting 2 sec were applied with

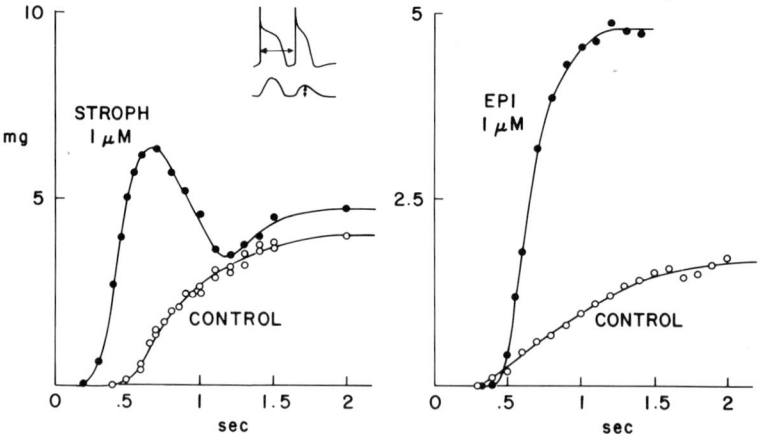

Fig. 11. *Influence of strophanthidin and adrenaline on the restitution of contractility. Each point represents a single test contraction following a regular train of about 8 beats. Twitch amplitude (ordinate) is plotted against the interval between the last regular stimulus and the test stimulus. (A) Control run (open symbols) and run beginning 14 min after exposure to 1 μM strophanthidin (solid symbols). (B) Control run taken after removal of strophanthidin (open symbols) and run beginning 3 min after exposure to 1 μM adrenaline bitartrate (solid symbols). Stimulus trains were applied at 0.5 Hz except in the adrenaline run where the rate was increased to 0.67 Hz to exceed the spontaneous frequency. Preparation R30-1. From Weingart et al. (1978).*

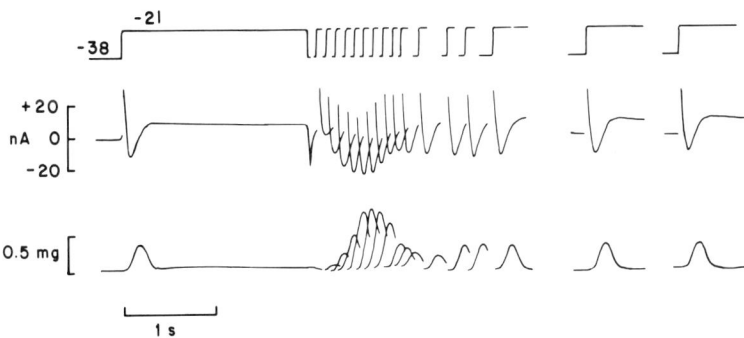

Fig. 12. Oscillatory repriming of slow inward current and twitch contraction in the presence of strophanthidin. A 2-sec depolarizing clamp pulse was followed after a variable interval by a second depolarizing step (A). Lower panels show superimposed tracings of membrane current (B) and contractile force (C). Records were taken in a run beginning 14 min after administering 1 µM strophanthidin. Preparation R13-2. From Weingart et al. (1978).

a variable interval separating them. The family of contractions shows the same oscillatory dependence on interpulse interval seen in the previous figure using action potentials. Furthermore, the envelope of the peak slow inward current also follows an oscillatory time course that increases, decreases, and then slightly increases again. Once again there seems to be good correlation between digitalis-induced changes in slow inward current and changes in the size of the contraction.

Fig. 13 shows that other mechanisms may be involved in the overall cause of positive inotropic effect. As in the other experiments, voltage clamp pulses were accompanied by twitches and surges of slow inward current. The horizontal dash indicates the level of the slow inward current in the absence of drug. Strophanthidin exposure for 4 min increased the slow inward current dramatically and also increased the twitch. After a little longer, up to 8 min, the slow inward current was no larger than at 4 min, but the twitch has continued to grow. Here, then, the correlation seems to begin to break down. And finally, after 14 min, the slow inward current became smaller, even though the tension increased even further, and one clearly sees an indication

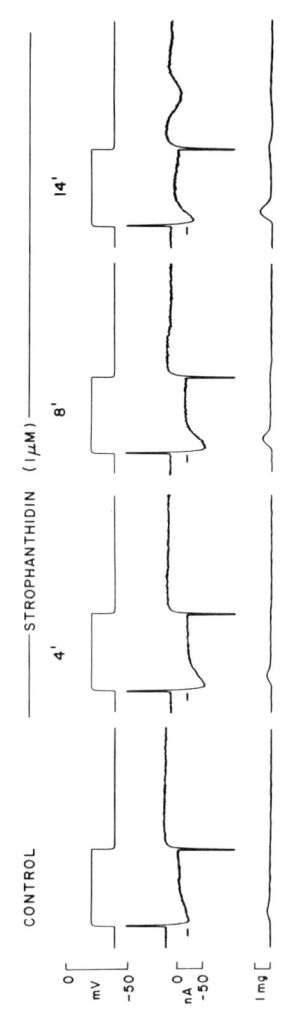

Fig. 13. Temporal separation between development of therapeutic and toxic effects at various times during continuous exposure to 1 μM strophanthidin (A), control; (B–D), 4, 8, 14 min after drug administration. Upper records show depolarizing clamp pulses from a protocal similar to that used in Fig. 1. Associated current and force records are shown below. Horizontal bars indicate level of peak slow inward current in control. Preparation R20-1. From Weingart et al. (1978).

of a toxic effect. Looking at the overall time course, the increase in calcium current comes in early and then seems to disappear. Presumably something else must be involved in producing the late positive inotropic effect.

We have looked from an electrophysiological point of view at the kinds of events that might underly both the undesirable and the desirable effects of cardiotonic steroids. The experiments raise a lot of new questions about how the surface membrane potential is related to the activity of the sarcoplasmic reticulum. Why, for example, does a repolarizing voltage pulse produce a delayed increase in calcium release which we can see just by measuring the contractions alone? How does repolarization trigger calcium release? What is the significance of the oscillatory repriming of the slow inward current? The kinetics are usually thought of in terms of Hodgkin-Huxley theory: will this description need modification? Is it possible that an oscillation in intracellular calcium somehow controls how much calcium the surface membrane lets in?

DISCUSSION

Rougier: I want to make some comment concerning the action of ouabain on the frog atrial muscle. Four years ago, we showed with Garnier that ouabain is able to increase the slow inward current in frog atrial muscle. After some time, the contraction always began to increase, and the slow inward current decreased, just as with strophanthadin.

Tsien: Thank you, I'd appreciate knowing the exact reference and I'm sorry I didn't mention the work. I should also add, since you mention the frog atrium, that Giles, Noble, and Brown have looked for an increase in slow inward current and they occasionally see very large and dramatic increases, with low concentrations of either ouabain or strophanthidin. But the effect seems to be rather variable

Reuter: If you measure the action potential under the same conditions, does the action potential then have an overshoot? Your reversal potential of the TI is 0 mV.

Tsien: The overshoot of the action potential is at least 20 mV, if not more, positive to where the apparent reversal potential is, but I haven't done it systematically.

Reuter: Do you think that there might be some sort of compartmentalization?

Tsien: As I understand it, your question concerns the nature of the pathway for the transient inward current, and in particular, its selectivity for sodium. The pathway is probably not the normal excitatory sodium channel since tetrodotoxin concentrations which abolish excitability do not block the TI. Another possibility is that the TI is carried by a rather non-specific leak channel, perhaps the same channel which generates the steady inward current underlying the normal pacemaker depolarization. The inward background current involves a tetrodotoxin-insensitive sodium influx, but its selectivity is unknown. It is possible that K ions are permeant also, since Prof. Carmeliet has found considerable K efflux even after cesium is used to block the inwardly rectifying K currents I_{K1} and I_{K2}. If both Na and K ions passed through the background current channels, their reversal potential might correspond to that found for the TI.

Edjtehadi: You said that you can use anesthetics in order to prevent the accumulation of calcium inside the cells. What sort of anesthetic did you have in mind?

Tsien: I was thinking of a local anesthetic called aprindine, an experimental antiarrhythmic drug. Elharrar and his coworkers (*Fedn. Proc. 36*, 416) report that aprindine abolishes transient depolarizations induced by acetylstrophanthidin intoxication without affecting slow response activity due to the slow inward calcium current. We find that the TI and the aftercontraction are reduced concomitantly by exposure to aprindine. Since the aftercontraction reflects intracellular Ca, it seems possible, although far from certain, that the antiarrhythmic action involves an inhibition of calcium release from the sarcoplasmic reticulum.

Morad: Perhaps you could clarify Noble's results for me. In that low concentration range that he's talking about, 10^{-9} to 10^{-7}, is digitalis in fact stimulating the pump, and is the small potentiation in tension that one sees there compatible with loss of potassium from extracellular space, which in turn could potentiate the tension?

Tsien: It's always dangerous to speak for other people who aren't here. As far as I know, Noble, Blood, Cohen, and Daut thought seriously about the external potassium hypothesis for about a year, and they liked it very much but when they finally got around to writing their 1976 papers, they decided that the facts just did not fit well enough. The main point of their results was not just that ouabain induced hyperpolarization, but the fact that the change in overall current-voltage diagram can be mimicked by lowering the bathing potassium concentration. This was their evidence that ouabain was stimulating the pump, and that the pump was depleting potassium in a restricted extracellular space. The problem with using this kind of result to explain the whole low-dose positive inotropic effect is a quantitative one. The time course of the stimulation of the pump that they observed was often transient, while the positive inotropic effects that Blood has seen in separate but parallel experiments are very much sustained. They've also looked at the effects of manipulations in external potassium, and I think the curve for potassium is not sufficiently steep to account for these results, although it may be a contributing factor.

Carmeliet: Coming back to the action of local anesthetics, is it not possible that the effect of tetracaine is not only on the sarcoplasmic reticulum but also on what one may call the background current, the sodium current, and thus in your scheme you can mark an arrow from transient inward current to oscillatory calcium release, and back, so that both influence each other?

Tsien: It is certainly possible that tetracaine and other local anesthetics block an inward background current carried by sodium. Evidence for such an action of lidocaine has been presented by Weld and Bigger (*Circulation Res. 38*, 203-208, 1976). In preliminary experiments we have found that the TI and the aftercontraction decrease together after exposure to aprindine or tetracaine. These results are consistent with action at the level of the sarcoplasmic reticulum, but an additional effect on a surface membrane pathway cannot be ruled out. What we really need is a drug which acts solely on the inward background current: this would help in identifying the inward current pathway and in testing the overall hypothesis.

Cleemann: Oscillatory phenomena are usually produced by a chain of events that couples back to where they start. Can you outline exactly what kind of event you have in mind?

Tsien: At the moment our ideas about the subcellular oscillatory mechanism are rather vague. Mechanical oscillations in intact cells and skinned preparations are well-known from the work of Endo, Fabiato and Fabiato, Glitsch and Pott and others. There is a strong similarity

between the force oscillations in skinned preparations
and spontaneous periodic fluctuations in force and membrane
current in intoxicated Purkinje fibers (*Biophys. J. 16*,
25*a*, 1976). Fabiato and Fabiato have shown that the
force oscillations in skinned fibers are in all likelihood
generated by the sarcoplasmic reticulum, and we think
the same conclusion applies to our experiments in intact
Purkinge preparations. It is believed that the
oscillation consists of a cycle of calcium uptake and
release but the presice mechanism needs further investigation.

REFERENCES

1. Almers, W. & Best, P.J. (1976). Effects of tetracaine
    on displacement currents and contraction of frog
    skeletal muscle. *J. Physiol. 262*, 583-611.
2. Baker, P.F., Blaustein, M.P., Hodgkin, A.L. &
    Steinhardt, R.A. (1969). The influence of calcium on
    sodium efflux in squid axons. *J. Physiol. 200*, 431-458.
3. Blood, B.E. & Noble, D. (1977*a*). Glycoside-induced
    inotropism of the heart--more than one mechanism?
    J. Physiol. 265, 76-77.
4. Blood, B.E. & Noble, D. (1977*b*). Two mechanisms for the
    inotropic action of ouabain on sheep cardiac Purkinje
    fibre contractility. Submitted for publication.
5. Bozler, E. (1943*a*). The initiation of impulses in
    cardiac muscle. *Am. J. Physiol. 138*, 273-282.
6. Bozler, E. (1943*b*). Tonus changes in cardiac muscle and
    their significance for the initiation of impulses.
    Am. J. Physiol. 139, 477-480.
7. Carafoli, E., Tiozzo, R., Lugli, G., Crovetti, F. &
    Kratzing, C. (1974). The release of calcium from heart
    mitochondria by sodium. *J. molec. cell. Cardiol. 6*,
    361-371.
8. Davis, L.D. (1973). Effect of changes in cycle length on
    diastolic depolarization produced by ouabain in canine
    Purkinje fibers. *Circulation Res. 32*, 206-214.
9. Endo, M. (1977). Calcium release from the sarcoplasmic
    reticulum. *Physiol. Rev. 57*, 71-108.

10. Fabiato, A. & Fabiato, F. (1975). Contractions induced by a calcium-triggered release of calcium from the sarcoplasmic reticulum of single skinned cardiac cells. *J. Physiol. 249*, 469-495.

11. Ferrier, G.R. & Moe, G.K. (1973). Effect of calcium on acetylstrophanthidin-induced transient depolarizations in canine Purkinje tissue. *Circulation 33*, 508-515.

12. Ferrier, G.R., Saunders, J.H. & Mendez, C. (1973). A cellular mechanism for the generation of ventricular arrhythmias by acetylstrophanthidin. *Circulation Res. 32*, 600-609.

13. Fozzard, H.A. (1973). Excitation-contraction coupling and digitalis. *Circulation Res. 47*, 5-7.

14. Gadsby, D.C., Niedergerke, R. & Page, S. (1971). Do intracellular concentrations of potassium or sodium regulate the strength of the heart beat? *Nature 232*, 651-653.

15. Glitsch, H.G., Reuter, H. & Scholz, H. (1970). The effect of the internal sodium concentration on calcium fluxes in isolated guinea-pig auricles. *J. Physiol. 209*, 25-43.

16. Godfraind, T. & Ghysel-Burton, J. (1977). Binding sites related to ouabain-induced stimulation or inhibition of the sodium pump. *Nature 265*, 165-166.

17. Greenspan, A.M. & Morad, M. (1975). Electromechanical studies on the inotropic effects of acetylstrophanthidin in ventricular muscle. *J. Physiol. 253*, 357-384.

18. Hogan, P.M., Wittenberg, S.M. & Klocke, J.F. (1973). Relationship of stimulation frequency to automaticity in the canine Purkinje fiber during ouabain administration. *Circulation Res. 32*, 377-384.

19. Kass, R.S., Lederer, W.J., Tsien, R.W. & Weingart, R. (1978). Role of calcium ions in transient inward currents and aftercontractions induced by strophanthidin in cardiac Purkinje fibres. *J. Physiol.* (In press).

20. Kass, R.S., Tsien, R.W. & Weingart, R. (1978). Ionic basis of transient inward current induced by strophanthidin in cardiac Purkinje fibres. *J. Physiol.* (In press).

21. Katz, A.M. (1972). Increased Ca entry during the plateau of the action potential: a possible mechanism of cardiac glycoside action. *J. molec. cell. Cardiol. 4*, 87-89.

22. Morad, M. & Greenspan, A.M. (1973). Excitation-contraction coupling as a possible site for the action of digitalis on heart muscle. In: *Cardiac Arrhythmias*, Ed. Dreifus, L.S. & Likoff, W. Grune & Stratton.

23. Langer, G.A. & Serena, S.D. (1970). Effects of strophanthidin upon contraction and ionic exchange in rabbit ventricular myocardium: Relation to control of active state. *J. molec. cell. Cardiol. 1*, 65-90.

24. Lederer, W.J. & Tsien, R.W. (1976). Transient inward current underlying arrhythmogenic effects of cardiotonic steroids in Purkinje fibres. *J. Physiol. 263*, 73-100.

25. McDonald, T.F., Nawrath, H. & Trautwein, W. (1975). Membrane currents and tensions in cat ventricular muscle treated with cardiac glycosides. *Circulation Res. 37*, 674-682.

26. Okita, G.T. (1977). Dissociation of Na^+, K^+-ATPase inhibition from digitalis inotropy. *Fedn. Proc. 36*, 2225-2230.

27. Reiter, M. (1962). Die Entstehung von "Nachkontraktionen" im Herzmuskel unter Einwirkung von Calcium und von Digitalisglykosiden in Abhangigkeit von der Reizfrequenz. *Arch. exp. Path. Pharmak. 242*, 497-507.

28. Reuter, H. & Seitz, H. (1968). The dependence of calcium efflux from cardiac muscle on temperature and external ion composition. *J. Physiol. 195*, 451-470.

29. Rhee, H.M., Dutta, S. & Marks, B.H. (1975). Cardiac NaK ATPase activity during positive inotropic and toxic actions of ouabain. *Eur. J. Pharmac. 37*, 141-153.

30. Rosen, M.R., Gelband, H., Merker, C. & Hoffman, B.F. (1973). Mechanisms of digitalis toxicity. *Circulation 47*, 681-689.

31. Tuttle, R.S., Witt, P.N. & Farah, A. (1961). The influence of ouabain on intracellular sodium and potassium concentrations in the rabbit myocardium. *J. Pharmac. exp. Ther. 133*, 281-287.

32. Withering, W. (1785). An Account of the Foxglove and Some of Its Medical Uses: With Practical Remarks on Dropsy and Other Diseases.

33. Weingart, R., Kass, R.S. & Tsien, R.W. (1978). Digitalis inotropy: Enhanced slow inward calcium current? *Nature* (In press).

TWO MECHANISMS FOR THE INOTROPIC ACTION OF OUABAIN
ON SHEEP CARDIAC PURKINJE FIBER CONTRACTILITY

*B.E. Blood**
D. Noble

University Laboratory of Physiology
Oxford University
Oxford, England

The mechanism by which the cardiac glycosides increase
the force of contraction of the heart has been the subject of
considerable discussion (Besch & Schwartz, 1970; Lee &
Klaus, 1971; Lüllmann & Peters, 1976; Blood & Noble, 1977).
On the one hand, there are reports that show a remarkably
close correlation between increased force of contraction
and inhibition of the Na^+-K^+ exchange pump, and there is a
strong school of thought that attributes this correlation
to a causal link (Akera, Larsen & Brody, 1970; Besch, Allen,
Glick & Schwartz, 1970). This idea gained further strength
from the demonstration that a reduction in the sodium
gradient by pump inhibition may lead to a rise in the
intracellular calcium, since calcium efflux is sensitive to
the sodium gradient (the "sodium lag" hypothesis) (Langer
& Serena, 1970).

On the other hand, there are also reports showing that,
under certain circumstances, pump inhibition and increased
force of contraction may be dissociated (Rhee, Dutta &
Marks, 1976; Okita, Richardson & Roth-Schechter, 1973; Eick,
Bassett & Okita, 1973; Peters, Raben & Wassermann, 1974; Ku,
Akera, Pew & Brody, 1974). We have further examined this

British Heart Foundation Junior Research Fellow.

question by studying the increased force of contraction produced in sheep cardiac Purkinje fibers exposed to ouabain concentrations between 10^{-9} and 10^{-6} M. The reason for choosing this preparation is that the influence of low doses of ouabain on the potassium and sodium gradients has recently been studied.

Cohen, Daut & Noble (1976) measured the potassium gradient by estimating the reversal potential for the K^+ specific current, i_{K2}. They found that at levels of ouabain below 5×10^{-7} M, the K^+ gradient may be increased rather than decreased. For a given dose of ouabain, the change in K^+ gradient also depended on $[K]_o$. A decrease in $[K]_o$ may convert a stimulatory effect into an inhibitory one. Similar findings have been reported for frog atrium by Loh (1975). These results also agree with the intracellular sodium activity measurements, which indicate that increases in internal sodium only occur with high doses (greater than 5×10^{-7} M) of ouabain (Ellis, 1977). By contrast, the threshold concentration for increased force of contraction in the Purkinje fiber is considerably less than 5×10^{-7} M. Blood (1975) has reported positive inotropic effects between 5×10^{-10} and 10^{-6} M. Fig. 1 shows a dose-response curve for a sheep Purkinje fiber. A 50% increase in force is produced at 5.5×10^{-9} M ouabain. At 10^{-8} M, a 100% increase in seen. This level of ouabain is close to the reported therapeutic level. Thus a maintained substantial increase in force of contraction may be obtained at ouabain levels that are well below the threshold for pump inhibition in this preparation. Given the evidence quoted above that, under at least some conditions, positive inotropic effects are strongly correlated with pump inhibition, it is important to ask whether the effects we have observed at low doses differ in any way from those which occur when the pump is inhibited. We therefore compared the inotropic responses produced by inhibitory doses with those produced by noninhibitory doses.

Purkinje fibers (length 5–10 mm, diameter 0.3–0.8 mm) were excized from freshly killed sheep cardiac ventricles and were placed in apparatus described by Blood & Spindler (1975). The Tyrode solution employed had the following composition: NaCl 140; KCl 5.4 (or as given); $NaHCO_3$ 12; NaH_2PO_4 0.4; $CaCl_2$ 2; $MgCl_2$ 1; D-glucose 11.12. (All in mM/liter.) The pH was 7.1. All experiments were carried out at a temperature of $37°$ C $\pm \frac{1}{2}°$ C. Fibers were stretched to between 150–160% of their unloaded length. Recent studies

(Thornell, Sjöstrom & Andersson, 1976) have shown that such
elongation straightens myofibrils and aligns them with the
long axis of the strand. It also produces maximal active
tension responses from the stimulated fiber (Blood, 1977a).
A low frequency of stimulation, 1/5 sec, was chosen in these
experiments because while the onset of inotropism at high
frequencies of stimulation is beat dependent (Moran, 1976),
this is reduced to a basal rate of development at low
frequencies (Vincenzi, 1976). In this way, it was possible
to achieve the greatest possible temporal separation between
the actions of low dose and high dose inotropism.

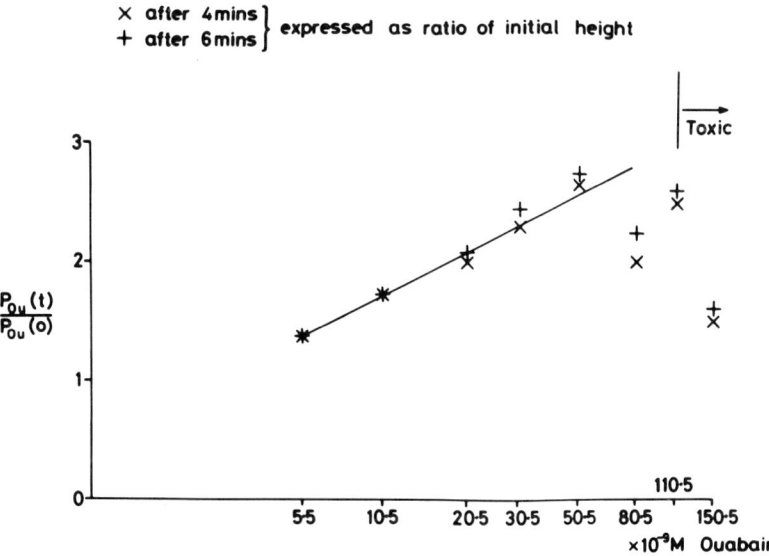

Fig. 1. *An increase in the ouabain concentration applied
to a Purkinje fiber in the range 5.5 × 10⁻⁹ to 5.05 × 10⁻⁸ M
produces a greater positive inotropic response expressed here
as the ratio of peak tension after time t ($P_{ou}(t)$) to peak
tension prior to application of ouabain ($P_{ou}(o)$). The
potassium concentration in the Tyrode solution was 4 mM.
The term toxic is used in this particular figure to denote
poor reversibility of the positive inotropic action of
concentrations of ouabain in excess of 1 × 10⁻⁷ M. The
threshold for the inotropic action of ouabain in this
particular experiment is between 1 and 5.5 × 10⁻⁹ M.*

A major characteristic of the low dose inotropism is
that it is rapid in onset and in decay. This is shown in
Fig. 2, which plots the time course of the inotropic
response to 10^{-8} M ouabain. A doubling of contraction
occurs within 3 min. The response is then constant until
the drug is removed, when it decays within a similar period
of time. Fig. 2 also shows the onset of the inotropic

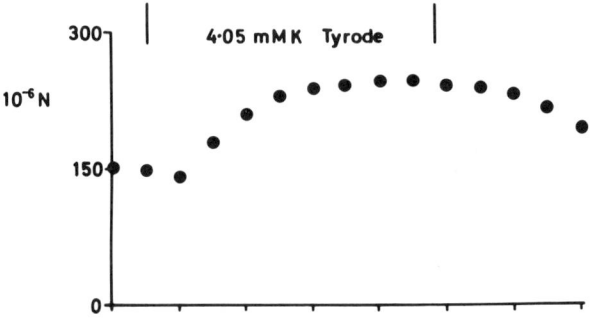

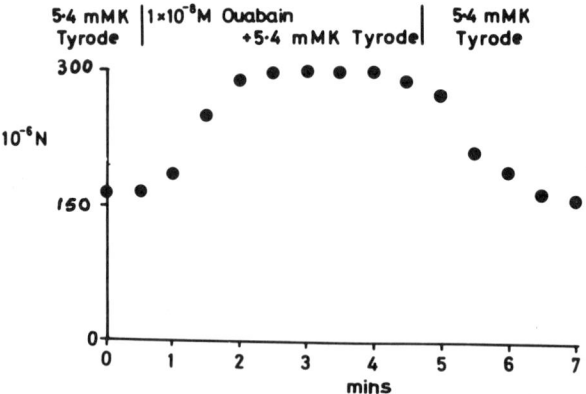

Fig. 2. *The time course of positive inotropic action of
either a reduction in external potassium concentration from
5.4 mM to 4.05 mM or the application of 1 × 10^{-8} M ouabain
in 5.4 mM K^+ Tyrode is similar. Both effects are rapidly
reversible and both produced approximately a 100% increase
in peak tension given here in absolute units. The same fiber
was used in both experiments.*

response to low K^+. The speed of onset is very similar to the ouabain response. This effect is also readily reversed. In Fig. 2 the reversal is somewhat slower than that of ouabain, but this is not a regular finding.

These results contrast with the well-established observation that inhibitory doses of ouabain produce an inotropic response that takes a considerable time to occur and is difficult to reverse (Straub, 1931). In Fig. 3 we have compared the response to 10^{-8} M and 10^{-6} M ouabain applied to the same preparation. The 10^{-8} M ouabain produces a 26% increase in force that develops within 3 min and is maintained at this level for a further 25 min.

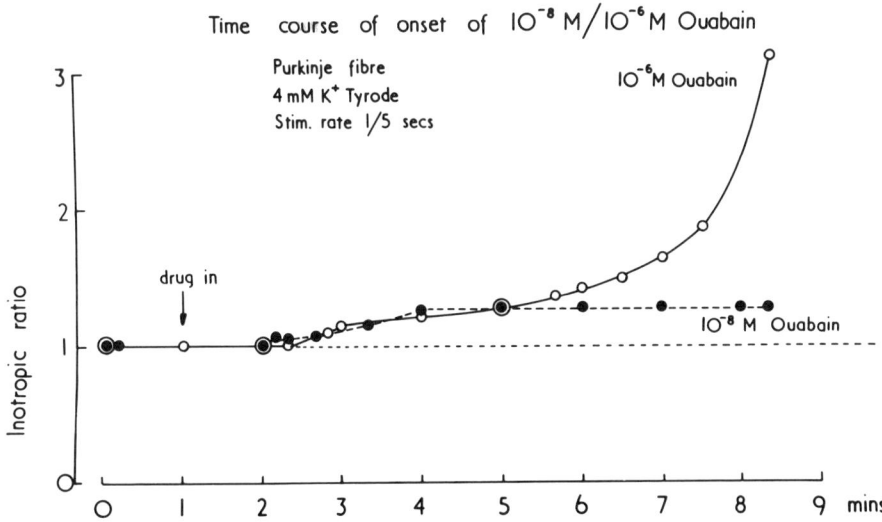

Fig. 3. *The positive inotropic action of 10^{-8} M and 10^{-6} M ouabain (expressed as an increase in the intropic ratio which is the ratio of the peak active tension at any particular time after drug application to peak active tension just prior to drug application) is similar over the first 4 min after drug application. However in the case of the lower concentration a sustained, readily reversible positive inotropic effect is observed while in the case of the higher concentration a progressive inotropic effect is observed over the next 20-25 min (see Fig. 4).*

Removal of the drug rapidly restores the initial level of contraction. Then 10^{-6} M ouabain is applied. A small inotropic response is seen for a few minutes, followed by a much larger inotropic response after 6 min. The full time course of the response to 10^{-6} M ouabain is plotted in Fig. 4. It can be seen that it requires 30 min to reach a steady-state level and is not readily reversible.

It is difficult to envisage a single action of ouabain that would explain these results. In general, we should expect the speed of approach to the steady state to be increased as the drug concentration in increased. Since the rapid onset effect is observed below the threshold for inhibition of the pump, while the slow effect is observed only at inhibitory doses, it is tempting to try to relate the two inotropic responses to the effects of ouabain on the ionic gradients. Thus, the low dose inotropism might be causally related to pump stimulation, while the high dose inotropism might be causally related to pump inhibition. However, the situation cannot be quite as simple as this.

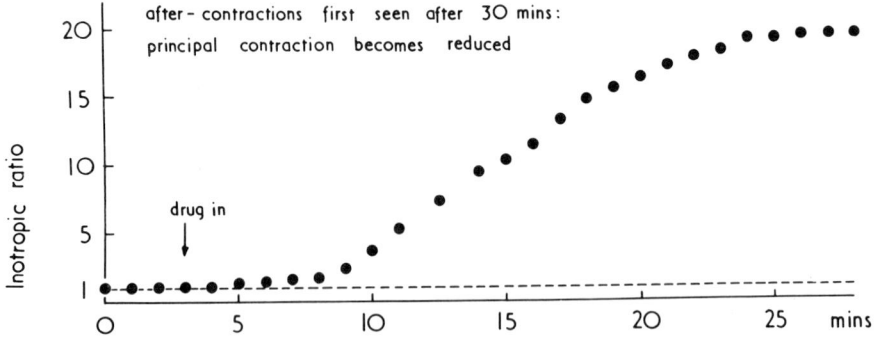

Fig. 4. *A progressive positive inotropic action of 10^{-6} M ouabain on a cardiac Purkinje fiber results in an almost irreversible peak after 20-25 min and after about a further 10-20 min after-contractions are recorded. This latter effect is associated with a loss of inotropism.*

The increase in K^+ gradient at low doses of ouabain continues to develop for much longer than the few minutes required for the inotropic action to reach its peak. It is not likely, therefore, that the low dose inotropism is causally related to the change in Na^+ or K^+ gradients. We are inclined to the view that this action, which occurs at therapeutic levels, is consequent upon some action of ouabain other than its action on the sodium and potassium gradients via the sodium pump. We have no evidence concerning the mechanism involved, although it may be related to the action of ouabain on sarcolemmal calcium binding reported by various workers (Lüllmann & Peters, 1976).

On the other hand, the high dose inotropism may be causally related to changes in Na^+ or K^+ gradients, since it occurs relatively slowly and, like the reduced gradients following pump inhibition, it is difficult to reverse. This action occurs at "toxic" levels of the glycoside, as judged by the appearance of after-contractions in the responses after prolonged exposure. Since in most preparations the high dose inotropism is very much larger than the low dose inotropism, it is easy to envisage a situation in which there may be a strong correlation between net inotropic action and pump inhibition, as observed by many workers.

It should be emphasized that, although our results suggest the possibility of more than one inotropic mechanism, this does not necessarily imply that there must exist more than one binding site for ouabain. One property of the ouabain receptor on the sodium pump is the antagonism between ouabain binding and potassium ions (Matsui & Schwartz, 1966). This antagonism is observed in the physiologically expressed potassium protection against high dose ouabain toxicity. If the effect of potassium is determined on the ouabain dose response curve it is found that increasing external potassium shifts the threshold concentration for ouabain induced inotropism and, both at high concentrations and at low concentrations, reduces the degree of inotropism induced by a particular concentration of ouabain. This is shown in Fig. 5. This similarity between the action of potassium on low and high dose inotropism may be taken as evidence for a common site of action. In the following article (Blood, 1977b) a model is presented which allows a single (inhibitory) binding site to generate net pump stimulation at low doses and net inhibition at high doses. We might suppose that the basic inotropic mechanism (such as calcium release—see above) may occur at all doses but that the effect is

secondarily enhanced when pump inhibition occurs. This
secondary enhancement may well involve the currently popular
"sodium lag" theory. A third contribution to a raised
internal calcium level which is associated with "toxic"
features of glycoside action may be sodium induced release
of mitochondrial calcium. The levels of increased
internal sodium demonstrated by Ellis (1977) could lead to
calcium release from these intracellular organelles
(Carafoli, Tiozzo, Lugli, Crovetti & Kratzing, 1974).

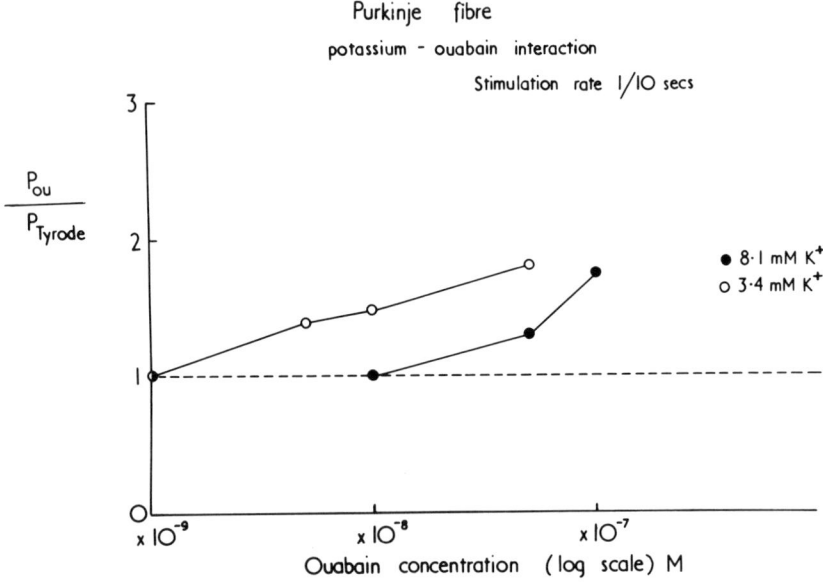

*Fig. 5. The dose-response curve for ouabain in two
different potassium containing Tyrode solutions demonstrates
that as the potassium concentration is increased so there is
a reduced inotropic effect at any particular ouabain
concentration and there is a shift in the inotropic threshold.
Each dose-response curve is normalized with respect to the
peak active tension recorded in the absence of ouabain in its
respective potassium containing Tyrode. An increase in
potassium concentration alone induced a negative inotropic
effect on cardiac Purkinge fibers. All inotropic effects
recorded in this figure were readily reversible.*

REFERENCES

1. Akera, T., Larsen, F.S. & Brody, T.M. (1970).
    Correlation of cardiac sodium- and potassium-activated
    adenosine triphosphatase activity with ouabain-induced
    inotropic stimulation. *J. Pharmac. exp. Ther. 173*,
    145-151.
2. Besch, H.R. jnr. & Schwartz, A. (1970). On a mechanism
    of action of digitalis. *J. mol. cell. Cardiol. 1*,
    195-199.
3. Besch, H.R. jnr., Allen, J.C., Glick, G. & Schwartz, A.
    (1970). Correlation between the inotropic action of
    ouabain and its effects on subcellular enzyme systems
    from canine nyocardium. *J. Pharmac. exp. Ther. 171*,
    1-12.
4. Blood, B.E. (1975). The influence of low doses of
    ouabain and potassium ions on sheep Purkinje fibre
    contractility. *J. Physiol. 251*, 69-70P.
5. Blood, B.E. (1977a). The action and interaction of
    potassium and ouabain on the contractility of cardiac
    muscle. D.Phil. thesis.
6. Blood, B.E. (1977b). Glycoside induced stimulation of
    membrane Na-K ATPase--fact or artifact? (this volume)
7. Blood, B.E. & Spindler, A.J. (1975). Apparatus for
    recording contractile activity in Purkinje fibres and
    other contractile preparations. *J. Physiol. 252*, 2-3P.
8. Blood, B.E. & Noble, D. (1977). Glycoside induced
    inotropism of the heart--more than one mechanism?
    J. Physiol. 265, 76-77P.
9. Carafoli, E., Tiozzo, R., Lugli, G., Crovetti, F. &
    Kratzing, C. (1974). The release of calcium from heart
    mitochondria by sodium. *J. molec. cell. Cardiol. 6*,
    361-371.
10. Cohen, I., Daut, J. & Noble, D. (1976). An analysis of
    the actions of low concentrations of ouabain on
    membrane currents in Purkinje fibres. *J. Physiol. 260*,
    75-103.
11. Ellis, D. (1977). The effects of external cations and
    ouabain on the intracellular Na activity of sheep
    heart Purkinje fibre. *J. Physiol. 273*, 211-240.
12. Ku, D., Akera, T., Pew, C.L. & Brody, T.M. (1974).
    Cardiac glycosides: Correlations among Na^+, K^+-ATPase,
    sodium pump and contractility in the guinea-pig heart.
    Naunyn-Schmiedeberg's *Arch. Pharmakol. 285*, 185-200.

13. Langer, G.A. & Serena, S.D. (1970). Effects of strophanthidin upon contraction and ionic exchange in rabbit vantricular myocardium: relation to control of active state. *J. mol. cell. Cardiol. 1,* 65-90.

14. Lee, K.S. & Klaus, W. (1971). The subcellular basis for the mechanism of inotropic action of cardiac glycosides. *Pharmac. Rev. 23,* 193-261.

15. Loh, C.K. (1975). Effects of cardiac glycosides on myocardial K in amphibian atrium. *Fedn. Proc. 34,* 434.

16. Lüllmann, H. & Peters, T. (1976). On the sarcolemmal site of action of cardiac glycosides. *The Sarcolemma.* pp. 311-328. Baltimore: University Park Press.

17. Matsui, H. & Schwartz, A. (1966). Kinetic analysis of ouabain K^+ and Na^+ interaction on a Na^+, K^+-dependent adenosine triphosphatase from cardiac tissue. *Biochem. biophys. Res. Commun. 25,* 147-150.

18. Moran, N.A. (1967). Contraction dependency of the positive inotropic action of cardiac glycosides. *Circulation Res. 21,* 727-740.

19. Okita, G.T., Richardson, F. & Roth-Schechter, B.F. (1973). Dissociation of the positive inotropic action of digitalis from inhibition of sodium- and potassium-activated adenosine triphosphatase. *J. Pharmac. exp. Ther. 185,* 1-11.

20. Peters, T., Raben, R.H. & Wassermann, O. (1974). Evidence for a dissociation between positive inotropic effect and inhibition of the Na^+-K^+-ATPase by ouabain, cassaine and their alkylating derivatives. *Eur. J. Pharmac. 26,* 166-174.

21. Rhee, H.M., Dutta, S. & Marks, B.H. (1976). Cardiac NaK ATPase activity during positive inotropic and toxic actions of ouabain. *Eur. J. Pharmac. 37,* 141-153.

22. Straub, W. (1931). *Lane Lectures in Pharmacology,* Vol. 3. Palo Alto: Stanford University Press.

23. Ten Eick, R.E., Bassett, A.L. & Okita, G.T. (1973). Dissociation of electrophysiological and inotropic actions of strophanthidin-3-bromoacetate: possible role of adenosine triphosphatase in the maintenance of the myocardial transmembrane Na^+ and K^+ gradients. *J. Pharmac. exp. Ther. 185,* 12-23.

24. Thornell, L-E., Sjöstrom, M. & Andersson, K-E. (1976). The relationship between mechanical stress and myofibrillar organisation in heart Purkinje fibres. *J. mol. cell. Cardiol. 8,* 689-695.

25. Vincenzi, F.F. (1967). Influence of myocardial activity on the rate of onset of ouabain action. *J. Pharmac. exp. Ther. 155,* 279-287.

GLYCOSIDE INDUCED STIMULATION
OF MEMBRANE Na-K ATPase--FACT OR ARTIFACT?

*Brian E. Blood**

University Laboratory of Physiology
Parks Road
Oxford, England

Recent work on guinea-pig atrium (Ghysel-Burton &
Godfraind, 1975), amphibian atrium (Loh, 1975), calf
ventricular muscle (Peters, Raben & Wassermann, 1974), sheep
cardiac Purkinje fibers (Cohen, Daut & Noble, 1976), dog
heart (Steiness & Valentin, 1976), skeletal muscle (Moore,
1972), and cultured HeLa cells (Lamb, 1977) has suggested
that at low concentrations some cardiac glycosides induce a
stimulation of the Na-K ATPase (EC 3.6.1.3) colloquially
termed the "sodium pump." The range of glycoside
concentrations over which this effect is observed corresponds,
in cardiac tissues, to in vivo therapeutic levels (Smith &
Haber, 1974), and in calf ventricular muscle (Peters *et al.*
1974), guinea-pig atrium (Ghysel-Burton & Godfraind, 1975),
sheep cardiac Purkinje fibers (Blood, 1975), and dog heart
(Steiness & Valentin, 1976), it is accompanied by positive
inotropism. The evidence for the stimulatory action of some
cardiac glycosides on isolated enzyme preparations is
contradictory; some workers consider it a well-established
phenomenon (Lee & Klaus, 1971), while others have been unable
to reproduce it (Schwartz, Lindenmayer & Allen, 1975; Glynn
& Karlish, 1975). Failure to observe the stimulatory action
of glycosides when applied to membrane fractions with a high
specific Na-K ATPase activity is not altogether surprising.
Increased mechanical or chemical disruption, which

**British Heart Foundation Junior Research Fellow*

characterizes modern procedures for the extraction of
material with a high specific Na-K ATPase activity
(Jorgensen & Skou, 1969; Matsui & Schwartz, 1966; Nakao,
Tashima, Nagano & Nakao, 1965; Schoner, vol Ilberg, Kramer
& Seubert, 1967), might irreversibly eliminate certain
properties which would be observed in less drastically
treated membrane material (Brown, 1966; Blood, 1977; Repke,
1963; Bonting, Hawkins & Canady, 1964). This phenomenon is
an example of allotropy (Racher, 1967). While it is possible
that excessive isolation of the enzyme has inactivated a
specific stimulatory site on the lipo-protein ATPase
(McClane, 1965), an alternative approach may be examined in
which it is assumed that neighboring enzymes interact to
modify total enzyme activity. Features of interactions
between plasma membrane constituents, the structural
proteins, lipids, and enzymes, have been demonstrated
experimentally (Coleman, 1973; Lüllmann, Peters, Preuner &
Rüther, 1975; Fiehn & Seiler, 1977) and treated theoretically
(Marcelja, 1976; Schindler & Seelig, 1975). Isolation
techniques that increase specific enzyme purity by extracting
apparently nonessential structural proteins and lipid, while
preserving the enzyme-lipid associations necessary for basic
activity (Roelofsen & Van Deenen, 1973), will destroy some
of the features associated with the integral membrane matrix,
including the basic liquid-crystal structure of membrane
lipids (Bresler & Bresler, 1974).

For the purpose of this model, it is assumed that pump
sites are spread homogeneously over the plasma membrane
surface, with a density of s per unit surface area, and that
the interaction between neighboring pumps is inhibitory.
This inhibition will be expressed as a steric factor $f(r)$
where r is the distance between pumps. The activity of a
pump infinitely separated from its neighbors is determined
by its local ionic environment. This activity, p, is
reduced, however, by the steric inhibition to a_p, where a_p
is given by the expression

$$a_p = p - 2\pi ps \int_0^\infty f(r) \cdot r \; dr \tag{1}$$

If A_t is the activity per unit mass wet weight of tissue and
R is the ratio of membrane surface area to unit mass wet
weight of tissue, then

$$A_t = Rsp - Rs^2pF \tag{2}$$

where $F = 2\pi \int_0^\infty f(r) \cdot r \, dr$. F may be derived from a statistical model assuming discrete sites nonhomogeneously spaced over the membrane, but this alternative approach does not affect the general conclusions made from the model.

If a drug is applied to the tissue which inhibits total pump activity by forming inactive drug-enzyme complexes, and if it can be assumed that these inactive complexes exert no steric inhibition upon their neighbors, it is possible to simulate low drug concentration net stimulation of activity and high drug concentration net inhibition of activity. It is generally accepted that cardiac glycosides are taken up by specific and nonspecific sites (Godfraind & Lesne, 1972). If K_a and K_b are the equilibrium constants for drug-nonspecific site and drug-specific site interactions, respectively, and a and b are the capacities of the two types of binding site, then U, drug uptake per unit wet weight of tissue, is given by

$$U = \frac{aG}{G + K_a} + \frac{bG}{G + K_b} \tag{3}$$

where G is the free drug concentration. Note that in this model b is identical to s. If $K_a \gg G$, then equation (3) reduces to the form derived experimentally by Godfraind & Lesne (1972). The result of drug binding to the specific site is to reduce the effective density of interacting sites. Thus the density of sites will be reduced from s to

$$\frac{s}{(1 + GK_b^{-1})}.$$ Most chemical data are expressed as

the ratio of activity in the presence of a drug concentration G to that in the absence of any drug. This ratio $A_r(G)$ is given therefore by the expression

$$A_r(G) = \frac{1 + GK_b^{-1} - sF}{(1 + GK_b^{-1})^2 \, (1 - sF)} \tag{4}$$

The condition that pump stimulation will be observed is given by $\left(\frac{dA}{dG}\right) r > 0$, which reduces to the form $sF > \frac{1}{2}(1 + GK_b^{-1})$. Fig. 1 shows a plot of equation (4) for different values of aF. Using an equation of this form, it has been possible to fit a curve to the data of Bonting, Hawkins & Canady (1964), for the action of erythrophleine on rat brain ATPase activity (see Fig. 2). The erythrophleum alkaloids are glycoside-like in their actions on Na–K ATPase (Schwartz *et al.* 1975; Bonting *et al.* 1964). From this curve-fitting exercize, it

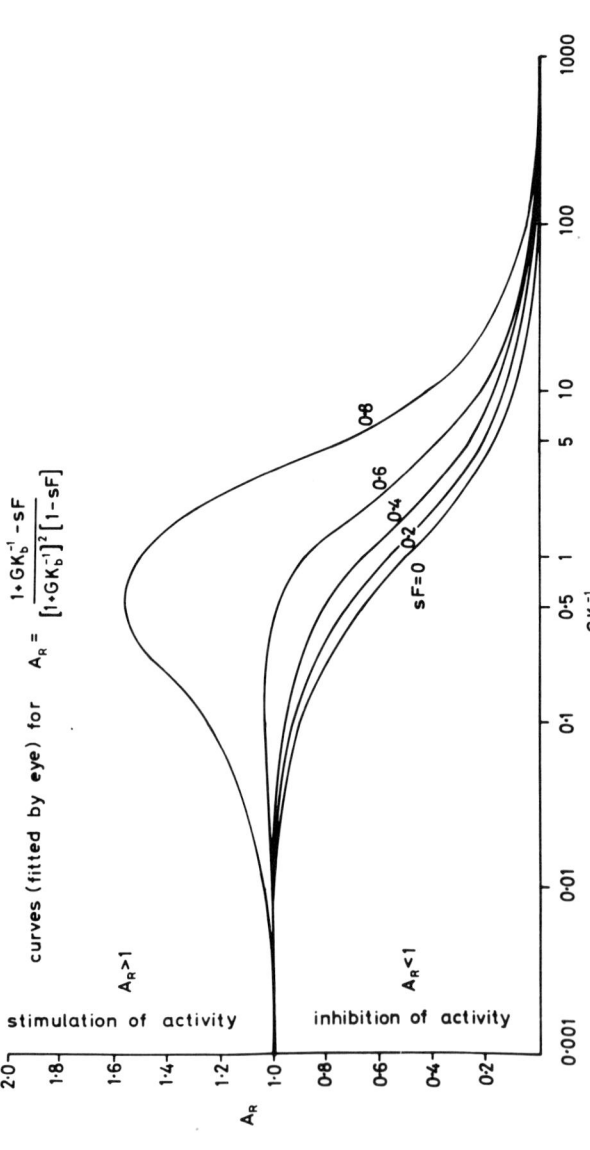

Fig. 1. *The variation of the activity–drug concentration profile with different degrees of site-site interaction, denoted by different values of sF, demonstrates the ability of this particular model to produce different amounts of low drug concentration induced increases in enzyme activity (Ar > 1) as well as higher drug concentration induced decreases in enzyme activity (Ar < 1). These effects are normally termed stimulation and inhibition, respectively.* GK_b^{-1}, *A_r, and sF are dimensionless numbers.*

is possible to determine the values 6.67×10^{-8} M for K_b and 0.595 for sF. It has not been possible to reproduce, so strikingly, the data of Bonting *et al.* relating to the action of ouabain and cassaine (another glycoside-like drug) on the same preparation, and this fact may identify simplifications in the model which cannot hold under all circumstances. It has been assumed that the steric factor $f(r)$ depends only on the distance between pumps and not on the concentration or structure of the applied inhibitor. Nonspecific binding of the drug to nonpump proteins could modify the steric interaction factor by altering interpump protein structures. The nature of this protein structure modification will depend on the structure of the applied drug. Similarly, the degree of relief of interpump inhibition by the inactivation of a pump may not be 100% and may significantly depend on the particular pump inhibitor applied. In addition, lipid soluble glycosides may modify lipid-enzyme interactions which a more sophisticated model will have to consider.

A major assumption of this model is that the stimulatory action follows the binding of the glycoside to the specific inhibitory site of the Na-K ATPase. It is well known that an increase in the potassium concentration in the incubation medium reduces the inhibition induced by high glycoside concentrations (Palmer & Nechay, 1964). The biochemical antagonism between potassium and ouabain finds a clinical correlate in the protection that elevated serum potassium gives against glycoside toxicity (Moe & Farah, 1968). The contention that the stimulatory and inhibitory sites are distinct (McClane, 1965) was supported by the lack of potassium influence on ouabain-induced Na-K ATPase stimulation and by the loss of the stimulatory effect after aging for 14 days at $-20°$ C (Palmer, Lasseter & Melvin, 1966). The inhibitory effect was not affected by aging. The latter distinction can be explained on the basis of structural modification that accompanies aging. The effect of potassium provides a more substantial test of the model developed above.

The lack of potassium effect on ouabain induced Na-K ATPase stimulation was parallelled by the earlier failure to observe any influence of potassium levels on the therapeutic, inotropic effect of ouabain on cat papillary muscle (Garb & Venturi, 1954). However, more recent evidence suggests that the potassium concentration is indeed a critical determinant of whether glycoside application is to induce a stimulatory or an inhibitory response. Reducing the external potassium concentration changes a stimulatory concentration to one

$$A_R = \frac{1 + AG}{[1 + BG]^2} = \frac{1 + 38 \cdot 4 \, G \, (\mu M)}{[1 + 15 \, G \, (\mu M)]^2}$$

× from data of Bonting, Hawkins and Canady (1964)
on activity of rat brain ATP-ase

•—• from equation

data for Erythrophleine

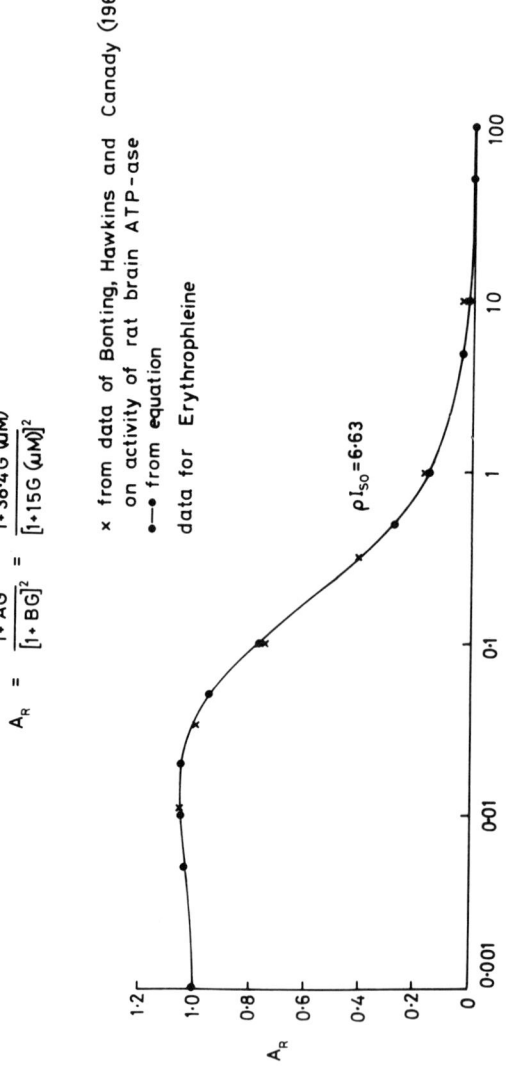

$\rho I_{50} = 6 \cdot 63$

$G \, (\mu M)$

A_R

Fig. 2. Bonting, Hawkins & Canady (1964) published data on the effect of erythrophleine, a glycoside-like alkaloid, on the activity of isolated rat brain ATPase. Their results are shown in this figure by crosses (x). Using an equation derived from the model described in this paper, it has been possible to reproduce the published data; theoretical results are shown by the filled circles joined by the curve (•—•).

which produces net pump inhibition in amphibian atrium (Loh, 1975) and sheep cardiac Purkinje fibers (Cohen *et al.* 1976). An increase in bathing potassium has also been shown to move the ouabain inotropic threshold to higher drug concentrations as well as reducing the inotropism measured in sheep cardiac Purkinje fibers exposed to any particular ouabain concentration (Blood & Noble, 1977*a*). A reassessment of the earlier evidence that stimulation is unaffected by variation of the external potassium level (McClane, 1965; Palmer *et al.* 1966) shows that experimental error was large enough to make the observation of Na-K ATPase stimulation--potassium interaction--difficult. The stimulation effect is rarely greater than 20%. In addition, the choice of ouabain concentration (1×10^{-10} M) by Palmer *et al.* (1966), one which produced close to the maximum stimulation of their Na-K ATPase preparation, was unfortunate, for it lay on a part of the dose-response curve relatively insensitive to an order of magnitude change in the ouabain concentration. This would correspond to a similar insensitivity to changes in the external potassium concentration. McClane (1965) used a larger ouabain concentration (1×10^{-8} M), and over short periods of exposure he could detect an increase in ATPase stimulation if the external potassium level was increased from 3.4 to 14 mM. After a longer exposure time, increased experimental uncertainty obscured this effect by reducing its statistical significance. This does not, therefore, appear to provide strong evidence for postulating two different sites to explain the biphasic action of glycosides on Na-K ATPase activity.

Considerable doubt has been expressed that glycoside-induced pump stimulation is directly responsible for the inotropic action at these concentrations. Thus, the inotropic response develops more rapidly than, and stabilizes sooner than, either the increase in enzyme activity (Peters *et al.* 1975) or the increase in the transmembrane potassium gradient which is produced by the increased inward pumping of potassium (Cohen *et al.* 1976; Blood & Noble, 1977*b*). The considerable changes in protein conformation that accompany the binding of ouabain, even at concentrations as low as 5×10^{-9} M, are thought to be accompanied by the release of calcium from the plasmalemma (Lüllmann *et al.* 1975). It is the increased availability of this calcium which may be responsible for the positive inotropism (Lüllmann *et al.* 1975). The calcium is displaced from many more sites than those which are directly associated with the Na-K ATPase. It is possible that the displacement of calcium from intrinsic specific Na-K ATPase binding sites, occupancy of which inhibits enzyme activity (Besch & Schwartz, 1970), leads to part of the observed Na-K ATPase stimulation. This effect must precede the establishment of a significant increase in internal potassium concentration. Increases in net enzyme activity will depend on the proportion of released calcium that has been displaced from the specific calcium-enzyme inhibition site. If efficient calcium chelators [e.g. EGTA (Blood, 1977)] have been used in the preparation of the Na-K ATPase fraction, much of the membrane bound calcium will have been removed. While this may increase the basal enzyme activity, it will also remove a mechanism by which glycosides can induce pump stimulation. In either case, a close temporal relationship between inotropism and pump stimulation need not exist. The onset of pump inhibition induced inotropism is well correlated with the loss of activity of enzyme preparations exposed to high concentrations of cardiac glycosides (Blood & Noble, 1977a).

The major feature of glycoside and erythrophleum alkaloid interactions with Na-K ATPase lipo-protein complex can be reproduced by this model without the need to postulate the existence of a specific stimulatory site on the enzyme. The failure to observe stimulation in highly purified membrane fractions is explained by the reduced integrity of the isolated material when compared with the native membrane matrix, and by the corresponding reduced membrane component interactions. Recent work showing the influence of potassium ions on low dose glycoside-induced inotropism and on Na-K ATPase stimulation strongly supports the contention that it is the action of glycosides on the Na-K ATPase receptor which is the initial interaction responsible for glycoside-induced pump stimulation, pump inhibition, and low and high dose glycoside inotropism. Further biochemical work is clearly necessary to clarify the mechanism by which low concentrations of certain glycosides induce Na-K ATPase stimulation. A clearer understanding of the way in which nonenzyme components are important in the drug--Na-K ATPase interaction through the use of techniques such as proton magnetic resonance (Kagan, Sibel'dina, Ritov, Kobeler, Kozlov & Kayushin, 1975), for example, is also vital. While this paper demonstrates that the existence of a stimulatory site is not essential to explain experimental evidence, it cannot be concluded that such a site does not exist, and that it is more labile than the well-studied inhibitory site.

REFERENCES

1. Besch, H.R. Jr. & Schwartz, A. (1970). On a mechanism of action of digitalis. *J. molec. cell. Cardiol. 1,* 195-199.
2. Blood, B.E. (1975). The influence of low doses of ouabain and potassium ions on sheep Purkinje fibre contractility. *J. Physiol. 251,* 69-70.
3. Blood, B.E. (1977). The action and interaction of potassium and ouabain on the contractility of cardiac muscle. D.Phil. thesis.
4. Blood, B.E. & Noble, D. (1977a). Glycoside induced inotropism of the heart--more than one mechanism? *J. Physiol. 265,* 76-77P.
5. Blood, B.E. & Noble, D. (1977b). Two mechanisms for the inotropic action of ouabain on sheep cardiac Purkinje fibre contractility. (this volume)
6. Bonting, S.L., Hawkins, N.M. & Canady, M.R. (1964). Studies of sodium-potassium activated adenosine triphosphate. VIII - Inhibition by erythrophleum alkaloids. *Biochem. Pharmac. 13,* 13-22.
7. Bresler, S.E. & Bresler, V.M. (1974). The liquid crystal structure of biological membranes. *Dok. Akad. Nauk SSSR 214* No. 4. 936-939.
8. Brown, H.D. (1966). A characterisation of the ouabain sensitivity of heart microsomal ATPase. *Biochim. biophys. Acta. 120,* 162-165.
9. Cohen, I., Daut, J. & Noble, D. (1976). An analysis of the actions of low concentrations of ouabain on membrane currents in Purkinje fibres. *J. Physiol. 260,* 75-103.
10. Coleman, R. (1973). Membrane-bound enzymes and membrane ultrastructure. *Biochim. biophys. Acta. 300,* 1-30.
11. Fiehn, W. & Seiler, D. (1977). (Na^+, K^+)-ATPase activity and sterol composition of membranes. *Arch. Pharmac. 297,* S21.
12. Garb, S. & Venturi, V. (1954). The different actions of potassium on the therapeutic and toxic effects of ouabain. *J. Pharmac. exp. Ther. 112,* 94-98.
13. Ghysel-Burton, J. & Godfraind, T. (1975). Stimulation and inhibition by ouabain of the sodium pump in guinea-pig atria. *Br. J. Pharmac. 55,* 249.
14. Glynn, I.M. & Karlish, S.J.D. (1975). The sodium pump. *Ann. Rev. Physiol. 37,* 13-55.

15. Godfraind, T. & Lesne, M. (1972). The uptake of cardiac
     glycosides in relation to their actions in isolated
     cardiac muscle. *Br. J. Pharmac. 46,* 488-497.
16. Jorgensen, P.L. & Skou, J.C. (1969). Preparation of
     highly active (Na^+-K^+) - ATPase from the outer medulla
     of rabbit kidney. *Biochem. biophys. Res. Commun. 37,*
     39-46.
17. Kagan, V.E., Sibel'dina, L.A., Ritov, V.B., Kobeler, V.S.,
     Kozlov, Yu.P. & Kayushin, L.P. (1975). An investigation
     of protein-lipid interactions in sarcoplasmic reticulum
     membranes by the technique of high-resolution proton
     magnetic resonance. *Dok. Akad. Nauk SSSR 222* No. 5.
18. Lamb, J.F. (1977). Unpublished observations.
19. Lee, K.S. & Klaus, W. (1971). The subcellular basis for
     the mechanism of inotropic action of cardiac glycosides.
     Pharmac. Rev. 23, 193-261.
20. Loh, C.K. (1975). Effects of cardiac glycosides on
     myocardial K in amphibian atrium. *Fedn. Proc. 34,* 434.
21. Lüllmann, H., Peters, T., Preuner, J. & Rüther, T.
     (1975). Influence of ouabain and dihydroouabain on the
     circular dichroism of cardiac plasmalemmal microsomes.
     Arch. Pharmac. 290, 1-19.
22. Marcelja, S. (1976). Lipid-mediated protein interaction
     in membranes. *Biochim. biophys. Acta. 455,* 1-7.
23. Matsui, H. & Schwartz, A. (1966). Purification and
     properties of a highly active ouabain-sensitive Na^+,
     K^+ - dependent adenosinetriphosphatase from cardiac
     tissue. *Biochim. biophys. Acta. 128,* 380-390.
24. McClane, T.K. (1965). A biphasic effect of ouabain on
     sodium transport in the toad bladder. *J. Pharmac. exp.
     Ther. 148,* 106-110.
25. Moe, G.K. & Farah, A.E. (1968). Digitalis and Allied
     Cardiac Glycosides. *The Pharmacological Basis of
     Therapeutics,* ed. Goodman, L.S. & Gilman, A. New York:
     Macmillan.
26. Moore, R.D. (1972). Effect of acetylstrophanthidin upon
     the sodium pump: Inhibition and stimulation. *Abs.
     biophys. Soc. U.S.A. FPM G.1.*
27. Nakao, T., Tashima, Y., Nagano, K. & Nakao, M. (1965).
     Highly specific sodium-potassium-activated ATPase from
     various tissues of rabbit. *Biochim. biophys. Res.
     Commun. 19,* 755.
28. Palmer, R.F. & Nechay, B.R. (1964). Biphasic renal
     effects of ouabain in the chicken: correlation with a
     microsomal Na^+-K^+ stimulated ATPase. *J. Pharmac. exp.
     Ther. 146,* 92-98.

29. Palmer, R.F., Lasseter, K.C. & Melvin, S.L. (1966). Stimulation of Na$^+$ and K$^+$ dependent adenosin triphosphatase by ouabain. *Arch. Biochem. Biophys. 113*, 629-633.

30. Peters, T., Raben, R-H & Wassermann, O. (1974). Evidence for a dissociation between positive inotropic effect and inhibition of the Na-K ATPase by ouabain, cassaine and their alkylating derivatives. *Eur. J. Pharmac. 26*, 166-174.

31. Racher, E. (1967). Resolution and reconstitution of the inner mitochondrial membrane. *Fedn. Proc. 26*, 1335-1340.

32. Repke, K. (1963). Metabolism of cardiac glycosides. *New Aspects of Cardiac Glycosides*, ed. W. Wilbrandt, pp. 47-73. New York: Pergamon Press.

33. Roelofsen, B. & Van Deenen, L.L.M. (1973). Lipid requirement of membrane-bound ATPase: Studies on human erythrocyte ghosts. *Eur. J. Biochem. 40*, 245-257.

34. Schindler, H. & Seelig, J. (1975). Deuterium order parameters in relation to thermodynamic properties of a phospholipid bilayer. A statistical mechanical interpretation. *Biochemistry, N.Y. 14*, 2283-2287.

35. Schoner, W., von Ilberg, C., Kramer, R. & Seubert, W. (1967). On the mechanism of Na$^+$- and K$^+$-stimulated hydrolysis of adenosine-triphosphate. 1. Purification and properties of a Na$^+$- and K$^+$-activated ATPase from ox brain. *Eur. J. Biochem. 1*, 334-343.

36. Schwartz, A., Lindenmayer, G.E. & Allen, J.C. (1975). The sodium-potassium adenosin triphosphatase: Pharmacological, physiological and biochemical aspects. *Pharmac. Rev. 27*, 1-134.

37. Smith, T.W. & Haber, E. (1974). *Digitalis*. Boston: Little, Brown.

38. Steiness, E. & Valentin, N. (1976). Myocardial digoxin uptake dissociation between digitalis-induced inotropism and myocardial loss of potassium. *Br. J. Pharmac. 58*, 553-560.

ANALYSIS OF CATECHOLAMINE ACTION
IN FROG ATRIAL FIBERS

Sally Page

Department of Biophysics
University College London
Gower Street
London WC1E 6BT, England

It is well known that the β-action of catecholamines in the heart induces enhanced membrane adenylate cyclase activity which, in turn, increases the levels of cyclic AMP and enhances protein kinase activity (Robison, Butcher & Sutherland, 1971). It has also been shown that catecholamines increase the slow calcium inward current (Vassort, Rougier, Garnier, Sauviat, Coraboeuf & Gargouïl, 1969; Reuter, 1974) crossing the cellular membrane during the action potential. However, there is still uncertainty about the way in which these processes are linked and the time course with which they occur. I present here some of the results from the analysis of the β-action of the catecholamines in the frog heart (Niedergerke & Page, 1977) relating to these questions.

The preparation used is a single atrial trabecula, 80-120 μm wide. The perfusion chamber allows the fluid composition around the trabecula to be changed within 0.2 sec, but equilibration of solutes at muscle cell surfaces is considerably slower than this, because access is impeded by a layer of endothelial cells surrounding the muscle fibers. Thus, diffusion equilibration at external heart cell surfaces occurs with a half-time of 1.5-4 sec for calcium ions, for example, and can be determined for a given preparation from the time course of the response to a step change of calcium concentration in the perfusing fluid (Lammel, Niedergerke & Page, 1975). The diffusion coefficient of adrenaline is probably similar in magnitude to that of calcium ions, and its diffusion time therefore should also be similar.

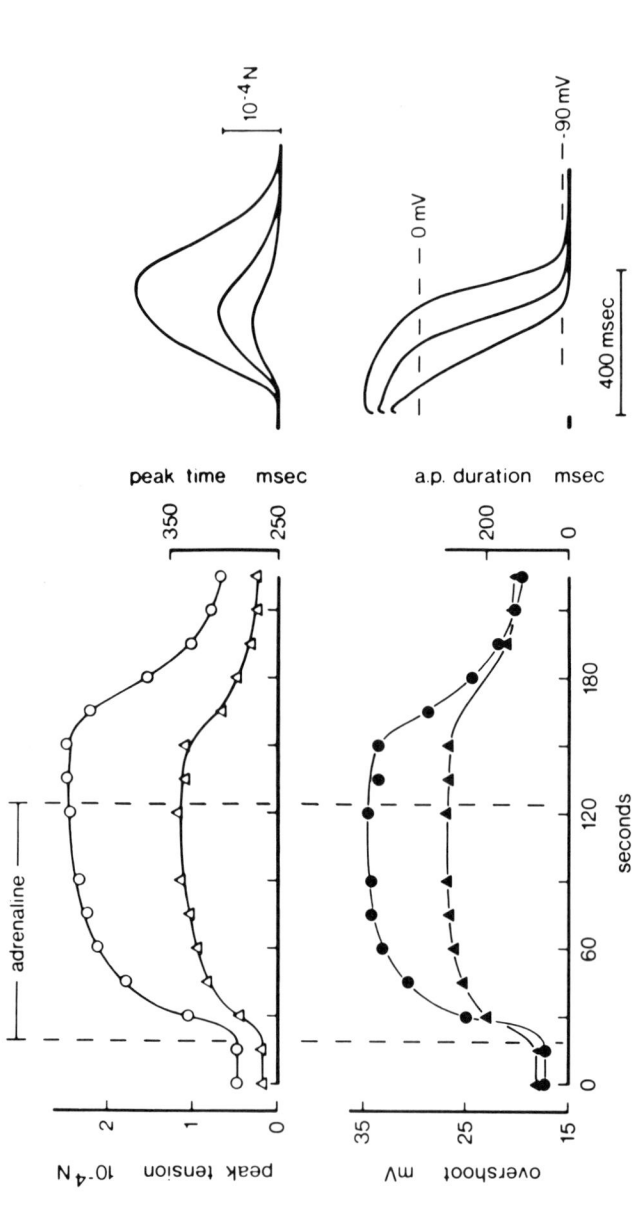

Fig. 1. Adrenaline response of a single heart trabecula. Upper graph: peak tension (open circles) and peak twitch time (open triangles), the latter estimated from the end of the stimulus artifact (not shown) to the peak of the twitch. Lower graph: overshoot of action potential (filled circles) and duration of action potential (filled triangles) determined at zero potential level. Right: superimposed tracings of three tension and potential records, obtained before, at the initial, and at a later, more fully developed state of adrenaline action. 1 mM $[Ca]_o$, 10^{-6} M adrenaline, stimulus rate 4 min^{-1}.

Fig. 1 illustrates the changes in the parameters we measure, the isometric twitch tension and the action potential, produced in this case by a just maximal dose of adrenaline. The most striking features are (a) the parallelism in time course of twitch and action potential changes, during both the buildup and the decline of the response, and (b) the relative slowness of the changes compared with diffusion equilibration.

The parallelism between electrical and mechanical changes suggested that tension increases as the immediate consequence of the increased calcium inward current. However, since frog heart cells contain a sarcoplasmic reticulum, albeit a rather sparse one (Page & Niedergerke, 1972), it seemed worth inquiring whether progressive accumulation of calcium within this cellular store contributes to the catecholamine response observed. In the experiments concerned with this question (Fig. 2), the likelihood of cellular calcium accumulation during development of the catecholamine action was much reduced by lowering the calcium influx either (a) by cessation of stimulation from the moment of drug application onwards (Fig. 2a), or (b) by lowering the external calcium concentration at the moment of drug application (Fig. 2b). In the first case, the catecholamine action was followed by applying, in successive runs, single test shocks at various times during drug exposure. Comparison with the adrenaline response obtained during continuous activity showed that neither the time course nor the final level differed greatly under these conditions.

In the second type of experiment, calcium was either withdrawn altogether, or reduced, during the first minute of adrenaline exposure, the time taken for steady tension levels to be attained at constant 1 mM calcium. As can be seen, the first twitch on restoring the original $[Ca]_o$ was already more than 90% of the final tension increment and was indeed of just the size expected from extracellular equilibration of calcium, whose time course is indicated.

The conclusion from both types of experiment is that calcium accumulation in a cellular store is not essential for the catecholamine action in the frog heart. This is in contrast to the situation existing in mammalian heart cells (Reuter, 1974).

Fig. 3 shows the time course of the catecholamine response in more detail. During maintained exposure to the drug, twitch tension rises in an S-shaped fashion after an initial latency, of about 8 sec in this case. If the drug

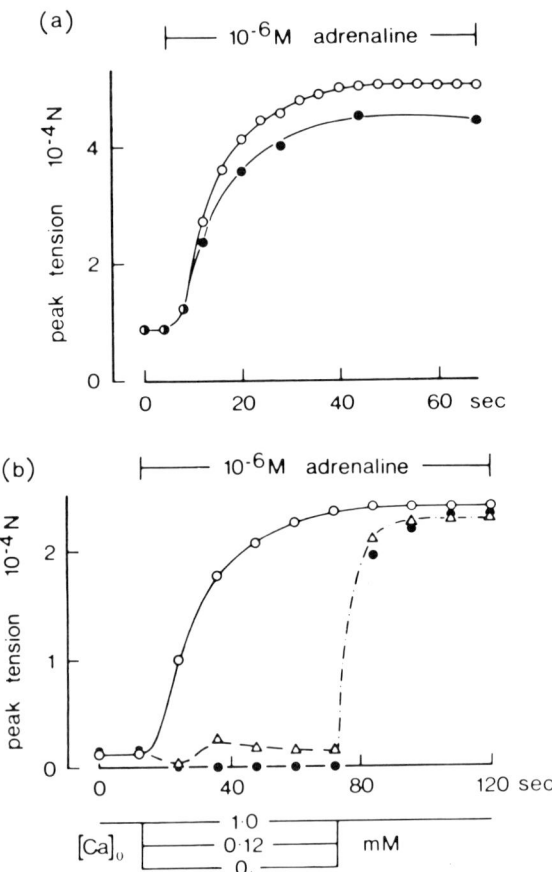

Fig. 2. *Adrenaline responses in conditions of varying*
stimulation pattern and external calcium concentration.
Different preparations in (a) and (b). (a) Tension buildup
either during regular stimulation at 15 min^{-1} (filled circles)
or at rest (filled circles), with single test shocks applied.
1 mM [Ca]$_o$. (b) Tension buildup in adrenaline either in 1 mM
[Ca]$_o$ throughout (open circles) or with [Ca]$_o$ reduced, for the
time indicated, to 0.12 mM (open triangles) or to zero (filled
circles), and then returned to 1 mM. Stimulus rate 5 min^{-1}.
Dot-dashed line: expected time course of tension change due
to extracellular diffusion equilibration of [Ca] from 0.12
to 1 mM [Ca]$_o$.

exposure is only brief (e.g., ≤ 5 sec) and the drug is
withdrawn before the end of the latency (Fig. 3b), a sizable
tension response can still be obtained. The buildup of this
response must have coincided with a decline of drug
concentration at the cell surfaces, suggesting that the short

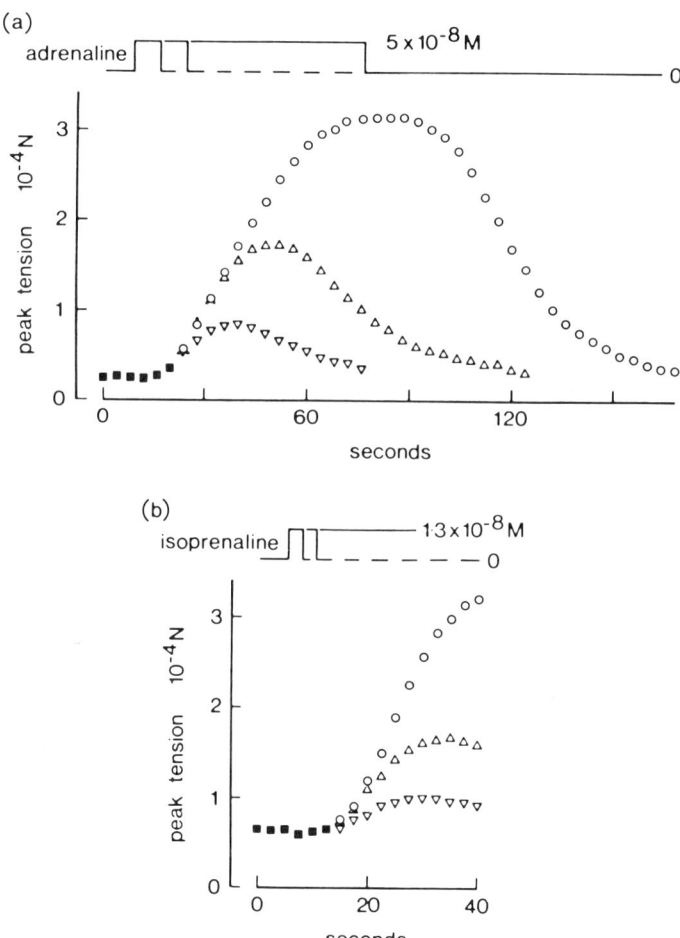

Fig. 3. *Catecholamine responses evoked by drug exposure
of different durations. Tension responses to adrenaline (a)
and isoprenaline (b) applied for the periods indicated.
Different preparations in (a) and (b). 1 mM $[Ca]_o$ in both,
stimulus rate 15 min^{-1} in (a) and 24 min^{-1} in (b). Initial
level of peak tension of twitches of the various responses in
(a) and (b), respectively, their means indicated by (■),
were practically identical.*

drug action had initiated cellular processes which then proceeded on a time course of their own. Similarly, during washout after more prolonged drug action, tension was maintained at a high level for some time, again suggesting the continuation of cellular processes after the original stimulus had ceased. We therefore attempted to analyze responses such as those of Fig. 3 in terms of the sequence of reactions, beginning with increased adenylate cyclase activity, which are known to result from β catecholamine action.

We considered first the latency in onset of the response, where again the mechanical event goes in parallel with the action potential, with a comparable latency in both before any change can be detected. The duration of the latency was found to be concentration dependent, decreasing with increasing doses down to a certain minimum value of about 3-4 sec. We could show that the latency did not arise from delays related to either the external diffusion process or a slow drug-receptor interaction. An alternative explanation accounting for the concentration dependence was therefore examined, according to which a threshold-like mechanism exists associated with one or several of the later intracellular processes. We tested this possibility by comparing the responses of the trabecula to a given catecholamine dose obtained with and without pretreatment to a lower dose of the drug, during which a small but sustained tension response indicated that the hypothetical threshold had been exceeded. Contrary to the prediction of this hypothesis, a latency period preceded both responses; and, indeed, the latency was little altered by the pretreatment.

Qualitatively, some of these results resemble those reported for isolated membranous adenylate cyclase preparations which, during hormone-induced activation, show a lag in onset of enzyme activity, and also a dose dependence of duration of the latency period (Kaumann & Birnbaumer, 1974; Rodbell, Lin & Salomon, 1974). By analogy with these findings and their interpretation, we ascribe the latency in our results to the time required for occupied receptors to induce the conformational changes of the adenylate cyclase molecule, the process responsible for its activation. This idea, although difficult to test directly, is supported by the results obtained with two different agonists. For example, when the time course of responses to equipotent doses of isoprenaline and adrenaline were compared, it was found that the latency in isoprenaline is about 50% longer than that in adrenaline, while the time course of the

subsequent increase in twitch tension did not differ. This
suggests that the latency is, indeed, associated with an
early reaction step that depends on the chemical structure of
the drug.

The subsequent phases of the catecholamine action are
most readily analyzed from the responses following sudden
withdrawal of the drug. Fig. 4 shows the time course of
the change of twitch tension either after steady tension
levels had been attained at various drug doses (Fig. 4a), or
after a given concentration of the drug had been present for
different exposure times (Fig. 4b). It can be seen that
during an initial period tension declined, if at all, only
quite slowly, and could even rise (i.e., after brief drug
exposure), but this is followed in all cases by a quite rapid,
final phase of exponential decline. A striking feature is
the constancy of the time course with which the final tension
decline occurred, regardless of previous dose or exposure.
Furthermore, it was found that the time constant of this
decline was identical with that of the fast phase of the
descending staircase, i.e. the response to a reduction of
heart rate, in the absence of catecholamines. However, it is
not regular stimulation itself which determines this time
course, as shown by experiments of the type illustrated in
Fig. 4c. In these, the transient response to a brief
adrenaline exposure during regular stimulation was compared
with the response to the same drug exposure for which,
however, stimulation was stopped at the moment of drug
application and single test shocks were applied, in
successive runs, at different times afterward. It is clear
that the modifying effect of regular activity on the
catecholamine action is only slight.

The simple shape of these responses and their
reproducibility encouraged us to attempt to interpret the
underlying events by means of rate equations describing
several consecutive steps. Fig. 5 illustrates the reaction
scheme used: the activated adenylate cyclase catalyzes the
formation of cAMP which, in turn, activates a protein kinase
and is itself destroyed by a phosphodiesterase. The
catalytic unit of the protein kinase phosphorylates specific
sites at the internal membrane surface, a process which
converts "latent" calcium channels into the "active" mode,
thereby enabling them to conduct inward current during a
subsequent action potential. Reconversion of the channels
to their former state is associated with dephosphorylation of
these sites, presumably by a protein phosphatase. A number
of greatly simplifying assumptions are made (for more
detailed discussion and a refinement of the hypothesis, see

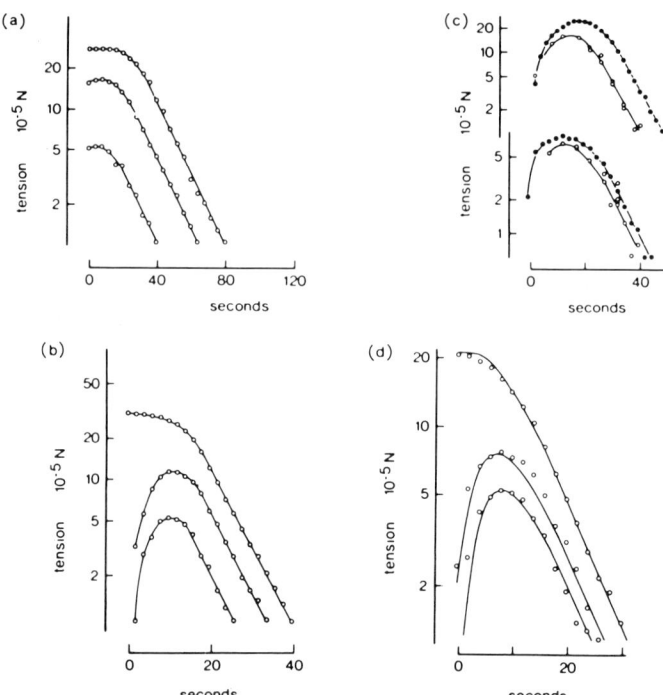

Fig. 4.   *Tension changes after short and long periods of adrenaline exposure. Logarithm of tension increments with respect to final steady tension level plotted against time after adrenaline withdrawal. (a) Tension decline after prolonged exposure and attainment of upper steady tension levels with (from top to bottom) 10^{-7}, 2.5×10^{-8} and 10^{-8} M adrenaline; 1 mM $[Ca]_o$, stimulus rate 15 min^{-1}. (b) Responses after exposure to 10^{-7} M adrenaline for various durations (from top to bottom): 80 sec, 8 sec and 4 sec; 1 mM $[Ca]_o$, stimulus rate 15 min^{-1}. (c) Two sets of results with two trabeculae after short exposures to adrenaline, with either regular stimulation (filled circles) or application of single test shocks (open circles); heart rate during regular stimulation 15 min^{-1} in upper graph, 24 min^{-1} in lower graph; adrenaline concentration 2.5×10^{-7} M (upper graph), 10^{-7} M (lower graph), 1 mM $[Ca]_o$ in both. Curves in (a), (b) and (c) fitted by eye. (d) Responses after exposures to 5×10^{-7} M adrenaline of (from top to bottom) 80 sec, 12 sec and 8 sec, fitted by theoretical curves with rate constants: $k_1 = 0.25$ sec^{-1}, $k_2 = 0.105$ sec^{-1} and $k_3 = 0.073$ sec^{-1}. (A 2-sec lag assumed between the time of drug withdrawal and commencement od decay of adenylate cyclase activity.) 1 mM $[Ca]_o$, stimulus rate 15 min^{-1}.*

Niedergerke & Page, 1977); in particular, steady tension responses to drug action are taken to vary in proportion to receptor occupancy, and three "back reactions," all first order, of rate constants k_1, k_2, and k_3 (as indicated), are proposed.

From the differential equations which can be set up on this basis, it is easily shown that after drug withdrawal, the time course of the tension response is given by the sum of three exponentials with rate constants k_1, k_2, and k_3, respectively. It is assumed that, as seems likely, the decline of adenylate cyclase activity, associated with rate

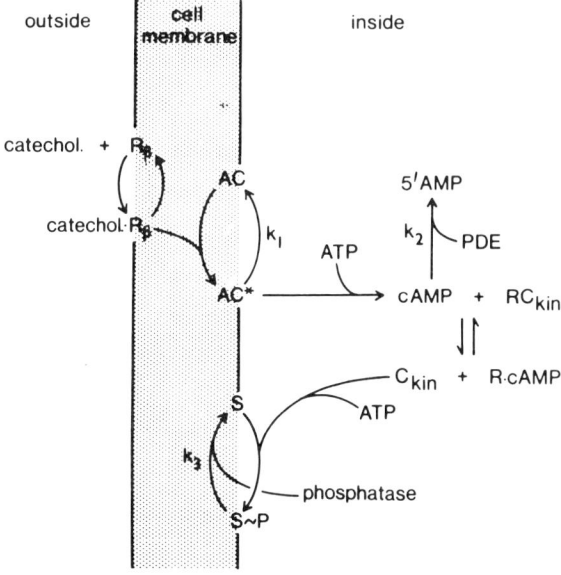

Fig. 5. Sequential processes associated with β-action of catecholamines. Processes at the cell membrane, i.e. the interaction of a catecholamine with the receptor (R_β) and subsequent activation of adenylate cyclase (active form: AC), are followed by formation of cAMP inside the cell. cAMP interacts rapidly with a protein kinase (RC_{kin}) and is destroyed, more slowly, by a phosphodiesterase (PDE). The catalytic subunit of the kinase (C_{kin}) promotes phosphorylation of membrane sites (S), which are reconverted to their original state by the action of a protein phosphatase. Rate constants for back reactions indicated.*

constant k_1, is rapid compared with the rates of the two
other reactions. On the other hand, if $k_2 > k_3$, as suggested
by results to be discussed later (in connection with Fig. 6),
then the final tension decline is controlled by the rate
constant k_3, that attributed to the dephosphorylation process
at the specific membrane sites; and k_3 can, therefore, be
obtained, approximately, from the slope of the exponential
tension decline.

In estimating k_2, which describes the rate of cAMP
hydrolysis, use was made of the observation that after short
drug action at a given concentration, the time to the summit
of the response was practically independent of the exposure
time used. For this condition, one can calculate k_2 from
values such as the summit tension and the time to summit,
which are read off several of these curves. Finally, k_1 was
taken as the parameter which provided the best overall fit to
a set of data such as those shown in Fig. 4d, where the points
are the experimental values and the curves calculated from
rate constants estimated, as above, for this preparation.

To investigate the assumption that $k_2 > k_3$, i.e. that the
de-phosphorylating step was rate determining, and at the same
time to test whether cAMP changes are indeed involved, we
applied the same type of analytical procedures to the
transient catecholamine responses obtained in the presence of
a phosphodiesterase inhibitor. In these conditions, k_2, the
rate constant for the decay of cAMP, should be decreased, and
it is easily shown that the amplitude of the transient
response is expected to be enhanced, its duration prolonged
while the final phase of tension decline is slowed. Fig. 6
shows that all these features were seen, but with lower doses
of inhibitor, the effects on amplitude and duration
predominate while the final decline is only slightly affected.
However, with higher doses of inhibitor, the slowing of the
tension decline is also clearly apparent, as expected if the
decay of cAMP is now largely rate determining (i.e., that
now $k_3 > k_2$) during this phase instead of, as is the case
normally and with low doses, the subsequent step associated
with the rate constant k_3. The theoretical curves of Fig. 5c
are constructed with the values of the rate constants
determined for this preparation (e.g. $k_2 = 0.11$ and 0.05 sec^{-1}
in the presence and absence of inhibitor, respectively), and
they demonstrate reasonable agreement of experimental points
with prediction.

We have made certain modifications to this scheme, in
particular to the original linearity assumptions, but these
do not affect the conclusions in any essential way. Rather

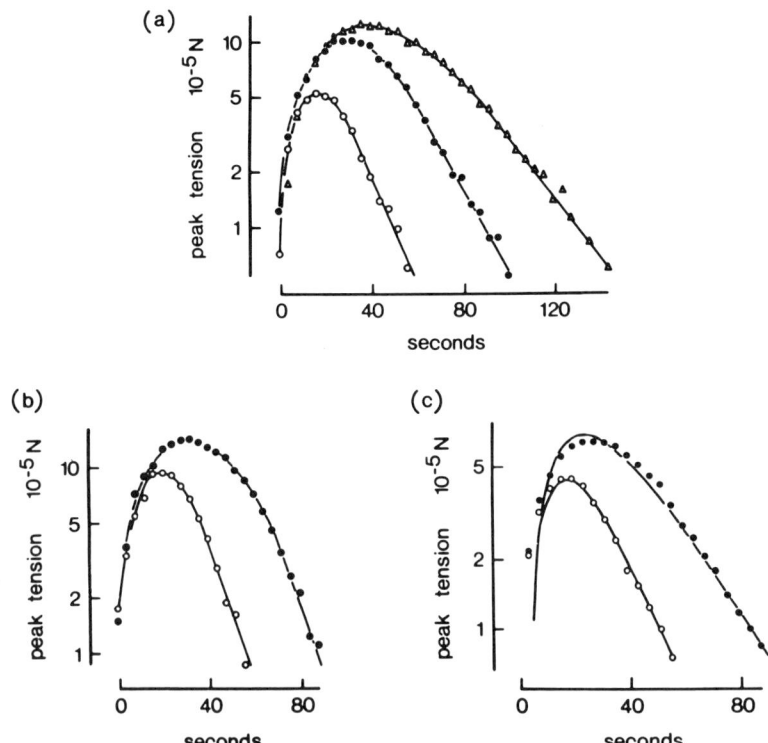

Fig. 6. Modification of adrenaline response by
phosphodiesterase inhibitors. Plot of tension changes as in
Fig. 4. Different preparations in (a), (b), and (c).
(a) Responses following 8-sec exposures to 5×10^{-7} M
adrenaline in the absence (open circles) or presence of 2.5 μm
(filled circles) or 10 μM (open triangles) papaverine; 1 mM
$[Ca]_o$, stimulus rate 15 min^{-1}. (b) asin (a) but with 10^{-7}
adrenaline in absence (open circles) or presence (filled
circles) of 7.5 μM ICI 63 197. Curves of (a) and (b) fitted
by eye. (c) as in (b), in absence (open circles) and presence
(filled circles) of 10 μM ICI 63 197. Curves constructed with
values for k_1 of 0.25 sec^{-1}, k_2 of 0.05 sec^{-1} (upper curve),
and 0.11 sec^{-1} (lower curve) and k_3 of 0.07 sec^{-1} (assumed
lag for start of decay of adenylate cyclase activity after
drug withdrawal: 3 sec).

than detail them, therefore, I want to finish by returning
to a point mentioned earlier, the finding that the final
exponential phase of tension decline from a catecholamine
response has the same time constant as that of the initial
exponential phase of the staircase, the phase that Chapman &

Niedergerke (1970) attributed to frequency-dependent changes
in calcium influx during the action potential. This finding
suggests a common step in the processes underlying the two
effects. Of the various possibilities for the location of
this step, the one most easily tested was the formation and
destruction of cAMP. This was done (Fig. 7) by examining the
staircase response in the presence of a high dose of a
phosphodiesterase inhibitor, which was known to slow the rate
of final tension decline after adrenaline by a factor of 2
or 3, but was found to have a negligible effect (not more
than 10-12%) on the decline of the descending staircase.
We therefore concluded that steps before and including that
of protein kinase activation by cAMP can be ruled out as
common to both catecholamine and staircase responses. Hence,
we propose that the sequential process underlying the two
responses might converge at the hypothetical membrane sites.
For example, phosphorylation of these sites could be
catalyzed either by a cAMP-dependent protein kinase, as
during catecholamine action, or by another, perhaps
membrane-bound kinase, cAMP-independent, whose activity may
in some way be linked to the membrane potential and be
stimulated by the repetitive depolarizations during enhanced

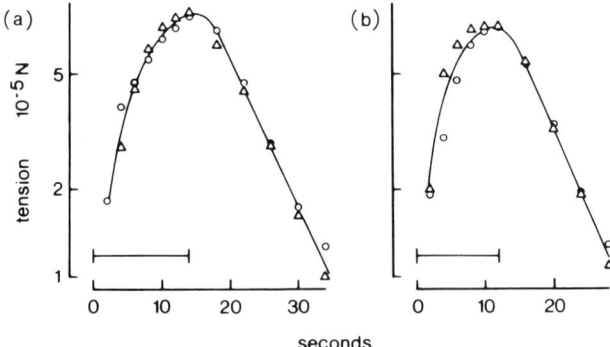

Fig. 7.   Short staircase and phosphodiesterase inhibition.
Tension response to change in heart rate from 15 to 30 min^{-1}
and vice versa, in absence (a) and presence (b) of 25 μM ICI
63 197.   Horizontal bars: periods of enhanced heart rate,
same trabecula in (a) and (b); each curve represents results
from two responses (open circles, open triangles).

heart rates. In both cases, the result would be the conversion of calcium channels to their "active" mode, while a single enzyme, a protein phosphatase, might restore the system to its original state.

In summary, it appears that the inotropic action of the β catecholamines in the frog heart is largely if not entirely due to an increase in calcium inward current during the action potential. Many aspects of the response are readily accounted for in terms of a sequence of intracellular processes, in agreement with biochemical studies. The initial phase, especially the latency, reflects the processes by which the occupied receptors activate the adenylate cyclase. A later phase depends on the size and kinetics of the changes in cAMP levels, and the final phase of tension change appears to be governed largely by the process which converts calcium channels between their "active" and "latent" states, a step attributed to the level of phosphorylation of membrane sites. Our results suggest that this final step might also be involved in the staircase response.

DISCUSSION

McNaughton: We've been examining two components of membrane current, when adrenaline is put on in a preparation which undoubtedly suffers much from the diffusion delays of Purkinje fiber, but we find the time course is very different on two different components. On the outward current and the plateau, the time course is quite slow, a couple of minutes, whereas on the diastolic potassium current it's very quick, of the order of the time course that you've mentioned here. Have you observed any very slow effects on the action potential?

Page: Not on the action potential. The relaxant effect on the potassium contracture to which Dr. Morad referred is much more slowly developing in our hands than the effect on the twitch, which is already maximal well before the relaxant effect.

Armstrong: Are there specific examples of phosphorylation of membranes?

Page: Yes, Wollenberger and his colleagues have been working with membrane fragments from frog cardiac muscle and

have shown that there is a cyclic AMP dependent
phosphorylation of these membranes.* They are working with
isolated membrane fragments, so they can only show that it
is phosphorylated, and we are making the step that it is the
phosphorylation which produces the increased calcium inward
current.

Tsien: The results are very elegant, but I have a
couple of qualms that I'd like to share with you. One is
that you used papaverine as a phosphodiesterase inhibitor,
and I'm 90% sure that you're aware of the fact that it also
blocks slow inward current, and I think that's been shown by
a couple of different groups.

Page: I'm sorry, I was not aware of that

Tsien: But you also used the ICI compound, and it would
be very important to show that that particular drug does not
have the undesirable side effect of blocking calcium currents.

Page: We have found that, particularly with papaverine,
with prolonged exposure there is indeed a depression of the
β catecholamine response. In fact, quantitatively, the
results with papaverine are slightly different from those
with either the ICI inhibitor or one of the Hoffman La Roche
inhibitors which we've used, which gives results
quantitatively the same as the ICI drug. And both of these
have a much smaller depressant effect with long exposure
times than the papaverine.

Tsien: The other question is that you've used a linear
analysis, and it seems to work remarkably well. Now, if
you think about all the possible nonlinearities, one of
them is the way calcium interacts with myofilaments, whether
it be troponin or the heavy meromyosin. The other is the
fact that you're not voltage clamping, and thus the change
in the height of the plateau is going to change the driving
force on the calcium entry. So even if you increase g_{Ca}
linearly, you'd expect I_{Ca} to increase in a nonlinear way.
Do you think that you have different nonlinearities that are
all cancelling out, or is it that the system just isn't so
nonlinear?

Page: As to your first nonlinearity, the relationship
between calcium binding to troponin and tension, we have a
pretty good idea of the tension range in our preparations
where we think we are in a linear part of that relationship.
If we work with high calcium and high doses of adrenaline,

*Wollenberger, A. & Krause, E.G. (1973). Pharmacology and
the future of man, Proc. 5th Int. Congr. Pha-macol., San
Francisco 1972, Vol. 5, 170-191. Basel: Karger.

such as μM, then we do indeed get up into a range where we begin to get the nonlinear response, and so all these analyses have been done in a tension range where we hope we're safely away from that. We have been surprised how reasonable the fit is with such a simple model. We improve the fit if we allow for saturation at certain stages such as the protein kinase step; in particular, the time course of the buildups is then accounted for rather well. I can only suppose that as far as the effects of the increased driving force are concerned, our results haven't sufficient sensitivity to detect these and other sorts of nonlinearities and saturations which obviously are there.

Morad: Did you work with the α stimulators at all?

Page: This is something I left out. These responses are only β responses. Forty percent of our preparations also show an α response which is dramatically different in its properties, time course, and so on. The results I have shown are chosen from strips which have no α response; we get the same type from the strips which do have the α response if we have phentolamine present, or, as you saw with some of them, with isoprenaline.

Morad: Adrenaline is a mixed drug.

Page: Yes indeed. In fact the response we can get from the α component in some preparations is even larger than the maximum β.

Morad: That's interesting. If I am allowed to quote some preliminary data, it seems that the α drugs, at any rate, the pure α drugs, have very little effect on the suppression of relaxation or enhancement of relaxation. That is, the potassium contractures don't seem to go down.

Page: Of the pure α drugs that we've used, some three or four give a very small response in the same preparation that gives us a large α response to adrenaline. So the efficacy of the pure α drugs in this particular preparation seems to be rather low. That may be one reason why you're not picking up much on the relaxant effect.

Tsien: Why are you so sure that the rate of drug coming off adenylate cyclase isn't rate limiting? If you look at TTX, it's supposed to come off its receptor pretty slowly.

Page: As I understand it, if we reach the steady level occupancy very rapidly, which we have shown, it implies that the back reaction cannot be slow. We do think there is a lag in deactivation, but it's a matter of a few seconds, and not on the time scale of these other responses, of tens of seconds.

REFERENCES

1. Chapman, R.A. & Niedergerke, R. (1970). Interaction
    between heart rate and calcium concentration in the
    control of contractile strength of the frog heart.
    J. Physiol. 211, 423-443.
2. Kaumann, A.J. & Birnbaumer, L. (1974). Studies on
    receptor-mediated activation of adenylyl cyclases. IV.
    Characteristics of the adrenergic receptor coupled to
    myocardial adenylyl cyclase: stereospecificity for
    ligands and determination of apparent affinity constants
    for β-blockers. *J. biol. Chem. 249*, 7874-7885.
3. Lammel, E., Niedergerke, R. & Page, S. (1975). Analysis
    of a rapid twitch facilitation in the frog heart. *Proc.
    R. Soc. B 189*, 577-590.
4. Niedergerke, R. & Page, S. (1977). Analysis of
    catecholamine effects in single atrial trabeculae of the
    frog heart. *Proc. R. Soc. B 197*, 333-362.
5. Page, S.G. & Niedergerke, R. (1972). Structures of
    physiological interest in the frog heart ventricle.
    J. Cell Sci. 11, 179-203.
6. Reuter, H. (1974). Localization of *beta* adrenergic
    receptors, and effects of noradrenaline and cyclic
    nucleotides on action potentials, ionic currents and
    tension in mammalian cardiac muscle. *J. Physiol. 242*,
    429-451.
7. Robison, G.A., Butcher, R.W. & Sutherland, E.W. (1971).
    Cyclic AMP. London: Academic.
8. Rodbell, M., Lin, M.C. & Salomon, Y. (1974). Evidence
    for interdependent action of glucagon and nucleotides on
    the hepatic adenylate cyclase systems. *J. biol. Chem.
    249*, 59-65.
9. Vassort, G., Rougier, O., Garnier, D., Sauviat, M.P.,
    Coraboeuf, E. & Gargouïl, Y.M. (1969). Effects of
    adrenaline on membrane inward currents during the cardiac
    action potential. *Pflügers Arch. ges. Physiol. 309*,
    70-81.